The Blue Hill
Meteorological
Observatory

The Blue Hill Meteorological Observatory: The First 100 Years—1885–1985

by John H. Conover

American Meteorological Society
45 Beacon Street, Boston, Massachusetts 02108

ISBN 0-933876-89-0

Library of Congress catalog card number 90-81186

Published by the American Meteorological Society, 45 Beacon St., Boston, MA 02108

Richard E. Hallgren, Executive Director
Kenneth C. Spengler, Executive Director Emeritus
Evelyn Mazur, Assistant Executive Director
Arlyn S. Powell, Jr., Publications Manager
Jonathan Feld, Publications Production Manager

Editorial services for this book were contributed by Linda Esche, Eileen Furlong, Lisa Greene, Greenhills Books, Corinne Kazarosian, Susan McClung, and Jay Talbot.

Printed in the United States of America by Salina Press, Manlius, New York

Table of Contents

To those who measure and study
the atmosphere and its effects
upon us

Preface

Long before I retired from the Air Force Geophysical Laboratory it had been suggested that I prepare a history of the Blue Hill Observatory, simply because I had worked there for more than twenty years and was quite familiar with the early investigations. In spite of similar suggestions after my retirement in 1977 and my continued interest in the Observatory, the thought was routinely dismissed on the basis that I never wanted to write another article or report in my life. Furthermore, my interest in painting and woodworking, which had been restricted somewhat, as well as the construction of a second home in the mountains of New Hampshire, were occupying every minute of my retirement.

However, in 1982 I assisted the Friends of the Blue Hills in the preparation of an application for nomination of the Observatory to the World Heritage List. This called for a bibliography of work performed at the Observatory, much of which was dutifully listed in detail. In the course of reading the old titles, scores of memories returned, prompting me to a sense of duty, especially to the scientific community; I decided then to make this work better known and, to a certain extent, to relate the work of the Observatory to that of other organizations and observatories. The thought was discussed with Dr. Kenneth Spengler, executive director of the American Meteorological Society at that time. He not only endorsed the idea, but graciously suggested that the Society handle the publication, thereby eliminating that expense on my part.

A basic problem has been whether to present the history in thematic or chronological form. Although the former usually makes easier reading, it was found that the multitude of projects demanded an excessive number of subject divisions, and the interaction of many of the projects became unwieldy. For these reasons, most of the writing and presentation of events are given in a strict chronological form. Also I believe this provides more scientific basis. The notation of each year provides an easy time reference.

This labor of love has taken far longer than anticipated, especially in the face of failing eyesight, not only due to the time required to look up and collate facts but also because of the difficulty in getting them down in presentable fashion. I make no pretense regarding my ability as a writer, but I trust the facts have been accurately documented for the future.

JOHN H. CONOVER

Acknowledgments

This book would not have been possible without the encouragement of Dr. Kenneth C. Spengler, executive director emeritus of the American Meteorological Society, and without the Society itself for sponsoring its publication. Also acknowledged are its staff members, in particular Evelyn Mazur, Corinne Kararosian, and Phebe Chace.

Miss Aimee and Miss Rosamond Lamb, nieces of the founder of the Observatory, generously supplied background material on the Rotch family and financial assistance toward the preparation and publication of the history.

Mrs. Marion Philpot volunteered to do the rough typing of the draft, and Mr. John F. Murphy copied a large number of pictures.

Dr. Richard Goody, fourth director of the Observatory, was of great assisstance in explaining the details of his appointment and his subsequent work at Harvard; he also made possible the loan of various annals and records.

I am deeply grateful to Mr. Abbott Lawrence Rotch, a grandson of the founder of the Observatory, who loaned Mr. Rotch's diaries and memorabilia, and to Nancy Rotch Magendanzt, who supplied drawings and photographs of the founder.

The following Harvard University libraries gave invaluable assistance: The Gordon McKay (of which the Blue Hill Collection is a part), the Houghton, the Godfrey Lowell Cabot Science Library, the Widner, and the Archives, where the associate curator, Dr. Clark A. Elliott, was of special help. Mr. Eben Gay at the Collection of Historical Scientific Instruments at Harvard assisted by showing old Observatory instruments, photographs, and reprints.

The technical library at the Air Force Geophysics Laboratory in Bedford, Massachusetts, enabled me to verify many references.

The Harvard controller's office furnished information regarding personnel.

Information concerning the geology of the Blue Hills was supplied by Mr. Clifford Kaye, its anthropology by Mr. George Horner, and antiquarian events by the Milton Public Library.

Barry Surman of MIT's paper, *The Tech,* and Warren Seamons of the MIT alumni office supplied information about Mr. Rotch's life as a student.

Personnel at the National Archives and Records Administration, Dr. Sharon Gibbs Thibodeau in Washington, and Mr. James K. Owens, in

Waltham, Massachusetts, assisted with the old records. Other information pertaining to the records was furnished by the National Records Center in Asheville, North Carolina, and by Dr. J. Murray Mitchell.

Dr. Ronald C. Taylor of the National Science Foundation offered comments and valuable reference material.

Mr. Frank Creedon graciously shared all his work preparatory to the writing of a thesis on the structure of the Observatory.

Mr. Dick Sizzer prepared maps of the areas discussed.

The Blue Hill Observatory Weather Club and Museum provided financial assistance toward publication and Dr. William E. Minsinger supplied information regarding the Club's hundredth anniversary celebration.

The many others who helped by sharing memories of the Observatory should be mentioned by name: Sally Wollaston, George Kenneth Thompson, Charles Pierce, Charles B. Pear, Jr. (for details regarding the radio-meteorograohs), Alexander A. McKenzie, Robert G. Stone, Paul Dalrymple, Richard Ashley, Ralph Newcomb, Dave Atlas, Fred Volz, and the present Blue Hill observer in charge, Robert Skilling.

And, lastly, my wife Ethel, who not only shared memories but assisted in innumerable and often tedious phases of the project.

Chapter I

Abbott Lawrence Rotch, Founder of the Blue Hill Observatory

Abbott Lawrence Rotch founded the Blue Hill Meteorological Observatory in his twenty-fourth year. This congenial young man possessed unusual energy; he was well educated, a frequent traveler abroad, spoke German and French fluently, was financially secure and properly trained in business. He became an indefatigable student of science, and was soon recognized as a world authority in his first interest—meteorology.

The attributes of such a man may relate to his forefathers. Was he a single progeny of unusual ability or one of a long line of worthy descendants? In this case Abbott Lawrence was a member of the fifth generation of a historic family that deserves further mention.

According to Bullard (1947), Joseph Rotch was the first member of the Rotch family to achieve great prominence. Joseph was born in 1704 in Salem Massachusetts. He went to Nantucket penniless as a youth, but immediately started to lay aside money earned from his occupation as a cobbler. He became a Quaker and married into a well-known Quaker family on the island. Soon, through frugal management of his resources, he purchased a small ship and joined in the commerce between Spain, the West Indies, and the colonies. He proved to be a good merchant, business prospered, and more ships were purchased. Of his large family, 10 died in infancy and 2 sons died early without issue, but William, born in 1734, married at 19 and lived to be 93.

When Joseph Rotch neared the age of 50, he retired from active participation in the navigation of his ships. Meanwhile, he had become one of the wealthiest men on Nantucket. After retirement, he invested in large tracts of land on the mainland in an area that became New Bedford and Fairhaven, Massachusetts. Anticipating the need for a deep-water port, which was not the case in Nantucket, he decided to settle in the area. It was first named Bedford; then it was recognized that a town in Middlesex County already bore the same name so the town became New Bedford. With mighty impetus, this energetic and pioneering businessman used his abundant means to set the wheels of industry in motion at the new seaport. His first new ship, the *Dartmouth*, was involved in the Boston Tea Party, and a son, acting as a managing owner, went through the trying experience of attempting to straighten out the affair. The family pioneered in the whaling industry, and soon New Bedford was recognized as the foremost whaling port in the world.

William, a more devout Quaker than his father, lived in Nantucket and ran his business from that port. The former counting house of William Rotch, later the home of the Pacific Club, still stands at the foot of the square in that quaint little village.

The Quakers, or Friends, who made up most of the population of Nantucket, stood neutral during the Revolutionary War years and firmly opposed all violence and bloodshed. Their ships were seized by both British and Americans and their seamen imprisoned. Fishing virtually came to a halt as lengthy negotiations were held in efforts to obtain whaling permits. Because of their beliefs, the island people nearly starved, as their ships were refused food supplies by the mainland. Distress was not confined to the island during the war years. New Bedford was burned by the British in 1778, and the financial loss was set at £105,000. At war's end the price of American sperm oil fell below the cost of its manufacture, but English prices held up. Because of this, William reluctantly went to England in an attempt to obtain reparation and to reestablish the business. Payment of damages on his fleet was refused, but eventually he became established in Dunkirk, France. His family joined him there, but misfortune struck again. As a result of the French Revolution they were forced to leave, so they returned to America, where William spent the last 30 years of his life in New Bedford. In 1821, a grandson, Samuel Rodman, started a diary and weather record in New Bedford, that later became part of the Blue Hill Observatory collection. The record is now famous for its length and homogeneity.

William Rotch, Jr., lived from 1759 to 1850. He followed the family tradition working in the business of whaling and had extensive interests in commerce and international trade. He too was very prosperous and amassed a large fortune in his lifetime.

Joseph Rotch, son of William Rotch, Jr., was not prominent when compared with his forefathers, but he did operate some of the first factories in New Bedford.

Joseph had two sons, Benjamin and William, both of whom went to Cambridge to attend Harvard University. The brothers, who were independently wealthy, could easily have squandered their resources, but they recognized their civic responsibilities early in life and became interested in cultural affairs. Benjamin married Annie Bigelow Lawrence, whose father was a prominent merchant in Boston, a United States ambassador to Great Britain, and one of the founders of the city of Lawrence, Massachusetts.

Benjamin and Annie had seven children; the youngest, Abbott Lawrence, was born in 1861. Both Benjamin and Annie were avid participants in Boston's social life and actively promoted the arts. Benjamin was a trustee of the Boston Athenaeum and the Museum of Fine Arts. He was considered a fine artist and businessman, and Annie was recognized as an accomplished harpist. Benjamin also filled posts, both political and other including that of representative in the Massachusetts legislature in 1843 and 1844.

The family first lived at 3 Commonwealth Avenue, Boston, across from the famous Boston Garden. In 1857 Benjamin purchased a farm in suburban Milton, just north of the Blue Hills. It is said that the first Jersey cattle in America were imported to this farm, an example of his interest in accepting and furthering new endeavors.

The family, including children, traveled abroad often in the company of Mrs. Rotch's father, Mr. Lawrence. Their tours were extensive, sometimes lasting for months as the family enjoyed the European culture and scenery. In the course of these trips the children were enrolled in various schools. Abbott Lawrence attended schools in Paris, Berlin, and Florence, thus providing an excellent background in foreign languages.

In 1877, at the age of sixteen, he started the first of a series of diaries, which he continued until his death. These show that, even at that time, he was already interested in science and mathematics. His inventive mind attempted to rearrange tables of logarithms to facilitate multiplication, an attempt which, incidentally, proved unsuccessful. He also prepared tables from which he could derive speed from odd measures of distance and time. He used these to compute speeds while traveling on trains.

His father encouraged his scientific interests and, on one occasion, while in Italy, he was taken to a demonstration of the burning of petroleum with the aid of asbestos.

Attendance at the theater and opera with his parents was a routine part of his life. In the summer, Abbott Lawrence traveled in Austria and Germany, where he became an enthusiastic mountain climber. He continued to indulge in this pastime throughout his life whenever time permitted. Prolific correspondence with friends and relatives clearly shows an early facility for writing. When his parents departed for America, to leave him in Europe for the first time, he was very homesick, but after

Christmas he was permitted to travel to Paris to be with his older brother, Arthur, who was studying architecture at the Beaux-Arts.

The date of Abbott Lawrence's first interest in meteorology is not known, but his small weather diary, which begins on 4 November 1878 at "Rear of 3 Commonwealth Avenue, Boston, Massachusetts," reveals that he was a proficient observer and meticulous recorder at age 17. He explicitly noted his "expression of the wind's velocity" in a table of eight categories and carefully defined his meaning of "fair" and "clear" skies.

Abbott Lawrence's activities were not all studious; on Beacon Hill he measured coasting speeds with other boys and fought in snowball skirmishes on Boston Common. Once he experimented in an attempt to cure a cold by not eating, which he found unsuccessful on the second day. He indexed interesting articles in *Scientific American,* studied calculating machines, and made a cardboard device, apparently similar to a circular slide rule, for multiplication. Frequent visits were made to Mr. Roper, a local inventor, to whom he showed the results of his experiments and talked about entering the "Institute" (the Massachusetts Institute of Technology, MIT). Lectures were attended several evenings a week, and over one short period of time he twice visited an exhibit of Edison's inventions. During this time in Boston he attended the Chauncy Hall School, a preparatory school for MIT.

As so often is the case with adolescents, interests suddenly change. In the spring of 1879, he took up walking in earnest and made careful notes of distance covered and the time required; in one instance he recorded an "average speed of 5 m.p.h. for 3.2 miles." The 12 km (7 3/8 mi.) walk to his father's farm in Milton and return he covered at 6.8 km hr^{-1} (4 1/4 mi. hr^{-1}). He apparently did not consider walking trips of this length excessive. In the summer he obtained a bicycle and after painstaking practice was soon riding back and forth to Milton. By fall, bike riding had become one of his chief interests and he drew maps of Milton and Quincy which were used to log his trips in a book.

While attending Chauncy Hall, Abbott Lawrence's life continued in this tenor, interspersed with parental trips around the Northeast, but with an added interest in girls. He was obviously a versatile individual, although he sometimes referred to his inability to make things with his hands. His father must have recognized this shortcoming, because he enrolled Abbott in MIT in a course in woodworking while Abbott was still in Chauncy Hall School. Abbott did quite well at cabinetmaking but occasionally noted frustration at having to make things over. Meanwhile, early in 1880, he

made more weather observations and obtained his first book in meteorology. After a concentrated period of cramming in June 1880, he
graduated from Chauncy Hall.

Abbott now spent much of his time with girls while vacationing at Bar
Harbor Maine and Newport, Rhode Island. In the fall he entered MIT
and continued to live at home. He kept diaries of the weather in 1881 in
Boston and Swampscott, a resort town on the Massachusetts Bay coast. At
MIT Rotch chose the mechanical engineering course. He wrote for *The
Tech,* the school paper, which was published every two weeks at that time.
In his junior year he presented a talk before the EME society titled
"Comparative Speeds of the Fastest Trains in Europe and America." An
abstract of his presentation, published in *The Tech* on 3 January 1883, shows
his fine writing ability, scientifically enhanced by supporting facts and
figures. The article was later listed as the best of that issue and he was
awarded a prize of $2.00. Rotch noted that this was probably the first
money he ever earned. In February 1883 he was elected president of EME,
and in the following May his article, entitled "Fast Trains in Europe and
America," was published in *Van Nostrand's Engineering Magazine.*

Rotch's past traveling experiences were put to good use when he
arranged an excursion for a group of mechanical engineers by steamer
from Fall River, Massachusetts to New York, which continued by train to
manufacturing plants in the New York–Philadelphia area. His civic interests were illustrated by his work on a committee that prepared a memorial
tablet for Professor Barton Rogers, founder and first president of MIT. In
his junior year he became interested in local weather forecasts and began
a separate diary of Boston Signal Service[1] observations to facilitate an
evaluation of their predictions. Shortly afterward he met Alexander
McAdie,[2] a private at the U.S. Army Signal Service Office in Boston.

Rotch's grades were poor at the end of his junior year, and at the start
of his senior year he wrote that he was tempted to give it up, as he did not
intend to practice engineering as a profession. Nevertheless, he was elected
class president almost unanimously and made an editor of *The Tech.* At this
time he recorded a brief note in his diary describing himself as weighing
"140 lbs. and height 5 ft. 8 in." (64 kg and 1.73 m).

In the spring he wrote a timely article for *The Tech* on the winter
weather of 1883–84 in Boston, including ample data illustrating his points.

---

[1] At that time the U.S. Army Signal Service was responsible for routine weather observations and predictions in this country.

[2] McAdie became director of the Blue Hill Observatory after Rotch's death.

In May 1884 Rotch graduated. His thesis, coauthored with H. F. Baldwin, was titled "An Application of the Steam Engine Indicator to a Locomotive." This apparently was a device that indicated the overall efficiency of the engine in pulling it along the track. A photograph, taken of Rotch for the *MIT Yearbook*, is shown in figure 1.

Figure 1. Abbott Lawrence Rotch (c. 1883) Courtesy Rotch family.

Rotch emerged from his college years a worldly man. His father died in 1882, leaving him with a more than adequate income, which he frugally monitored in detail. His formal education was complete, and he wrote and spoke with dexterity. In addition, he had traveled at home and abroad, was conversant in French and German, and could be at ease in both formal society or outdoor sports. He presided over and participated in numerous activities, which illustrates his outgoing and popular character as well as his leadership ability. Very few people, especially in those times, were well educated and wealthy enough to pursue their own interests at such an early age.

References

Bullard, J.M. 1947. *The Rotches.* Milford, N.H.: The Cabinet Press.

Chapter II

Through the Ages

The hill and bedrock on which the Blue Hill Observatory is firmly anchored, according to Kaye (1983), has its origin in the Ordovician Period, about 450 million years ago. Preceding that period, about 630 to 525 million years ago (late Proterozoic to Cambrian periods), a great thickness of sediments accumulated under shallow seas, lakes, and floodplains in the area that is now eastern Massachusetts. The sediments were mostly shale and mudstone, but also contained some gravel rock (conglomerate) and several thick layers of volcanic rock from volcanoes that were occasionally active in the area.

About 450 million years ago, an igneous mass squeezed up magma into these sedimentary rocks. This mass, now known as the Quincy Granite stock, makes up the Blue Hills of eastern Massachusetts. The granitic intrusion within the sedimentary rock was a large tack-shaped body, or pluton, the uppermost part of which appears to have broken through the surface at that time, forming a volcano.

Millions of years passed, during which time the upper part of the Quincy Granite stock was unearthed by surface erosion of the surrounding sedimentary rock. At that time the pluton must have stood high, probably several thousands of meters above the surrounding plains, forming a formidable range of mountains. Erosion of the pluton carried sediments, including many large, well-rounded cobbles and boulders down from the higher, steep slopes during Carboniferous time, about 250 million years ago. In this area, the earth's crust was then squeezed and, as a consequence, the rocks were faulted and deformed into great folds.

The final geologic event of consequence to the Blue Hills was the Ice Age, the million or so years during which ice sheets, several kilometers thick, covered all of New England, scraping, plucking, and abrading the ice-buried rock surface. Rock and soil eroded away; softer rocks eroded more and harder and more massive rocks eroded less. It is mainly for this reason that now the granitic Blue Hills rise 100-150m (300-500 ft.) above the softer sedimentary rock terraces to the north and east. Moreover, the outcrop of granitic rocks has been smoothed by the glacial abrasion into a series of rounded hills of which Great Blue Hill is the largest. The topography of the hills was exposed to the light of day for the first time about 15,000 years ago, after the melting of the ice. At first, the newly exposed land appeared as an arctic landscape on which mastodon and caribou roamed. At the time, the shoreline was far out at Georges Bank because the glaciers had not melted sufficiently to raise the sea to present levels.

Archaeological events from the time of the early natives to the time of their assimilation by the European settlers are abstracted from Horner (1985). The first migrants to the Blue Hills came from the Georges Bank coastal area in search of game about 11,000 years ago. The hunters found stone suitable for spear weights and projectile points in the hills, another cause for frequent trips.

About 6,000 years ago, the local climate warmed more rapidly and the ice melt caused the sea to rise close to its present levels. In consequence, the arctic tundra and animals of the area receded northward and the early hunters followed.

The warming climate gradually changed the flora and fauna to temperate types, and a new group of people slowly migrated north-eastward from the present-day Carolinas. The more persistent bands moved into New England. Family units established seasonal villages along the waterways. One such camp was on Green Hill, just west of Blue Hill near the Neponset River. Meanwhile the quarries, which were located on the north side of the Hills, continued to supply points and tools.

A third wave of people migrated to the area from southeastern United States beginning about 2,500 years ago. These people were agriculturists and manufactured pottery utensils and carved soapstone bowls and pots. The increase of food production allowed a population increase: ". . . kin-group loyalty, obligation rights, allegiance and duty" developed, all of which were controlled by the village heads-of-family. In time, federations were formed and ruled by a chief called a *sachem*. The Blue Hills fell within the bounds of the Massachuseuck Federation. Activity at the lithic quarries, which has proved to be the most extensive in New England, increased around 400 A.D., thus benefiting the villages at Green Hill and Ponkapoag, about 1.5 km (1 mi.) south of Blue Hill. Horner writes:

> The significance of the Blue Hills as a lithic source is reflected in the Anglicized word "Massachusetts . . . which is probably a composite of three Algonquin words, Massa-chus-et. The term has been translated variously as 'Blue Hills,' 'A Hill in the form of an Arrowhead,' 'the Place of the Great Rock,' and 'Big Rock.' However, each could be ascribed and confused with other 'Blue and Great' Hills in southeastern Massachusetts.

Horner offers a possible compromise considering the hill's location, function and size as "The Great Arrowhead Hill Place." Thus it appears that the State name "Massachusetts" was derived from the Blue Hills.

The more recent history of the Blue Hills, and in particular the site of the Observatory atop Great Blue Hill, has been set forth in considerable detail by Teele (1884).

In 1614 Capt. John Smith caught sight of the hills, which can be seen from far out at sea. (The general location of the hills in relation to other landmarks in Massachusetts is shown in figure 2.) When he entered Boston Harbor, whose waters reach within 11 km (7 mi.) of Great Blue Hill, he named the hills "Massachusetts Mount" (see figure 3, showing the Blue Hills Range). At a later date he had Prince Charles place new names on his map and the name "Chevyot Hills" replaced "Massachusetts Mount."

Further reference to figure 3 shows that the hills lie in a west-south-

Figure 2. Map showing the location of most southern New England sites referred to in the text. Courtesy J. H. Conover.

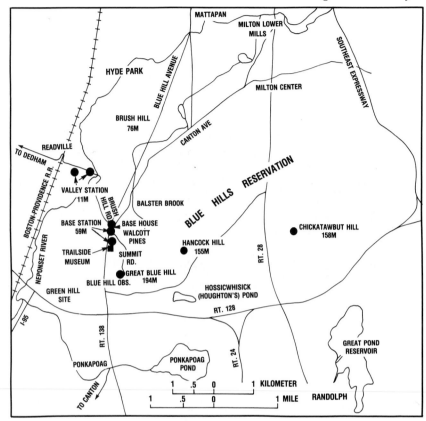

Figure 3. Map showing the area within the square in figure 2. Heights above sea level in meters. Courtesy J. H. Conover.

west/east-northeast line about 11 km. (7 mi.) long. The highest hill, Great Blue Hill, [194 m (635 ft.) above sea level], lies near the western end of the range and about 16 km (10 mi.) south of the center of Boston. The hill as seen from the south is shown in figure 4.

The leader of the Massachuseuck tribe, Chickataubet, which is also the present name of the second highest hill in the range, died of smallpox in 1633. His brother, Kitchamakin, gathered the tribe near the head of the tidewater on the Neponset River (which is now Milton Lower Mills) and in 1636 deeded the whole Uniquity Territory to Richard Collicot of Dorchester. Boundaries were to "the utmost extent" while the tribe was allowed 40 acres "where I (Kitchamakin) like best."

Manning (1895) in his account of the vegetation over and around the Hills, quotes from a letter by Master Graves, which was appended to Rev. Francis Higginson's account of New England's plantations in 1629:

it is very beautiful in open lands mixed with goodly woods, and again

Figure 4. Great Blue Hill, the Observatory, and TV tower as seen from the south. Circa 1957. The land remained essentially unchanged from 1885 to 1985 when the Codex Corporation erected a large two-story structure on the flat land shown on the right side of the picture. Courtesy J. H. Conover.

open plains, in some places five hundred acres; some places more, some less, not much troublesome for to cleere for the plough to goe in; no place barren but on the tops of the hills.

However, by 1669 invasion of the Hills for timber had commenced. Mr. Balster, a ship builder from Boston, bought and cut standing timber, probably white oak and white pine, in the section near Balster Brook, a few kilometers north and east of Great Blue Hill.

Late in the seventeenth century, hardwoods were taken for firewood, and probably all the virgin trees had been cut before the turn of the century. Fire was often used to clear the land in the valleys and fertile slopes and undoubtedly occasionally got out of hand and swept over the hills. This prevented the development of a healthy growth and resulted in a primary cover of scrub oak over the hills. The summit of Great Blue Hill was virtually barren except for a few scrub pine at the time the Observatory was built in 1885. Since then scrub oak and pine have gradually returned, with the pine achieving a more rapid growth in the last fifty years.

In Teele (1884) mention is made of the winter of 1780, which was very cold. Great suffering resulted among the destitute in Boston. John Hancock had a large quantity of wood cut from his lot in Milton—probably

Hancock Hill, 1 km (1 mi.) east of Great Blue Hill—and sledded down the river (Neponset) and over the harbor ice to Boston to distribute the wood among the poor.

The summit of Great Blue Hill, being the highest elevation in the region, always seems to have been a point of interest and recreation. The following example is from Rev. Peter Thatcher's journal:

> October 18, 1681. Brother Claff and his wife, brother Paul and his wife, and we, went upon Blue Hill to the pillar of stones, and Quartermaster (Thomas Swift) came to us there, and divers others; there we dined; we came home by Brush Hill, they came into our house and drank and smoked it.

The significance of "the pillar of stones" of that time is not known. Such outings, in general, were most unusual in the seventeenth century; in fact family outings in America were not common until late in the nineteenth century. Throughout the Revolutionary War the top of Great Blue Hill was occupied by soldiers as an observation post.

By an act of the Legislature in 1776, the summit was named Great Blue Hill and beacons were established. From that vantage point an enemy by sea or by land could be detected, at least in clear weather, and beacons could be used to warn surrounding townsfolk of dangers. On many occasions the hill was illuminated, presumably by huge bonfires. Such fires were lit for the repeal of the Stamp Act by the British, the promulgation of the Declaration of Independence and the surrenders of Burgoyne and Cornwallis.

According to Teele (1884) on 30 May 1798 at the summit of Blue Hill,

> . . . a foundation of heavy stones was built, twenty-one feet square, and ten feet high. Upon this foundation was erected a structure of wood three stories high, each story ten feet, with substantial flooring and with plank seats and railing securely fixed around the outside, accessible by stairs on the inside.

> The work was devised by the proprietor of the "Billings Tavern" a hostelry located near the hills, famous after the war as a resort for fancy dinners, parties, balls, and summer boarders. The old tavern was built about 1684 and was among the oldest buildings in Milton. The neighbors joined in the work of building the Observatory (not to be confused with the meteorological observatory), and the patrons of the tavern freely contributed to it. The passage up the hill was at the same time repaired and greatly improved, so that carriages could reach the top. This was accom-

plished by means of a "Bee"—a favorite method among neighbors, in olden times, of joining hands to secure a much desired object.

The Observatory was built for the purpose of opening a wider range of vision, and of affording an easy and comfortable position from which to take in the magnificent view. Mr. Billings' carriages were passing up and down the hill with his guests almost daily.

Four years later the structure was blown down, and a second staging was erected. In 1822 it was again repaired by Dr. E. H. Robbins, and remained many years, contributing greatly to the pleasure of the numerous visitors on the hill.

In 1828 Moses Gragg, former owner of the Norfolk Hotel in Dedham, took a new hotel in Milton. He planned to call it the Blue Hill Hotel (Austin 1912) but it continued to operate under the old name of the Billings Tavern. It was located on Canton Avenue about 1/4 mile north of what became the Base House. In those days the law required display of a swinging sign in front of every tavern. Gragg's sign showed a table of victuals on one side and on the other a picture of the road leading up Blue Hill to the three-story lookout at the summit. The sign shows a large tree at the entrance to the road. This probably depicted the "great oak," a landmark frequently referred to in the early Observatory history.

About 1834 the Hill was also used as an enticement to dine, this time at the Cherry Tavern, which was located at the base on the south side. During the cherry season "cherry parties" were held and guests were led to the summit of the Hill by a foot path.

In 1836 authorities of the Harvard College astronomical observatory in Cambridge erected a column 105 m (345 ft.) due west of the summit of Blue Hill. This was to serve as a meridian line marker from their observatory, then in the Dana la mer House, at the present site of the La Mont Library in the southeastern corner of the Harvard Yard. The circular column was 6.7 m (22 ft.) high, 3.6 m (12 ft.) in diameter at the base, and about 1.8 m (6 ft.) at the top. It was built of stone, the outer course laid in mortar, and filled solid to the top. It is interesting that, at some point in the natural course of development, another building, presumably in Boston, was erected which obstructed the view. A compromise was reached when the owner of the new building consented to cut a valley through the roof to open the line of sight between the Observatory and the stone column. Not long afterward sightings using the column were no longer needed and the column was destroyed by vandalism and the natural effects of time.

   In 1830 the state trigonometrical survey was authorized by the Legislature. Simeon Borden conducted the survey work, starting from a precise baseline near Deerfield. From there the network grew to about 100 hilltops which were designated as primary stations. A position on Great Blue Hill, which was one of these stations, was determined by sightings from Marblehead and Nahant. This position, 42° 12' 44"N, 71° 06' 33" W, and 193.69 m (635.05 feet) above sea level was marked by a copper bolt which became known as the "Borden bolt." A brass plate in the floor of the original meteorological observatory and later in the floor of the new tower marks the site. The bolt, however, has been missing since before the construction of the original observatory.

   In the summer of 1845 the Corps of Engineers of the U.S. Coast Survey occupied the summit of Great Blue Hill. They not only carried up surveying apparatus, but also cooking utensils and tents. In order to make the summit more accessible, a new road was opened from the junction of Canton Avenue and Blue Hill Avenue. Supplies and apparatus were carried easily by carriage over this road. Professor Boche, of the survey party, spent a part of the summer there and at times was accompanied by his wife. The U.S. Coast Survey marker was 8.00 m (26.25 feet) and on a bearing of 15° 37' E from the Borden bolt site.

   In 1875 the station was again occupied by the state and in 1885 and 1886 by the U.S. Geological Survey in connection with the State Topographical Survey. Since then numerous survey markers and various towers suitable for observers and instruments have occupied the site.

References

Austin, W. 1912. *Tale of a Dedham Tavern*. University Press, p. 56.

Horner, G. 1985. The geology, the geography - the people. In *Braintree, Massachusetts, its history*, edited by H. Holy. Braintree (MA): Braintreee Historical Soc.

Kaye, C. 1983. Verbal communication regarding the geology of the Blue Hills.

Manning, W. H. 1895. Notes on the vegetation of the reservations. *Rept. to the Board of Metropolitan Commissioners, Pub. Doc. No. 48. Han. 1895.* Boston, MA. Wright and Potter Printing Co. State Printers.

Teele, A. K. 1884. *The History of Milton, Mass. 1640-1887*. James M. Robins, Charles Breck, and Edmund J. Baker on the committee for writing and publications. Boston: Press of Rockwell and Churchill.

Chapter III

Conception and Construction of the Observatory

In the summer of 1884 Abbott Lawrence went through a period of restlessness following his graduation. He tried to unwind by visiting relatives and friends, fishing in New Brunswick, visiting Ottawa, and buying clothes in Montreal. On one occasion he climbed Blue Hill from his home in Milton and made special note of the good visibility. About two weeks later, perhaps with the climb still fresh in his mind, on 5 August 1884 he wrote in his diary (see figure 5), "The idea has struck me that a station might be established by me in connection with N. E. Met. Soc. (New England Meteorological Society) on Blue Hill and wrote to Prof. Niles and Mr. Teele regarding it."[1] The blossoming of this idea would end Rotch's unrest and set in motion a project that would carry him through the remainder of his life.

Teele was minister of the Congregational Church in Milton and town historian. Niles was professor of geology and geography at MIT and president of the New England Meteorological Society. Niles apparently referred Rotch's letter to Professor Winslow Upton, who taught classes in mathematics, meteorology, and astronomy at Brown University, and who had conducted meteorological and other observations at the Ladd Observatory for the federal government.

Although it is not known exactly what Rotch asked of the two gentlemen, Teele's response (Harvard University Archives), dated 7 August 1884, was very encouraging. He gave advice concerning the land, stating that the summit road had been open to the public for so long that it did not require a special right of way for use. He also advised discretion in revealing plans while negotiating purchase of acreage on the summit, for fear of a sudden increase in prices. He suggested two springs for drinking water; one beyond the turn half way up the road was thought to exist year round. This, however, was not the case. He also proposed the collection of

---

[1] In McAddie's 1914 memoir of Rotch (see bibliography) he notes that Rotch "conceived and executed plans for the Observatory on Great Blue Hill" in his junior year at MIT. However, the Rotch diaries make no mention of such plans. To the author it appears that Rotch may have indicated to McAddie at that time that he would have liked to have his own observatory, but had not seriously contemplated the endeavor.

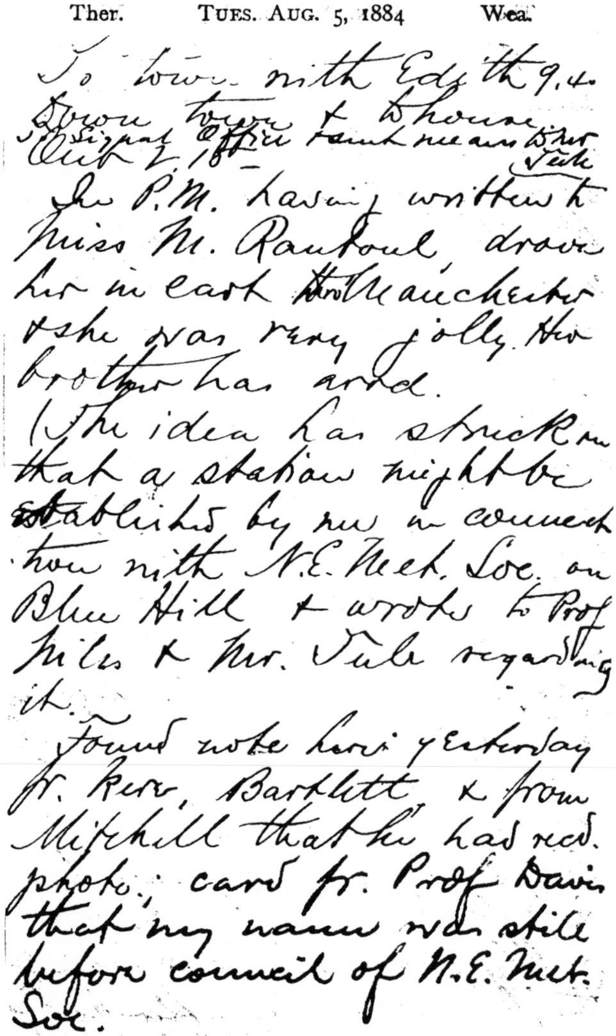

Ther.   TUES. AUG. 5, 1884   Wea.

[handwritten diary entry]

Figure 5. Copy of Abbott Lawrence Rotch's diary, 5 August 1884. Courtesy Rotch family.

rainwater in a cistern for general use. In a letter dated 16 August 1884, Upton expressed enthusiasm about the proposal.

Rotch had originally proposed to study the winds atop the hill and elevation related changes in precipitation. He was also eager to observe from a location unobstructed by buildings or trees, as had been the case with all his earlier observations and those made by the Boston Signal Station. Upton admired the location for these studies, and emphasized the need for continuously recording instruments, since records of this type were almost nonexistent in the United States. Because Upton provided Rotch with a list of the instruments and their cost, it is believed that Rotch planned to outfit the observatory with instruments identical to those in use at a first-order U.S. Signal Service Station. On 18 August 1884 Rotch wrote to the chief of the U.S. Signal Office (Harvard University Archives) and outlined his proposal to build the new station. In his letter he added the study of the paths of thunderstorms. He stated that he planned to connect the Observatory to the Boston Signal Office by telephone or telegraph, and that Professor Niles of Cambridge and Sergeant Cole of the Boston Signal Office had approved a plan of the observatory. He also proposed that the Signal Office supply a staff of observers who, as a result of their military training, would be more reliable and accurate. Rotch then would operate the building and conduct research with his own staff. The chief of the Signal Service replied on 1 September 1884, saying that the army would not enter into an agreement without full control. Upton advised Rotch against this and the proposal was dropped after a round of firm but elegantly written letters between Rotch and the chief of the Signal Service.

In further correspondence Upton suggested a building with a heated room for the staff and instruments. He did not recommend an enclosed tower but merely poles for the wind instruments. He had a strong preference for an isolated thermometer shelter of the Hazen design.[2] Nevertheless, he outlined the construction and method of installation of a window shelter if Rotch was to insist on this design. He also suggested corresponding with Alexander McAdie at the Boston Signal Service Office.

Rotch proceeded with his plan by purchasing an acre of land on the south side of the summit from Deacon John A. Tucker, and another parcel, 139 m^2 (1,500 ft.^{2}), which included the highest point on the hill and the ruins of the old lookout under which was the site of the Borden bolt. The latter was obtained from Mrs. E. Lewin after lengthy negotiations on the price. The town line between Canton and Milton ran roughly northwest–southwest through a point just south of the Borden bolt, putting the tower of the new building in Milton and the living quarters in Canton. At the time there were two foot paths leading to the summit from Canton Avenue; these joined about 0.4 km (1/4 mi.) from the top. It was decided to improve the one that had its beginning at the "Great Oak" opposite Brush Hill Road. A contract was awarded to James McLaughlin and work began on

---

[2] The Hazen design of shelter corresponds to that presently used at Blue Hill and is similiar to most shelters used in the United States.

1 September 1884. In ten days, with the work of 20 men and horses, a good carriage road to the summit had been built. At this time local residents were speculating about the building to be erected on Blue Hill, some saying it would be a fort, others a monastery, but most believed it would be a new Signal Service Station.

No stations existed in the country that could be used as models, so the design was left mainly to the architects, Arthur Rotch (Abbott Lawrence's brother), and George T. Tilden, his partner in a Boston architectural firm. They fixed the cost at $3,500. Construction began on 18 October 1884 by the J. H. Burt Company, and favorable weather during the fall allowed rapid progress.

Even in those days there were concerned environmentalists who made their presence known. E. G. Chamberlain of Auburndale, an Appalachian Mountain Club member, amateur surveyor, and authority on the Blue Hills wrote to Rotch (Harvard University Archives) and politely suggested the observatory be placed over the brow of the hill so it would not interrupt the 360° view. Rotch replied that from such a location the winds could not be measured properly, but he would gladly allow Chamberlain and his friends to enter the Observatory to enjoy the view from the tower. Chamberlain later became a good friend of Rotch, and was well known for his panoramic map of the view as seen from the Observatory.

By 6 December the building was roofed, but some of the outside masonry and pointing had to be left until spring. Plastering was begun on 22 December. A floor plan of the Observatory is shown in figure 6.

The Observatory was constructed of broken stone found on the hill, and granite trimmings. The building consisted of a two-story circular tower, 3.6 m (12 ft.) in diameter and 7.6 m (25 ft.) in height, with a flat roof and parapet. Extending southward from the tower was a one-story hip-roof addition which contained two bedrooms, a dining room, and kitchen. A wooden shed was attached. The walls were 51 cm (20 in.) thick and a metal-covered roof was securely anchored to them. The walls of the living rooms were plastered and the floors were of hard pine. A 3,400 liter (900 gallon) wooden tank was placed in the cellar for the storage of rainwater that drained from the roof. A large coal stove in the lower room of the tower heated the upper room through a register and its flue pipe, and a kitchen stove heated the other rooms. The windows were double paned throughout. A safe in the brick chimney was used to store records.

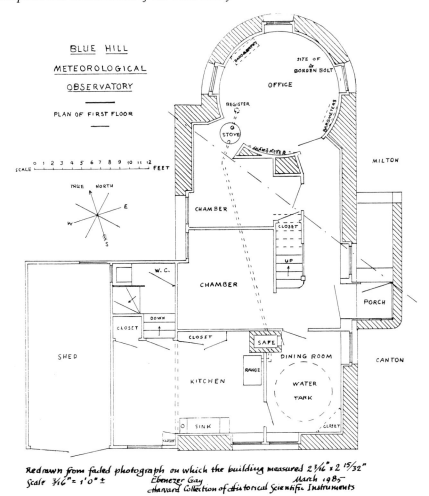

Figure 6. Floor plan of the original Observatory.

A telephone line was strung up on the south side of the hill, connecting the Observatory with the outside world. It cost $120 per year plus a one-time charge of $160 for the new line.

On 26 December Upton sent forms that he had specially prepared for recording the new observations, and at this time he suggested that Rotch become a volunteer observer for the Signal Service. An application was filed and Rotch was officially designated a volunteer observer on 6 January 1885 (Harvard University Archives).

Throughout the construction period, Rotch continued his busy schedule. He was often in his downtown office in the morning, on Blue Hill in the afternoon checking on the building, and at the theatre or some social event in the evening. On 29 December 1884 he and John Lowell departed for Mount Washington where they planned to usher in the New Year along with the Signal Service observers. They went to Fabyans, New Hampshire on that day and on the 30th continued to the Cog Railway base by sleigh. From there they climbed to the summit along the railroad tracks. The climb took 2 1/2 hours, and the weather was very mild with a strong southwest wind. On the summit Rotch gained experience observing under windy conditions and on New Year's Eve he noted in his diary, "games and music by Cahill (observer) and quite a jolly time. Highest people in New England to see '84 out." Thus began a rapport between the Mount Washington Observatory and Blue Hill, even before Blue Hill was officially opened. Before the group departed the temperature fell to -34°C (-30°F) with a 32 m sec^{-1} (70 m.hr^{-1}) wind, but they descended safely on the 3rd under less severe conditions.

Upton followed up his observation forms in January with tables for the reduction of the barometric readings to sea level.

At midnight on 31 January 1885 a red fire and rockets announced the official opening of the Observatory. Rotch and Willard P. Gerrish, an observer, moved in, although the lower room of the tower and adjoining sleeping apartments were the only rooms ready for occupancy. Gerrish came from the Harvard College Observatory where he had gained experience in observing. For the first two days they did their own cooking— and Rotch noted what a dismal failure he was at it—but on the third day the steward, Frank Brown, arrived to begin his duties.

References

Harvard University Archives: Blue Hill Observatory office files, 1884-1896l UAV.221.4.

Chapter IV

Initial Detailed Observations: 1885-1886

1885

In spite of the discomforts experienced during the first days at the Observatory, Rotch must have felt a sense of pride to see his new observatory atop the barren, windswept summit of Blue Hill (figure 7), while Gerrish dutifully began the long series of observations.

Conforming to the system of the Signal Service, observations were made three times daily: at 0700, 1500, and 2300 EST. Observations for February 1885 were entered in a hand-ruled notebook; those for the first day are shown in figure 8. These observations were made with thermometers, a barometer, a rain gage, and a "Hahl" anemometer whose dial had

Figure 7. Blue Hill Meteorological Observatory, looking toward the southwest, 1885. Note the lack of vegetation. Copyright President and Fellows of Harvard College. Center for Earth and Planetary Physics.

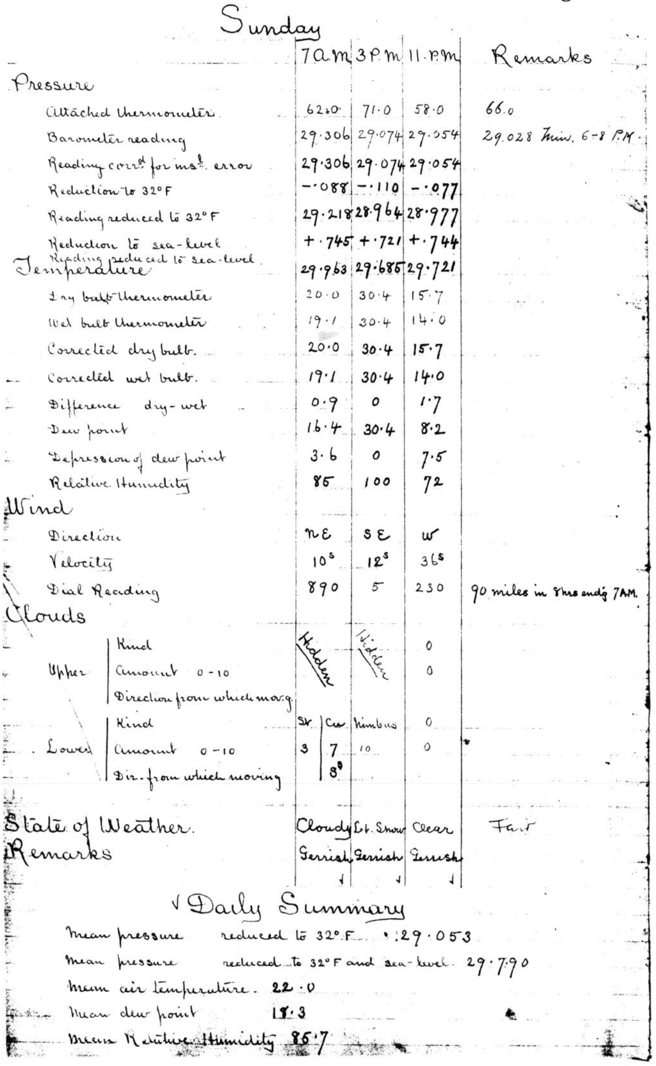

Sunday

	7 a.m	3 p.m	11 p.m	Remarks
Pressure				
Attached thermometer	62.0	71.0	58.0	66.0
Barometer reading	29.306	29.074	29.054	29.028 min. 6-8 P.M.
Reading corr.t for m.t error	29.306	29.074	29.054	
Reduction to 32°F	-.088	-.110	-.077	
Reading reduced to 32°F	29.218	28.964	28.977	
Reduction to sea-level	+.745	+.721	+.744	
Reading reduced to sea-level	29.963	29.685	29.721	
Temperature				
Dry bulb thermometer	20.0	30.4	15.7	
Wet bulb thermometer	19.1	30.4	14.0	
Corrected dry bulb	20.0	30.4	15.7	
Corrected wet bulb	19.1	30.4	14.0	
Difference dry-wet	0.9	0	1.7	
Dew point	16.4	30.4	8.2	
Depression of dew point	3.6	0	7.5	
Relative Humidity	85	100	72	
Wind				
Direction	NE	SE	W	
Velocity	10⁵	12⁸	36⁵	
Dial Reading	890	5	230	90 miles in 8 hrs ending 7AM.
Clouds				
Upper — Kind	Hidden	Hidden	0	
Upper — Amount 0-10			0	
Upper — Direction from which mov'g				
Lower — Kind	St.	Cu.	Nimbus 0	
Lower — Amount 0-10	3	7 / 8⁸	10 / 0	
Lower — Dir. from which moving				
State of Weather	Cloudy	Lt. Snow	Clear	Fair
Remarks	Ferrish	Ferrish	Furrish	

√ Daily Summary

Mean pressure reduced to 32°F. ·.29.053
Mean pressure reduced to 32°F and sea-level. 29.790
Mean air temperature. 22.0
Mean dew point 18.3
Mean Relative Humidity 85.7

Figure 8. The first day's observations at Blue Hill Observatory, 1 February 1885. Blue Hill Meteorological Observatory.

to be read to determine wind passage. Wind speed was calculated by dividing the distance noted by the time interval between observations. The speeds, however, were much greater than later standards, which makes standardization of the record difficult.

Only five days after the start of observations, instrumental problems were encountered as 5 cm (2 in.) of rime ice slowed the anemometer. On the 15th the Observatory was tested by a severe winter storm with heavy rain driven before a southeast gale. The cups of both anemometers were lost and the stone masonry leaked—a problem which proved to be ever-lasting.

In March the first recording instruments were set up. These instruments, an anemoscope that recorded the wind direction and an anemograph that recorded wind movement, were designed by Dr. Daniel Draper of the Central Park Observatory in New York. They were constructed by Black and Pfister, also of New York. The anemoscope consisted of a vertical chart-covered cylinder attached through shaftwork directly to a wind vane. Through clockwork, a pen that fell down over the cylinder in 24 hours marked the oscillations of the vane on the chart. The coordinates were rectilinear.

Wind movement was measured by a four-cup "Draper" anemometer. This instrument was unique in that the distance along the cup arms from hub to cups was adjustable, thereby changing the number of rotations per unit of wind movement. Cup rotation was geared down and transmitted to the anemograph on the upper tower floor with a lightweight rod. The rotating shaft turned a cam one rotation for every 32 km (20 mi.) of wind passage. The cam was arranged to lift a pen resting on a chart-covered drum, which rotated once every 26 hours or so. The cam was cut so that after 32 km (20 mi.) of wind passed the pen dropped to zero, ready for the next cycle of 32 km (20 mi.) of wind movement. The slope of the trace, therefore, corresponded to wind speed. Earlier models of these wind recorders were used by the Signal Service in Washington as early as 1872. These two Draper instruments, which are shown in figures 22 and 95, remained in use until 1959 with only minor alterations.

In May a Draper barograph, shown in figure 9, was set up. The instrument consisted of a fixed glass tube that had been enlarged and sealed at the top. This was filled with mercury, as in any mercury barometer, but the lower end was immersed in a test tube, or cistern, of mercury that was supported by long springs. As the pressure fell, mercury flowed from the fixed tube to the test tube cistern, thus causing it to sink with additional weight. Rising pressure caused the converse action. Since changes of mass were measured, no correction for changes in density of the mercury due to temperature was necessary. Clockwork pulled a board, on which a chart was affixed, across a pen attached to the mercury cistern. This system provided a daily trace of the pressure changes, magnified three times, on rectilinear coordinates; it, too, saw continuous service until 1959.

From the opening day, a daily exchange of observations and forecasts took place by telephone between the summit and the Boston Signal Service

Figure 9. Draper mercurial barograph. Blue Hill Meteorological Observatory.

Office, and arrangements were made to fly cold wave warning flags based on the official forecasts. The Observatory staff experimented with different size flags, settling on an overall size of 2.7 x 2.7 m (9 x 9 ft.), in white with a black 1.2 x 1.2 m (4 x 4 ft.) center. These flags flew from a 9.1 m (30 ft.) mast atop the tower of the Observatory. A circular explaining the meaning of the flags was posted in twenty post offices and twenty-six railroad stations within 8 km (5 mi.) of the Observatory. These flags generated considerable interest in the surrounding populace, many of whom employed telescopes to look for the warning flags.

In the spring Rotch made first use of the new data in studying the diurnal wind variations; results were published in May 1885 (see bibliography). In 1887 he had to correct his first conclusions and write that nighttime maximum winds at Blue Hill only occurred during stable conditions at that level, and that they were not the general rule as had been deduced from the original small sample. He also experimented with short term (2h) rain warnings, using flags and a lantern, and the Observatory participated in "term days" of the New England Meteorological Society.

On these days, which were spaced at 7 day intervals, detailed thunderstorm observations were made at numerous stations.

Rotch occasionally took the evening and following morning observations to relieve Gerrish. He apparently arrived late at night because on 14 March he notes, "purchased a revolver for protection in walking the hill at night." In spite of this busy period from midwinter to summer, Rotch continued his social whirl in Boston and pursued the necessary matters of business at his Boston office. He attempted, but failed, to purchase more acreage on Blue Hill in order to protect his investment and at one point he wrote to Roger Wolcott, a neighbor at the base of the hill, about preserving the top of the hill and offered to subscribe $1000 for the cause.

An increasing procession of visitors representing local officials and scientists came to the Observatory to mark the beginning of a long list of distinguished visitors. Among those who made the climb were Rotch's old inventor friend, Mr. Roper, Reverend Teele, Professor Davis of Harvard, Professor Niles of MIT and the New England Meteorological Society, and Professor Upton. Percival Lowell, well-known astronomer, Professor E. C. Pickering, director of the Harvard College Astronomical Observatory, and Professor Vose, with MIT students, also made visits. By early winter two Japanese meteorologists and a French scholar and writer had also toured the Observatory.

Early in July, Rotch, armed with letters of introduction from Draper, departed for Europe to visit observatories and acquire new instruments. This was the first of many trips abroad which quickly raised the Observatory's instrumentation to a status equal to those of Europe. He visited Germany, Austria, Switzerland, France, Belgium, England, and Scotland and observatories on Pic de Midi, Puy de Dome, Ben Nevis, and other hilltops. In 1886 Rotch described in detail his visits to these observatories (see bibliography). He met numerous scientists, including Assmann, Buys-Ballot, and Hildebrandson.

In his absence, operation of the Observatory was placed under the guidance of the New England Meteorological Society. During this time, Upton and McAdie helped with the observing and in one instance McAdie (1885) spent a week on the hill making from the tower atmospheric electricity measurements with a kite. These measurements were made under the auspices of the Signal Service. Two tin foil-faced kites were flown some 260 m (850 ft.) above the hill; a fine wire was wound around the restraining line. At the ground potentials of over 500 volts from a vertical height of 61 m (200 ft.) were measured in clear and cloudy weather, and sparks to ground were clearly observed. Modern measurements show this potential to be far too low. However, the observations did lead McAdie to believe that condensation particles need not be present to cause a potential. In the fall, McAdie was ordered to a new Signal Service post at Washington, D.C.

Great crowds visited the hill during the summer. On 4 July the traditional bonfires, rockets, and Roman candles were set off, and on one August weekend it was noted that 180 people and 17 sporting teams came to the summit.

Rotch returned from Europe in October, bringing with him the first of three new standard Hicks barometers (number 818), from London for the Observatory, and gold watches for the Signal Service staff in Boston. This and other Hicks barometers are shown in their case in figure 10. The instruments were still in service in 1985.

Throughout the fall and early winter many other instruments arrived from Europe. Two "rain band" spectroscopes were received and used at

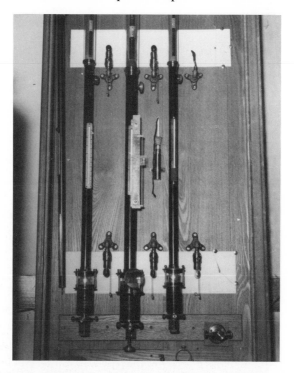

Figure 10. Hicks barometers numbered 1019, 872 and 818, left to right. Small electric lights illuminate the opaque glass panels to provide background light when making the adjustments at the cisterns and verniers. © 1950. Courtesy J. H. Conover.

observation times, but no analysis of the data seems to have been made. These pocket-sized spectrometers were popular at the time for estimating the total water vapor in a column extending to the top of the atmosphere. Water vapor absorption lines near 5893 A and 6563 A were observed; from their intensity total water vapor was crudely estimated. These instruments may be considered forerunners of the powerful techniques of remote sensing that have been developed for use in satellites.

A Campbell–Stokes sunshine recorder (figure 11) also arrived from Europe. This was the first of its kind in the United States consisting of a clear glass sphere having curved slots around it at a distance corresponding to the focal length of the sphere. Cardboard strips, marked with a time scale, are inserted in these slots. When the sunshine is bright, a narrow line is burned on the card. Cards are replaced after a day showing sunshine. On bright, clear days the card starts to burn 30 min. after sunrise and ceases to burn 30 min. before sunset; thus 100% sunshine actually corresponds to bright sunshine amounting to 1 h less than the sunrise-to-sunset interval. A Campbell–Stokes sunshine recorder has been in use since January 1886,

Figure 11. Campbell–Stokes sunshine recorder. Courtesy J. H. Conover.

making the record of "bright" sunshine the most homogeneous of all the records at the Observatory.

Another sunshine recorder, called a "Jordan," was also received and used experimentally. It employed light-sensitive paper that was housed in a cylinder light-tight, except for pinholes that allowed sunlight to focus on the paper. Duration of sunshine could be read from the traces on the paper, corresponding to the sun's movement, after it was developed. These instruments never suffered from clock problems since the apparent movement of the sun provided the timing element.

Other new instruments received were a "cloud reflector," or black mirror that could be used to follow the motion of the clouds, a Hicks standard anemometer, a Hicks standard thermometer, Negretti and Zambra "turn over" thermometers, and a hygrometer. A Gibbons single register was purchased to record 1-mi. electrical contacts from the Hahl cup anemometer. This was standard Signal Service equipment. This instrument completed the complement of self-recording instruments that were set up in 1885.

At the end of the year Rotch designed and had fabricated a pressure plate anemometer, thus bringing the complement of anemometers to four models. This anemometer consisted of a 0.09 m^2 (1 ft.^2) circular plate that was kept oriented into the wind by a vane. The plate moved against a spring, leaving an index at the maximum pressure exerted, thereby indicating a maximum wind gust. It was necessary to climb the anemometer mast to obtain a reading. Rotch also constructed a snow gage that was larger at the base than at the orifice thus aiding the retention of snow under windy conditions (figure 12). Water equivalent was determined by a simple weighing scale.

With the start of open scale (one page per day) autographic records the Observatory became one of the first stations in United States to obtain continuous records of the chief meteorological elements. The Central Park Observatory in New York City had been operating along these lines but after the stereotyped manner of the older European observatories (Waldo1901). Other open scale recordings of atmospheric pressure, precipitation, temperature, and wind were made by the Signal Service in Washington, D.C. as early as 1870 (Records of Experimental Self-Registering Instruments, 1870-88).

Figure 12. Snow gage designed by Rotch, 1885. Courtesy American Meteorological Society, Blue Hill Collection.

1886

The new year started with the addition of a cyclostyle, a form of copying machine, which was put to use immediately for duplicating the monthly weather summaries. Copies were sent to the Signal Office, to meteorological organizations at home and abroad and to interested friends.

On 28 January a severe ice storm occurred. Ice collected on a 1.6-cm (5/8-in.) diameter guy rod to a circumference of 33 cm (13 in.), and a vertical instrument support of 15-cm (6-in.) circumference increased to 46-cm (18-in.) circumference. The large 15-cm (6-in.) Draper anemometer cups were elongated to 30 cm (12 in.), and each cup carried a 2.7-kg (6-lb.) weight of ice.

Later in January, a Hicks black bulb thermometer in vacuum was brought up. This was used intermittently as a measure of solar radiation.

At the end of the month, Gerrish completed a busy year of observing with few breaks. Without his detailed journal of Observatory affairs, the weather and three-times-daily observations, all set down in meticulous handwriting, little would be known of the first year's operation. He helped in preparing the January monthly summary, then departed for work at the Harvard College Observatory in Cambridge.

At about this time Rotch said it cost him $2500 a year to run the Observatory. Frank Brown, the steward, was allowed $120 a month from which he supplied the food for Gerrish, himself, and visitors. It appears that Gerrish was given board and room and no salary.

The first year of operation ended with instruments capable of recording, on open time scales, atmospheric pressure, wind direction and speed and the duration of bright sunshine. Although many instrumental problems arose, resulting from the severity of the hilltop weather, progress was being made in their reduction. Anemometers, wind vanes, and their supports were strengthened, and observational techniques were improved. A host of new instruments, some of which bore certificates of calibration, were acquired as standards, spares in reserve, and for general use. The program of observational meteorology was rapidly becoming the best in the country. The data were distributed in the form of monthly summaries, some of which had already been analyzed, and results were published by Rotch.

References

McAdie, M. A. 1885. IV Contributions from the Physical Laboratory of Harvard College, XXV- Atmospheric electricity at high altitudes. *Proc. Amer. Acad. Arts & Sci.* 48: 129-134.

Waldo, F. 1901. The Blue Hill Meteorological Observatory. *Pop. Sci. Mon.* 59: 290-304.

Records of Experimental Self Registering Instruments. 1870 - 88 Washington, D.C.: National Archives, Records Group 27, p138, entry 15.

Chapter V

Cloud Studies: 1886–1894

1886 (continued)

This period, beginning about mid-1886, is classified as "cloud studies" because these have proved to be the most important of the time; nevertheless, weather forecasting, improvement of instruments, and many other activities continued to take place.

Henry Helm Clayton was placed in charge of the observing program on 1 February. Clayton who came from Murfreesboro, Tennessee, was 24 years old at the time. Because of his delicate health, he acquired his early education in private schools and by study at home. His interest in meteorology developed early in life with the study of local storms, and in 1882 he helped organize the Tennessee Weather Service. In 1884-85 he was an assistant at the University of Michigan and associate editor of the *American Meteorological Journal*. Before coming to Blue Hill he spent three months at the Harvard College Observatory, where he published numerous articles about the weather.

The cloud studies for which the Observatory became internationally recognized, can be traced to Clayton (see bibliography 1889). While in Tennessee he made his first cloud observations and later, at Ann Arbor, his attention was drawn to their forms. While there he set out to determine what existing system of nomenclature was best adapted to the clouds of this country. Ley's suggestion that clouds be classified as cumuliform or stratiform was adopted and the nomenclature of Hildebrandsson and Abercromby was used, except for nimbus, which Clayton omitted. Hildebrandsson and Abercromby defined "nimbus" as a cloud from which rain fell and since Clayton had seen rain fall from nearly all cloud forms, he declined to use the name.

For the first months of observership under Clayton the work continued as before, except that only one check observation, the evening one, was made per day. Data for the other observation times were obtained from the recording instruments.

Rotch continued his heavy schedule in Boston and on the hill. In February he traveled to Washington with Professors Upton and Davis to attend the Conference of State Directors of the Signal Service. He met with his old friend McAdie, and attended an army and navy reception at the White House. Before returning to Boston he dined with General Hazen, chief of the Signal Service and, in his words, felt "cordially received."

The comforts of home must have come to mind when on 1 March, a subzero day at the Observatory, Clayton noted "water left warm in a bowl on stand in observers' room, tho' not only a few feet from the stove in the adjoining room, froze over within an hour."

In March a Richard self-recording rain gage was received from Paris and set up. The gage worked on the principle of a weighing scale, but without springs; the traverse of the pen from base to top of the chart corresponded to 7.6 cm (3 in.) of precipitation. The instrument is shown in figure 13.

In April, Rotch procured a calculating machine designed by Thomas deColman of France. He notes that it saved "mental fatigue" but was "little gain in time by its use."

On 1 May the Boston Signal Office started to print a daily weather map by means of a cyclostyle apparatus furnished by Rotch. This was the first printing of a synoptic chart outside the Central Office in Washington, and the Signal Service soon extended the method of issuing maps to several of its other stations. A copy was received on the hill by 1700 EST via train and special messenger from nearby Readville. With the use of the map and local observations, the Observatory's first routine program of rain predic-

Figure 13. Richard Freres weighing precipitation gage with wire wind shield, left, and with cover removed, right. American Meteorological Society. Blue Hill Collection.

tions was inaugurated on 1 July 1886. The predictions were made at sunset for the following day with a revision the next morning at 0800 EST if necessary. Again, flags were used: a red one for rain, and none for no rain. The cold wave flags were flown as before, using the Signal Office forecasts.

Also on 1 July 1886 a base station, shown in figure 14, at an elevation of 61 m (200 ft.) above mean sea level (MSL) was established. It was located in the apex of land between Canton Avenue and Blue Hill Avenue, about 1.2 km (3/4 mi.) northwest of the Observatory. It was provided with maximum and minimum thermometers and a thermograph inside a Hazen shelter and a rain gage similar to that used on the Hill.

Meanwhile, it became necessary to protect the rain gages on the summit from spectators. This was accomplished by erecting a fence around them. This fenced-in area, which became known as the "enclosure," was virtually the same as the present enclosure. At the time, the Hazen shelter remained outside the enclosure at its original location east of the tower.

A note in Rotch's diary on 3 June states that Mr. Draper advised him to experiment with kites and to study the force and directions of the wind. This may have been the initial seed of thought that led to the extensive program in which kites lofted recording instruments some years later.

By 17 July the observation books show that Clayton was beginning to record from time to time cloud amounts, their types, positions in the sky,

Figure 14. Base station. Thermometer shelter and rain gage are shown. The house, owned by Rotch, was located at 1793 Canton Avenue. Copyright President and Fellows of Harvard College. Gordan McKay Library — Blue Hill Collection.

and their motion. Occasionally he did this for each hour of a day from 0800 to 2300 EST. His journal notes pertained mostly to clouds and at times an entire page was used to carefully depict changes during an interval of the day.

In September Rotch went to Burnside, Connecticut, to observe the effects of a tornado that had crossed the countryside. He prepared an article for the *American Meteorological Journal* (see bibliography) in which a detailed map illustrated counterclockwise winds. The directions of fallen cornstalks and trees and relative estimates of the winds' strength were carefully noted to illustrate the phenomena. Such observations of tornadoes were not new (Espy 1837). In this case, however, Rotch went on to give Davis' explanation of tornado formation. He assumed that cold air behind a cold front (not known as such at the time) had overrun warm, humid air in advance of the front, causing instability. A local breakthrough of the warm air presumably followed. Circular motion was imparted and the pressure then fell at the core of the circulation.

In the fall of 1886 Rotch was made an associate editor of the *American Meteorological Journal* and placed on the Council of the New England Meteorological Society.

Clayton designed a new "cloud mirror" which Rotch had Lowe of Boston construct (see bibliography, Fergusson 1934). This was a form of nephoscope that allowed accurate determinations of the direction of cloud movement and angular velocity. In a few cases, Clayton determined actual cloud speed by shadow movement on the ground and then cleverly determined cloud height from the angular velocity measured on the mirror.

The Observatory tower continued to leak in spite of marine varnish which was applied about every three months. An effort to reduce the dampness and cold was made by sheathing the walls of the upper room in the fall of 1886. The room photographed sometime after sheathing is shown in figures 15 and 16. Figure 15 looks toward the southwest. Two anemo-cinemographs are shown on the left; the upper operated from electrical impulses; the lower was mechanically connected to the windmill directly above. A pendulum clock was housed in the central box. It was used for sounding off seconds when nephoscope observations were being taken by using the mirror and eyepiece stand (made by Fergusson in 1895) shown beside the window. The Draper anemoscope that was connected directly to the wind vane above is shown on the right.

Figure 15. Second floor of the tower after sheathing, looking southwest. Copyright President and Fellows of Harvard College. Harvard collection of historical scientific instruments.

Figure 16 looks toward the northeast. A spare windmill for the cinemograph is shown at the lower left. The Draper anemograph is shown on the right. It was connected directly to the four-cup anemometer above.

Among the visitors of the year was Cleveland Abbe, chief forecaster for the Signal Service. After working almost straight through for eight months, Clayton was allowed a few days off, followed by a ten-day vacation over Christmas. Rotch and his friends substituted over this period.

The observations for the first complete year of record, 1886, were published by Rotch (see bibliography).

1887

Although the weather predictions from the hill were primarily of interest to Rotch, the records show that Clayton pursued them with as much enthusiasm as his cloud studies. Predictions were issued to the Boston evening newspapers, *The Transcript* and *The Record,* during the early part of 1887. Then, for budgetary reasons, the telegraphic reports from

Figure 16. Second floor of the tower after sheathing, looking northeast.
Copyright President and Fellows of Harvard College. Harvard collection
of historical scientific instruments.

the Signal Service, on which the Observatory depended for synoptic
reports, were suspended. This, naturally, made forecasting more difficult,
but Rotch and Clayton carried on by plotting "synopsis" data from the
papers. On 2 May telegraphic reports were resumed, largely through the
efforts of Rotch and the Boston Board of Trade. This was the first of a series
of incidents over many years where outside influences, including those of
the Blue Hill Observatory, were brought to bear on the weather agency at
Boston to improve the local service. Daily forecasts, with the exception of
Sundays, to the Associated Press in Boston were resumed on 2 May.

On 1 January the regular weather observations at the Signal Service
were changed to 0700, 1500 and 2200 EST. Those at the Hill were changed
on 1 February to conform.

In April Rotch proposed to Professor Pickering of the Harvard
Astronomical Observatory that, upon his death, he leave the Observatory
to Harvard with a $40,000 endowment. At that time the corporation
declined.

Clayton had proposed that clockwork and a chart be built into the tail of the pressure plate anemometer to provide a continuous record of wind pressure. This was done under Rotch's direction and the new instrument was set up for use on 22 June 1887. It is shown among a collection of anemometers in figure 25.

In July Rotch left for Iwanowa, Russia, to participate with Upton in the making of weather observations during a solar eclipse on 19 August. Köppen, whose expenses Rotch paid from Germany, also helped with the work. Data were obtained, but the sky was cloudy during the eclipse. Rotch then toured Europe, climbed numerous mountains, visited many meteorologists, and mountain observatories, and purchased new instruments. Meanwhile, at Blue Hill, a horse and buggy had been obtained and stabled on the hill to help Frank Brown in the delivery of food supplies.

On 27 September 1887 Sterling Price Fergusson arrived to take the position of assistant to Clayton. At the time he was nineteen years old. Fergusson came from Riddleton, Tennessee, and undoubtedly was known to Clayton before he came to the hill. Fergusson was quick to learn and, best of all, he was a genius when it came to solving mechanical problems. Such difficulties had plagued the observers since the beginning. Breakdowns were frequent and repairs had to be made off the hill, which led to long delays and considerable expense.

Throughout the year Clayton had been busy observing clouds on an hourly basis from 0700 to 2300 EST with few breaks. He adopted a nomenclature with abbreviations as follows:

	Stratus Type	**Cumulus Type**
Upper	Cirrus (C)	Cirro-cumulus (CK)
	Cirro-stratus (CS)	
Middle	Strato-cirrus (SC)	Cumulo-cirrus (KC)
Lower	Stratus (S)	Cumulus (K)
	Strato-cumulus (SK)	Cumulo-stratus (KS)

Fog

In addition to noting cloud amounts, he recorded their motion, position in the sky, direction of radiation of bands, if any, and remarks. Most of these observations were published in 1889 (see bibliography) with extensive tables of hourly values of the other meteorological elements measured on the hill during 1887.

On 7 December 1887 the following appeared in the journal: "A tramp was found in entry-way last evening and for fear his presence might give rise to trouble during the night, he was escorted to the base. At bedtime he was again found in the entry and again escorted to base; but as he again soon returned it was thought well to make the best of circumstances, so he was given an old overcoat and told to make himself comfortable. In the morning he peaceably departed."

Beginning in 1887 the practice was to use two thermometer shelters at different locations. From April or May to October a shelter located 20 m (66 ft.) east of the tower was used. During the remaining months a shelter housed the thermometers outside a window of the unheated second floor. It was located 4.6 m (15 ft.) above the ground, on the north-northwest side of the tower. This arrangement prompted a small study of comparative readings from the shelters. It was concluded from data obtained in a February-to-August sample that very little difference occurred in the average daily temperatures from each site. However, when the sun was high and clear, the temperatures in the isolated shelter were higher than in the window shelter.

In mid-December Rotch returned from Europe with a new lot of instruments and a change in his social status. He had become engaged. He had known this lady for a long time in Boston and reunited with her as she was traveling with her parents in England.

The new instruments were a Fineman nephoscope, Koppe hair hygrometer, self-recording Richard actinometer, solar black bulb thermometers, a Hicks barometer (number 872) still in daily use, charts for the Richard instruments, and a new pair of Negretti and Zambra turnover thermometers. These thermometers could be arranged to retain the temperature at preset times.

The year 1887 ended with the most severe southeast gale, with snow changing to rain, that had been experienced to that time. An indicated speed of 33 m sec^{-1}(74 mi. hr^{-1}) was recorded for one hour with a maximum speed of 39 m sec^{-1}(87 mi. hr^{-1}). Several panes of glass were blown in and a room was flooded.

1888

Early in 1888 the meteorological observational program at Harvard College was terminated because the new detailed Blue Hill record was proving to be far superior. Pickering (1886) of the Harvard College Astronomical Observatory agreed to publish the Blue Hill observations in detail as part of their annals and as long as the expense was shared by

Rotch. These publications, which commenced in 1889 with the 1887 observations, included twenty-four hourly values of pressure, temperature, relative humidity, wind direction and speed, precipitation, bright sunshine, and cloudiness from 0700 to 2200 EST. Discussions accompanied the summary tables and articles on current research were included. At this time Rotch discussed, once again, a possible bequest to Harvard at the time of his death, but he did not elaborate on the details.

Another cold spell in late January caused 2.5 cm (1 in.) of ice on the cistern below the dining room, and on the 28th the maximum and minimum temperatures for the day were -18°C (-1°F) and -23°C (-9°F) respectively.

In March Rotch traveled to St. Paul, Minnesota, to attend a conference on weather predictions and their verification. Forecasts to the Associated Press in Boston ended in April.

Meanwhile, animosity in connection with the weather predictions and their verification had been building between individuals in the Signal Service and Blue Hill. For some time Clayton had published in the *Bulletin of the New England Meteorological Society* verification comparisons of forecasts by The Signal Service and Blue Hill. A series of articles in the *American Meteorological Journal* followed. GAN (1888), an anonymous author, but undoubtedly associated with the Signal Service, claimed that by forecasting average weather the verifications would be better than Clayton's. He referred to them as "child's forecasts." Davis (1888), president of the society, wrote a strong rebuttal saying that the "object of the comparison seems to be to show up Mr. Clayton's work in a ridiculous light, making it appear to be no better than that of a child or ignorant person." Clayton also replied, saying that only persistently dry months and warm summer months had been used in GAN's comparison. At one point the *Boston Transcript* published a letter from Professor Hazen of Washington in which he "depreciated" the Observatory forecasts. Clayton, who used the weather map supplemented by local observations, a technique used in Europe since the advent of the exchange of observations by international telegraph, said, "It seems to me better to combine both methods: that is, to have scientific experts at Washington to study and predict the movements of areas of high and low pressure, and leave the local predictors to fill in the details of weather for their section. For the greatest efficiency, these local predictors should have reports from a net-work of secondary stations in their locality, such as those of our voluntary observers." The idea did not set well with the Service because all their forecasts originated in Washington, but clearly the idea was similar to modern methods that are now in use by many national weather services. Another round of articles by GAN and Clayton followed, but the arguments could not be resolved, primarily because the forecasts were issued at different times. General Greely, chief of the Signal Service, was not aware of this exchange and later he wrote to Rotch explaining that the comments did not represent the official policy of the service. In fact he praised the Blue Hill forecasts and viewed them as useful experiments in the research to improve forecasting which he could not pursue because "Congress stood in the way of reforms."

A valley station was opened on 1 July, located near Paul Bridge over the Neponset River in Readville, 2.8 km (1 3/4 mi.) northwest of the summit station. The station, shown in figure 17, was 15.2 m (50 ft.) sea level and operated by Mrs. Dean, who lived in the adjacent house. She was paid five dollars a month to make the observations. A shelter identical to those at the base and summit was supplied; it housed thermometers and a thermograph; a rain gage was located nearby.

Rotch sent up 22 kg (45 1/2 lbs.) of rockets, 24 roman candles, and 1.8 kg (4 lbs.) of red and blue fire to be set off on the hill on the 4th of July. Three lanterns shown from the tower and each tower window was illuminated. The observers and Rotch spent much of that memorable evening setting off the fireworks.

Even at this time in the life of the Observatory it had become so well known that other establishments sought to use it as a model. E.T. Turner

Figure 17. Valley station looking toward the southeast. Thermometer shelter and precipitation gage are shown. The location is near the present junction of Truman Highway and Neponset Valley Parkway. The house no longer stands. Copyright President and Fellows of Harvard College. Gordon McKay Library—Blue Hill Collection.

of Cornell University visited in July to observe the procedures and instrumentation with the view of opening an observatory at the university.

Leaks continued to be a great source of annoyance; in fact, they became so bad in the tower masonry that it was decided to coat the surface with a cement wash in August. Unfortunately, this too failed to stop the water during the next severe storm.

In the course of Rotch's travels abroad and at home he arranged the exchange of publications from nearly every organization he visited and individual that he met. Rotch's books and the rapidly growing collection of publications prompted the construction of an addition to the Observatory. This addition, which was started on 18 October 1888, consisted of two floors attached to the dining room and extending to the east. After completion in early 1889 the upper floor, which was entirely finished in natural wood, served as the library with attached fireproof vault. The library is shown in figure 18, a photograph taken about 1900, after the cloud photo collection had been expanded. The downstairs, which was reached by a cast iron spiral staircase, was partitioned to form a bedroom and small shop. An external view of the addition is shown in figure 19.

Rotch journeyed to Colorado in the fall to visit Pike's Peak and the observatory on its summit. He revealed his artistic ability by returning with a set of very fine, small pencil sketches showing the peak from various

Figure 18. Interior view of the east wing which served as the original library. Photo c. 1900. Copyright President and Fellows of Harvard College. Harvard collection of historical scientific instruments.

Figure 19. East wing addition looking toward the northwest. 1889. The high mast was used to mount and compare anemometers. Copyright President and Fellows of Harvard College. Gordon McKay Library —Blue Hill Collection.

distances with the observatory near the summit. At this time Rotch wrote of the closing of the observatories on Pike's Peak and Mount Washington (see bibliography). He noted "great regret to meteorologists the world over" and went on to say that the reports were received daily in Washington D.C. However, it appears that no one understood how to interpret the reports, although a brief report by Hazen (1888) noted that the direction of movement of cirrus above the stations was about the same as the direction of the wind at the summits. The speed of the cirrus he found to be about twice that of the summit winds. Rotch suggested that the stations might defray costs, which was the reason given for the closings, by feeding and lodging tourists as was done on several European mountains. It is interesting to note that at this writing, 53 years after the reopening of the Mount Washington Observatory, that observatory obtains about sixty percent of its necessary income from a museum that it operates at the summit.

In November, Rotch received notice of his appointment as Assistant in Meteorology at Harvard College dated 14 May 1888. He held this appointment until 1891.

Rotch returned west again in mid-December, this time to participate in the Harvard College Astronomical Observatory's observation of a total eclipse at Willows, California, on 1 January 1889. Fergusson made a new anemoscope for the expedition to obtain a detailed record of the wind direction. A good set of data was obtained which was later analyzed and reported.

The first pole star recorder at the Observatory was set up on the tower in December; it had seen previous use at the Harvard College Observatory. This instrument was essentially a camera that made use of the fact that the pole star completes a small circle every twenty-four hours when viewed from the rotating earth. When the motion is amplified through a lens and focused on a photographic plate that is exposed throughout the nighttime hours, a partial circle is clearly scribed. This trace can then be used to estimate night cloudiness as the trace is interrupted, much as with the bright sunshine record during the day. Later the instrument was set up inside the enclosure and still later it was moved to a position just west of the tower shown in figure 20.

On New Year's Eve Mr. Pfister arrived from New York to attach battery-driven electric motors to some of the recording instruments. Apparently this experiment proved unsatisfactory because the motors were soon removed.

Among the many papers and articles published during the year was a long report by Rotch (see bibliography) on the organization of the meteoro-

Figure 20. Pole star recorder. c. 1935. Copyright President and Fellows of Harvard College. Harvard collection of historical scientific instruments.

logical services in some of the European countries. In this report the operations in each country were described in detail. Central observatories administered outlying stations, performed weather analysis, issued forecasts, and calibrated instruments of all kinds. Many performed magnetic observations. Autographic instruments were used and hourly values published in most cases. At that time the meteorological instrumentation at Blue Hill was comparable with that in the central observatories of Europe, although the observatory at Pawlowsk, in Russia, probably exceeded any in the world in that respect. Research at the observatories consisted of a few discussions of the observations and an occasional reference to aids in forecasting.

Rotch strongly favored the use of mountain observatories and within the year wrote several separate articles about them.

Clayton published the first hourly cloud observations in the western hemisphere and from these showed diurnal and seasonal variations at Blue Hill (see bibliography).

1889

At the beginning of the year the observations were entered in a new form of record book. One book was used for each year. Pages separate from those used for the routine observations were provided for meteorological and journal notes.

The daily weather predictions were continued and arrangements were made to supply forecasts to the Boston Office of the Associated Press for thirty dollars per month.

A seismograph was set up in the basement in January, but later notes indicate that it frequently malfunctioned and, in one case, did not record a heavy blast that had been set off at the base.

Detailed cloud observations continued as through the previous year, except that hourly detailed cloud observations were made from 0700 to 2200 EST.

Clayton published an eighty-page paper in French on the formation of clouds (see bibliography). He also attempted to construct a few pressure maps for the 5 km level of the atmosphere. He used surface pressure and temperatures and applied lapse rates given by Glaisher, Hann, and others.

At this time Clayton received a salary of $100 per month, plus room and board at the Observatory.

Early in 1889 Rotch noted that he had spent less than a third of his annual income in the past year; this apparently prompted him to become more active in the stock market. He made many transactions; then, on 26 March, after losing $1,100, he noted, "shall keep to speculating on the weather."

In April Rotch attended a meeting of the board of directors of the Draper Company in New York City. At this time arrangements were made to manufacture the Fergusson weighing precipitation gage. This gage was the predecessor to a weighing gage used widely by the weather services of the United States (see bibliography for 1890).

For the first twenty days in July of 1889, experiments were made in sending time signals from the Observatory by means of magnesium flashes. Former observer Gerrish telephoned the correct time from the Harvard College Observatory; then a flash was set off at 2215 EST most nights. Some flashes were set off at the midpoint of the carriage road when the summit was in cloud. The flashes were reported as having been seen in Lynn, Marblehead, and Princeton, the most distant location, some 67 km (42 mi.) west-northwest of the Observatory.

During the year, short papers were published on the average temperature differences and temperature inversions between the base and the summit, and a set of illustrations showed meteorological phenomena as depicted on the open scale autographic records (see bibliography).

When it came to counting on an autographic chart, the ticks corresponding to the passage of each mile of wind, Clayton became tired, so he bridged the space between the 9th and 10th mile pins in the anemometer to cause a long electrical contact. This arrangement facilitated counting by tens, thus speeding the work. The idea appealed to the Signal Service and they immediately modified all their anemometers. The scheme remains, to this day, on anemometers of that type.

In June Rotch sailed for Europe to attend the Paris International Exposition where he was to serve as a juror to the meteorological exhibit. While in Europe he also visited weather stations throughout France, most of which were located on hilltops or mountains. He met frequently with J. Richard of Richard Freres, meteorological instrument makers in Paris, Teisserenc deBort, and other meteorologists. He made numerous ascents in captive balloons. Temperature, humidity, and pressure observations were made on these ascents. On one occasion he ascended to 700 m (2,300 ft.) in a free balloon as he drifted across France.

Rotch traveled around France and England, seemingly meeting friends and relatives from the United States as if he were in the small confines of Boston. Meanwhile his engagement was broken, but he found time for other American girls who were traveling with their parents. In England he visited Dines and observed his whirling machine that was used to calibrate anemometers. The instrument to be calibrated was mounted at the end of an arm that was rotated in a horizontal plane. Speed through the air at the end of the arm could be calculated easily and compared with the anemometer indication. When the device was used outdoors, days with very light wind movement were chosen. In December, Rotch went on a trip to Algeria with de Bort and a Dr. Raymond. The primary purpose of the trip appears to have been a series of magnetic observations that were made over a large area. Rotch carried a barometer, thermometer, and hygrograph from which he made many observations. The three researchers traveled by train, mule, and camel, camping many nights in tents in desert valleys between high mountains. When clear, it was cold at night and hot by day. These conditions, combined with chronic diarrhea presumably caused by the food or water, caused Rotch great discomfort throughout much of his trip.

1890

Rotch left Algeria for Paris and England on 9 February. While in England he met Margaret Randolph Anderson who was traveling with her parents from Savannah, and became quite infatuated with her. This began an intermittent romance culminating in marriage three years later. On 20 February Rotch departed for the United States with twelve pieces of baggage, and thoughts of soon returning to live in England. Most of this luggage (which was 180 kg, or 400 lbs. overweight) consisted of new instruments that arrived on Blue Hill on 8 March. The instruments were:

Anemo-cinemograph (Richard Freres)
Barograph (Hotlinger)
Thermo-hygrograph (Richard Freres)
Ombrograph (Hotlinger)
Vertical wind anemograph (Richard Freres)
Two Mohn cloud theodolites (Olsen)
Air meter (Richard Freres)
Polaroscope (Cornn)
Anemometer (Lind)

Conjugate black and bright bulb thermometers

Evaporimeter

A Sunshine recorder by Marvin of the United States was added to the inventory.

The anemo-cinemograph, which was invented in France by Richard, 1886-1894, recorded wind speed directly. The instrument sensor consisted of a six-bladed windmill that was kept facing the wind by a split-tail wind vane. Rotation of the "mill" was originally transmitted by electrical contacts to an elaborate recorder that was basically a disk and roller type tachometer. A roller at the end of a shaft was pressed between two disks rotating in opposite directions at constant speed. A worm on the opposite end of the shaft engaged a gear that turned at a rate proportional to the wind speed. Motion from the wind tended to move the shaft and roller out from the center of the disks. When the outward motion was exactly balanced by the rotary motion imparted by the two disks, the roller and shaft were in equilibrium. This position was linked to a pen-arm and chart where wind speed was continuously graphed. Lag response of the windmill was low, but that of the recorder, about two minutes, so the resulting trace represented a smoothed wind speed. This instrument took the place of the pressure plate anemometer devised by Rotch. It was in use at the Observatory, with only a few modifications, until 1959.

The tower and its instruments are shown in figure 21, a photograph taken from the Observatory roof toward the northeast some years later. A weather forecast flag was flying. The split-tail wind vane, on the left, was attached to the Draper anemoscope. The windmill anemometer was attached to the anemo-cinemograph below and is shown in figure 22. The large four-cup Draper anemometer is shown in the background. What appears to be a vertical dipole antenna was part of McAdie's atmospheric potential sensor. Details of the imported ombrograph are not known. It may have been a predecessor to the ombroscope, an instrument that records the duration of rain. The vertical wind anemometer, designed by the Reverend Fr. Dechavrens of China, consisted of a four-bladed windmill mounted on a vertical spindle. Rotation in either direction, which corresponded to the up- or downward vertical component of the wind, was transmitted by electrical contacts. These, in turn, were graphed versus time in such a way as to show the direction and magnitude of the vertical component.

Figure 21. The Observatory tower and instruments looking toward the northeast, about 1890. Copyright President and Fellows of Harvard College. Harvard collection of historical scientific instruments.

The Richard air meter consisted of a delicate windmill several centimeters (inches) in diameter whose shaft ran in fine bearings. Attachments indicated meters of wind passage on a dial and a stopwatch showed the time interval of exposure.

In May Rotch journeyed to Ann Arbor, Michigan, where the *American Meteorological Journal* was published. He notes, "paid share of deficit of the Journal for the year, $218," thus indicating his interest by financial help in furthering the science of meteorology

In the spring the Mohn theodolites, shown in figure 23, were set up for the purpose of determining cloud heights by triangulation.

A few cloud height measurements had been made in United States as early as 1837 (see bibliography, Fergusson 1933). At that time Epsy used data obtained by the Franklin Kite Club of Philadelphia to verify his

Figure 22. Draper anemoscope left, and anemo-cinemograph, right, as set up in the 1900s. Copyright President and Fellows of Harvard College. Harvard collection of historical scientific instruments.

calculations of cloud heights. Pickering (1886) reported that in the spring of 1885 about 300 pairs of simultaneous measurements of clouds were

Figure 23. Mohn cloud theodolite. Copyright President and Fellows of Harvard College. Gordon McKay Library—Blue Hill Collection.

made by Davis and McAdie at Harvard. The stations were connected by telephone, and wooden altazimuth instruments, which apparently allowed the measurement of elevation and azimuth angles, were used. Clouds ranging in height from 610 m (2,000 ft.) to 7,620 m (25,000 ft.) were measured. Systematic measurements from double theodolites were already in progress at Uppsala, Sweden, but no such program was in existence in the United States.

The instruments were first developed by Mohn for the measurement of auroras in Norway and later used to triangulate on clouds. They contained no optics but cross-hairs at the eyepiece and cross-wires in the open end for aligning the direction of sight.

A large field of view was made possible by the construction, as shown in figure 23. The wide field was necessary when viewing clouds and describing them to a distant observer in order to make the sightings as close as possible to the same point. Azimuth and elevation scales with verniers were attached. One theodolite was set up on the roof of the Observatory tower and the other on the roof of the base house at 1793 Canton Avenue; a private phone line connected the two stations. The horizontal base line between stations was 1,178.4 m (3,865 ft.). Since the lines of sight presumably to the same point on a cloud seldom intersected, due to small errors, laborious calculations were required to find the height and position where they were closest. This work was greatly reduced by a plotting machine, which was designed by Clayton and constructed by Fergusson (also shown in figure 24). Two miniature theodolites were positioned in height, latitude, and longitude on a scale proportional to the actual positions of the cloud theodolites. Lines of sight to the cloud were then set up with threads from each miniature theodolite. The point along the lines of sight, or threads, where they were closest could then be determined easily. A plumb was dropped from this point and the three coordinates of the point noted. When successive observations were made on the same cloud, the velocity could be determined from the vector in the x and y plane. A complete discussion of the procedures and error analysis was given by Clayton and Fergusson (see bibliography for 1892).

These cloud observations, which became routine in May 1890, were the first of their kind in the United States and they became the second most extensive compilation of heights, directions, and speeds in the world.

Arthur Kendrick, teacher and director of the meteorological observatory at a school in Leicester, Massachusetts, helped with reduction of the data in the summer.

Figure 24. Cloud plotting machine. Copyright President and Fellows of Harvard College. Gordon McKay Library—Blue Hill Collection.

Also in midsummer a screw-cutting lathe powered by a foot treadle arrived. This valuable asset helped Fergusson immeasurably in building new instruments and making repairs.

On 12 July, Rotch again departed for Europe, with an invitation from Monsieur Vollot to attend the dedication of the Mount Blanc Observatory. The dedication actually took place on the first of August at the cabin of Vollot's in a col at 4,400 m (14,432 ft.) above sea level. Rotch suffered from mountain sickness, but oxygen revived him overnight. The next day they climbed to the 4,810 m (15,777 ft.) high summit and then descended amidst snow and lightning.

In August Rotch attended a meteorological congress at Limoges, France, and in September the International Meteorological Congress at Paris (see bibliography). At these meetings observational meteorology and instrumentation were discussed, but no mention was made of theoretical work. A visit to the Uppsala Observatory in Sweden followed. There Rotch observed the cloud measurement program that was already underway. At Kew Observatory in England he observed Whipple's new method of measuring clouds with photography and a special cloud plotter.

Rotch returned early in December with the following new instruments:

Fineman nephoscope
Registering statoscope
Cinemograph transmitter and two wheels (windmills)
Hair hygrograph
Small aneroid barograph
Clocks and drums
Two tripods for a Crova actinometer

Assmann psychrometer

The hygrograph used a bundle of hair as the humidity sensing element. The stretching and contraction of the hair was transmitted through a cam mechanism to make the motion of the pen arm linear with changes in relative humidity. The statoscope measured pressure change from a prefixed level. At this time Rotch presented Clayton and Fergusson with gold watch chains.

Soon afterward Rotch made one of his rare notes regarding his directorship of the Observatory. At this meeting forms were decided upon for the next five years and metric units were adopted for all measurements except for the surface observations of temperature and precipitation.

Late in December Rotch traveled to Washington to show Marvin, Abbe, and members of the Hydrographic Office the Fineman marine nephoscope, thinking that the instrument would be useful in their programs.

Clayton published a brief paper on cirrus formation, which had been given by Rotch while in London at a Royal Society meeting. In this he illustrated his skillful observation of the formation of filaments and their curving form, but he misinterpreted these as representing large changes of wind speed at the same level, rather than at different levels, through which the ice crystals fell. At year's end Rotch recognized the value of Clayton's and Fergusson's work by giving each of them a $5 a month raise.

1891

The year started with another change in the check observation times to 0800 and 2000 EST and the detailed cloud observations were scheduled for 0800, 1300, and 2000 EST.

Since the road to the summit was not considered private, it was often used by individuals or groups who made the ascent in buggies or sleighs. It was noted that on 11 January the sleighing was particularly good and many sightseers rode to the summit. In July an ascent ended in near tragedy. A horse pulling a buggy up the hill became frightened at the halfway point and ran up without the occupants. The horse then made a complete turn around the thermometer shelter on the summit, then ran all the way back down to Canton Avenue. At that point the buggy was all but destroyed but the horse was merely scratched.

A severe ice storm occurred on 18 January. Excerpts from the journal state that the circumference of ice accumulation on a wire measured 25 cm

(10 in.) and 3.5 g (1/8 oz.); a 30 cm (1 ft.) long twig increased in weight to 340 g (12 oz.), nearly a hundredfold change in weight per unit length.

In March, Rotch delivered the first of three lectures at the Lowell Institute. His subject was "Climatological Elements and Meteorological Stations." His talk was considered very successful. Pickering of Harvard was among the 200 attendees.

In June Rotch was awarded an honorary A.M. degree at the Harvard University commencement exercise.

In the summer, as noted by Koelsch (1985), McAdie returned to Blue Hill with a "wagon load" of apparatus to resume his atmospheric electrical potential measurements.

In midsummer Clayton, followed by Rotch, left for Europe to attend the International Meteorological Congress in Munich. At the conference Rotch raised a question concerning the need for a universal time for reading maximum and minimum thermometers. This poses a problem when these readings are used to compute average temperature because readings at different times result in slightly different averages even when taken over periods of a month. He proposed that cloud observations should pertain to a "definite zone around the zenith" to reduce problems caused by haze or other obstructions near the horizon. Although this proposal was not adopted the method was later employed by Brooks at Blue Hill in the 1931-59 period. Before adjourning, the congress, owing to the successful cloud measurement programs at Uppsala and Blue Hill, decided to start an international cloud measurement program at seventeen stations, one of which was Blue Hill, not later than 1 May 1894.

Rotch stayed in Europe after Clayton's return home. He renewed old acquaintances, expanded the Observatory's publication exchange list, and climbed in Austria.

Harry G. White of the U.S. Department of Agriculture came to the hill in the fall and stayed on through September and October to observe and record bird migrations.

On 16 September 1891, Clayton was appointed "local forecast official" at the Boston office of the newly formed U.S. Weather Bureau. He started in this position one month later by preparing routine forecasts at 2000 EST in Boston while continuing his work at the Observatory during the day. At that time forecasts made by Clayton and Fergusson at the Observatory ceased, since there was no further need to demonstrate that better forecasts could be made utilizing local data, including cloud observations, in conjunction with the usual widespread synoptic data.

In spite of this extra work and travel time, Clayton published an article on the features of the diurnal and annual periods of the measured elements at Blue Hill over the first five years of observations. He also published a summary of the most significant work at the Observatory. In this article the first results of the double theodolite measurements of cloud heights and their velocities were given. Tables of the average and extreme heights for each cloud genera were listed. These data, the first of their kind in the western hemisphere, were compared with the values found at Uppsala, Sweden. In this article Clayton also noted, based on his cloud measurements, that wind speeds at Blue Hill and Mt. Washington were probably higher than those in the "free air" at the same level. This appears to be the first revelation of the well-known effect of mountains or hills on the speed of the wind.

The vertical component wind-measuring equipment that Rotch had brought from Europe was set up in October. The windmill was mounted on a pole 8.5 m (28 ft.) above the tower and 15.5 m (51 ft.) above the ground. Slightly more upward than downward motion was found to prevail, presumably because of the hilltop location where winds were forced up and over the summit. Over the initial November to December period it was noted that upward motion prevailed from 0900 to 1700 EST with downward motion over the remaining portion of the twenty-four hours.

The Gibbons recorder, which contained an electro-magnet and attached pencil to record contacts from an anemometer, always gave trouble because the pencil became dull; this lead to indistinct marks and illegible records. Attempts to alleviate the problem by adding a pen failed because each time the pen arm snapped over against the electro-magnet the ink flew out of the pen. Fergusson cleverly solved this problem by arranging the mechanics so the pen arm moved gently, due only to gravity, when the magnet was activated. This improvement was not only adapted to other Gibbons recorders but later to the Weather Bureau's so-called triple register recorders that made use of several electro-magnets.

In November, Mark W. Harrington, the newly appointed chief of the U.S. Weather Bureau, visited the Observatory. He missed seeing Rotch, who did not return from Europe until mid-December, but was afforded the opportunity of seeing the numerous observing techniques, the library, and the facility in general.

1892

At the start of 1892 the afternoon check observation was changed from 1300 to 1400 EST, while the other observations continued at 0800 and 2000 EST.

Fergusson's brothers, F. K. and W. H., worked as assistants helping with the routine and detailed cloud observations.

In the spring Rotch proposed to Harrington, to no avail, that the Boston station be moved to Blue Hill. He emphasized the excellent site for observations, but it is suspected that he also felt saddled with the routine observational program and would have liked to have another organization take it over, along with publication of the data, thus saving funds and time for additional research.

McAdie came to the hill again in August. He set up electrometers at the base station and Observatory; from these observations he was able to show that the patterns of atmospheric potential were similar at stations separated by this distance of about 1.6 km (1 mi.). He used this finding in his proposal to the Weather Bureau to set up a network of stations from which he could prepare synoptic charts of atmospheric potential and plot movements of the patterns.

In midsummer S. P. Fergusson began extensive comparisons of anemometers in the free air. He used the Dines helicord air meter, that had been calibrated on a whirling machine by the inventor, as a standard. Fergusson constructed a four pen-chart recorder on which he recorded electrical contacts from the anemometers mounted on a steel pipe 8.5 m (28 ft.) above the tower.

Clayton and W. H. Fergusson analyzed sudden temperature changes as recorded at the Observatory. Through a smoothing process and much averaging they claimed to have found periods of six or seven and twenty-six to twenty-seven days. They also noted that after varying lengths of "well-behaved" cycles, they were interrupted only to start off again, but in a different phase. These cycles and their relation to solar rotations intrigued Clayton and later he devoted most of his time to studies of this type on a global scale.

Rotch went to Europe in July to attend scientific meetings in Edinburgh, a Geographical Congress in Italy, and an Aeronautics Congress at Seville, Spain. Again he traveled extensively and found time to climb Ben Nevis in Scotland, Mont Blanc in France, and Monte Cimone in Italy. He returned home in November with additional notes and pictures of these mountain stations and a new Hicks mercurial barometer, number 1019.

In an effort to improve the forecasts, Clayton prepared, from the Blue Hill data, composite charts showing cloudiness in relation to a cyclone center. In his presentation of these data he also noted that dynamic methods had not helped forecasting but they most certainly would aid eventually.

Meanwhile Rotch had promised to increase Clayton's salary if he stayed at the Observatory, and also promised that he would put the base station house in order for Clayton and his future wife when they married in September.

At year's end Rotch retired from the Cadets after nine years of active participation. During this time he regularly attended drill and summer camp at Hingham. Also of interest at year's end, under the influence of Rotch, was the movement of the editorial offices of the *American Meteorological Journal* from Ann Arbor, Michigan, to Boston. Publication of the journal continued into 1896 from Boston, but due to excessive costs it was terminated with the April 1896 issue.

1893

The year started off with a severe southeast gale and rain on 2 January. Gusts of 45 m sec^{-1} (100 mi. hr^{-1}) were measured early in the morning. This storm, that produced the highest winds in the eight years of observations, caused no damage to the instruments or Observatory, indicating that the problems caused by similar storms had finally been conquered. However, on 13 February severe damage did occur. The halyard to the flag became iced and the flag could not be lowered. Under increasing wind and ice the flagpole broke, smashing the wind instruments and pipe supports.

On 29 January Rotch departed for South America. He planned to observe a solar eclipse on 16 April in Peru with Pickering and other scientists. Instruments for recording the surface weather conditions had already been shipped from Blue Hill. Prior to reaching the site of the eclipse, Rotch visited Lima, bought a horse and saddle, and traveled in the mountains. At one point he ascended to 4,450 m (14,600 ft.) above sea level without undue suffering. After the eclipse he continued south, visiting many cities in Chile where he was treated royally through his new friendship with the head of the Chilean Weather Service. He then sailed through the Straits of Magellan to the Falkland Islands, Uruguay, Brazil and northward, finally reaching Boston 26 July.

In the spring, two severe brush fires swept the hills. The first started near Canton Avenue and spread to the Observatory where the wooden shed attached to the west side of the building was saved only by extraordinary efforts. The second fire burned from the halfway point on the road to Chickatawbut Hill, a distance of 4.8 km (3 mi.).

Beginning in August, S. P. Fergusson or his brother was to sleep at the Observatory and take his meals at the base with Mr. and Mrs. Clayton. Frank Brown, the steward, performed the janitorial duties at the Observatory.

In August Clayton, Fergusson, and Rotch presented papers at the Chicago Conference on aerial navigation. Rotch read abstracts of papers by Von Bibber, Hann, and Koppen, that he had translated into English.

In the fall, officers on a French frigate accepted an invitation, that Rotch had issued some time earlier while in Paris, to visit the Observatory.

After a concentrated period of traveling, shopping, and attending social functions, Rotch married Margaret Anderson of Savannah, Georgia, on 22 November 1893.

The records show that Clayton had been suspended from the Weather Bureau in November 1893 and on the 16th of December a public hearing, presided over by Harrington, chief of the Bureau, was held in Boston. The problem can only be surmised from the fact that numerous "business men" were involved, indicating that they may have been unhappy with Clayton's forecasts. Rotch and others testified on behalf of Clayton's scientific abilities and at the end, although no details can be found, Harrington remained friendly with both Rotch and Clayton.

The double theodolite measurements of clouds continued throughout the year to produce a very significant collection of data.

Rotch prepared papers on the highest meteorological station in the world, Charcham, Peru, 5,076 m (16,653 ft.), and the meteorological services of South America (see bibliography).

During the year the Metropolitan Park Commission, constituted by an Act of the Massachusetts Legislature, took the whole of the Blue Hills for a public reservation. The taking included the sixty-seven acres on the summit and slope of Great Blue Hill that were owned by Rotch.

1894

On 1 January Clayton resigned from the Weather Bureau and once again devoted full time to the Observatory. He soon started to publish weekly forecasts, based on periodicities, in the *Norfolk County Gazette*. Later these forecasts appeared in the *Blue Hill Weather Bulletin*.

Rotch was appointed as Visitor in the Department of Physics at Wellesley College on 2 February.

Two notable snowstorms occurred in the winter of 1893-94. On 13 February 34 cm (13.5 in.) fell, making a total of 51 cm (20 in.) on the ground. Once again a 0.6 m (2 ft.) drift of snow was found in the vestibule behind the storm door in the morning. A late storm on 13 April left 30 cm (1 ft.) on the hill with 0.8 m (2-1/2 ft.) drifts on the summit road.

In April Fergusson mounted a new pressure plate anemometer on the tower. Movement of the plate in response to wind gusts was transmitted to the floor below by a chain that moved a pen up and down on a recording chart. This represented a great improvement over the earlier pressure plate where the recording mechanism was inside the tail, necessitating climbing the mast every day to change the chart.

In midsummer the Park Commission reconstructed the lower part of the summit road. The new entrance was located at the junction of Canton and Blue Hill Avenues. This new road joined the old road a short distance up the hill. The old road that started at the great oak, across from Brush Hill Road on Canton Avenue, was walled off.

Rotch, in the process of breaking up the old home at 3 Commonwealth Avenue in Boston, moved much of his book collection to the base house in June. This growing collection contained many old volumes by famous astronomers, physicists, and literary writers.

Mr. and Mrs. Rotch sailed for Europe in June. They visited the north cape of Norway, Sweden, and England. While in Sweden, as a member of the International Meteorological Committee, he worked with Hildebrandsson and deBort on the selection of cloud pictures for the International Cloud Atlas and instructions for the observation of clouds. Batavia and Norway promised to start a double theodolite cloud measurement program beginning 1 May.

1896

On 29 July a violent hailstorm occurred. Stones up to 5 cm (2 in.) in diameter were measured. No damage to the Observatory was mentioned.

At year's end Arthur Sweetland was appointed to assist W. E. Fergusson in the observations and work of reducing the cloud observations.

During the year the Dines pressure tube anemometer was set up. It transmitted the dynamic pressure of the air stream via tubing to a curve shaped manometer. When small pressure changes corresponding to low wind speeds occurred, the water in the manometer still moved a measurable distance in the sloping tube. As pressures increased, they were measured in a portion of the tube that was more nearly vertical. In this way Dines was able to measure low wind speeds as well as the higher speeds.

The anemometer comparisons that were started in 1892 were completed and later reported by Fergusson (see bibliography for 1896). The pole on which the anemometers were exposed is shown in figure 19. It supported one instrument at its top and just below the top a movable crossarm supported anemometers at each end. During tests the arm was positioned perpendicular to the wind, thus avoiding interference between the anemometers. The instruments that were compared are shown in figure 25.

Starting speeds, sensitivities, and comparisons in the free air were determined. A unique method of obtaining the response time or what was called "sensitivity" was employed. It consisted of mounting the anemometers in a windward windowsill of the tower with the roof trapdoor open.

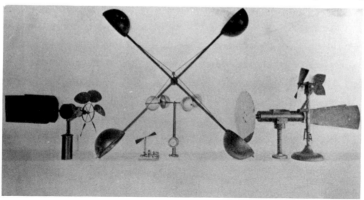

Figure 25. Anemometers compared, 1892-1894. Left to right: cinemograph, helicoid, large 4-cup of Kew, U.S. Weather Bureau 4-cup, Blue Hill pressure plate and brass helicoid. Copyright President and Fellows of Harvard College. Gordon McKay Library—Blue Hill Collection.

The window was quickly opened to allow a sudden inflow of wind on the anemometers and upward through the trapdoor. Times to achieve equilibrium speeds were measured by various ingenious methods contrived by Fergusson. In the case of the cup anemometer, he wound a thread around its spindle and recorded on a fast moving chart the rate at which the thread was unwound as the cups accelerated to reach equilibrium speed. The free air comparisons yielded the degree of over-running, at different average wind speeds, in the natural wind. These were the first measurements of that kind. Included in Fergusson's article were a long list of references, diagrams that showed the instrumental setups, and copies of traces showing gustiness of the natural wind (see bibliography).

The comprehensive observations of clouds that had been made over this period were documented in four summary articles that appear in the annals (see bibliography). Some of these articles were not completed until well after the next chapter in the Observatory's history had started because of the necessary time required for analysis and publication. The first article, by Clayton and Fergusson, published in 1892, summarized the methods and errors of observation and symbols used for the cloud types. A complete table of the actual observations was included.

The second article, by Clayton, and completed in 1896, consisted of 227 pages in 12 chapters plus 15 plates. He included a historical sketch of cloud nomenclature, beginning with the first published classification by Lamarck in 1801. He then discussed the most recent international nomenclature and additions of his own, both of which considered cloud height in the cloud definitions. The importance of height was obvious as the Observatory program of height measurement evolved. Clayton then presented tables showing annual and diurnal periods in cloud amount and wind at different levels as shown by the cloud movements. He also tabulated their relationship to rainfall and to the centers of cyclones and anticyclones.

The composite (figure 26) showing cloud type in relation to a cyclone, is of interest when compared with the Bjerknes model of 1919. Clayton's model shows the first cirrus and cirrostratus 2,400 km (1,500 mi.) ahead and to the northeast of the cyclone center as compared to about 1,500 km (930 mi.) in the Bjerknes model, a difference to be expected in view of the higher winds normally found over New England compared with Europe. Proceeding toward the cyclone center, altostratus and altocumulus were encountered, followed by a brief zone of stratocumulus just before the nimbus, as in the Bjerknes model. To the rear the sequence passed more rapidly from nimbus to stratocumulus and cirrus. Clayton's cross section

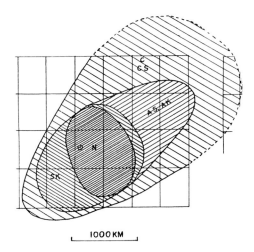

Figure 26. Composite of cloud types in relation to a cyclone center. The grid represents five degree squares at the latitude of Blue Hill. Copyright President and Fellows of Harvard College. Gordon McKay Library—Blue Hill Collection.

model, shown in figure 27, also bears resemblance to the Norwegian model, even to the uncertain lowering of cloud tops near the storm center shown by the dashed line; in fact, the Norweigan model may have copied Clayton's in this detail. Clayton's reason for the low cloud tops in that region was based on a few free balloon ascents that had been made through nimbus. For these ascents to have been successful, the writer postulates that they were made through weak systems and therefore showed the absence of high clouds.

Clayton must have been conscious of the sharp temperature changes associated with frontal passages and clearly revealed by the autographic

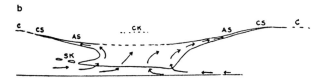

Figure 27. Clouds depicted along a southwest–northeast cross section through a cyclone. Copyright President and Fellows of Harvard College. Gordon McKay Library—Blue Hill Collection.

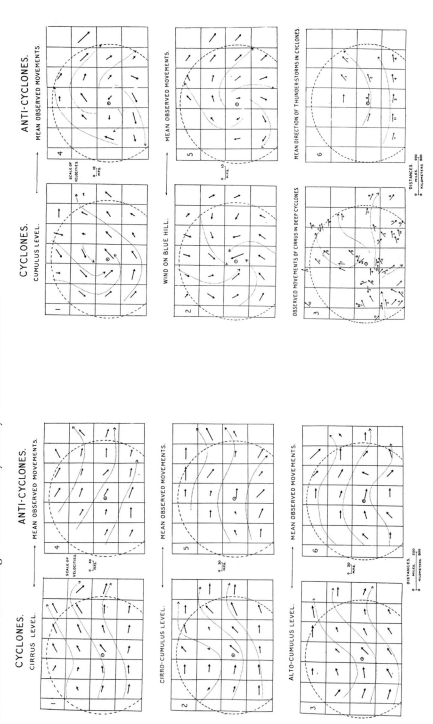

Figures 28a and 28b. Composite of cloud motions at different levels in relation to cyclone and anti-cyclone centers. Copyright President and Fellows of Harvard College. Gordon McKay Library—Blue Hill Collection.

records, but he never tied these to the clouds to derive a model. Therefore, it appears that the development of the Norwegian frontal model had to wait until the advent of synoptic reports in which cloud types were included from a large area.

Clayton's composites of cloud motions in relation to cyclone and anticyclone centers at the cirrus, cirrocumulus, altocumulus, and cumulus levels plus the wind at Blue Hill, (figures 28a and 28b) clearly show the well-known average patterns. These patterns, the first derived for any location in the western hemisphere, compared well with European models except at midlevels, where a somewhat sharper trough over the cyclone and sharper ridge over the anticyclone were depicted over Berlin.

Clayton also showed composites in which the mean motion of the atmosphere, derived from all the cloud motions, was subtracted. At cirrus levels the resulting patterns showed a slight anticyclonic motion above cyclones and cyclonic motion above anticyclones. Such observations and those from mountain tops helped to fuel discussions regarding the theories of cyclones by Ferrel, Hann, and others. Ferrel had postulated that cyclones formed by convective action while anticyclones were strictly dynamic in origin. Clayton, with little theoretical background, postulated that anticyclones formed from cold air at the surface to cause an increase in pressure at low levels and a decrease in pressure aloft, as he illustrated with the cloud motions. He presumed that cyclones were maintained by air that was forced into them from surrounding anticyclones. This work was first published in 1893 (see bibliography), in that paper he also concluded that with increasing height the wind speed increased at a rate proportional to the decrease in the air density. Later, a Frenchman named Engell confirmed the idea and at times it was referred to as the Clayton–Engell law. The idea was not far from the truth when dealing with average winds at temperate latitudes, as was the case at the time of the formulation. However, Clayton stuck with the idea in his book *World Weather* (1923). He then noted that the law only applied up to the level of the highest cirrus or about 11 km (36,000 ft.), and made no mention of its fallacy when it was applied to day to day conditions or over other latitudinal zones of the world.

Clayton was aware of the basic laws of thermodynamics but wanted to test the relationship of surface temperature gradient to wind speed and direction or cloud motion at high levels. This information would help in his daily forecasts. The data were tabulated and tables were published, showing the frequency and amounts of rises and falls of temperature following cirrus from different directions.

Clayton then presented data on the speed of storms and variability of the weather as related to the general flow of the atmosphere. He also presented tables showing the frequency of cirrus bands in different directions and their relation to the wind velocity. Other data were given on the frequency of cloud undulations from each direction.

The final chapter Clayton devoted to the use of cloud observations in forecasting. In this he gave tables showing the frequency of rain following various cloud genera and their motions and rain following the direction of the densest clouds.

The third summary article, by Sweetland in 1897, discussed temperature changes and rainfall probabilities following specific cloud forms.

The fourth article, by Clayton in 1900, included cloud theodolite measurements through 1897. Tables showed the difference between cumulus base heights determined by dew point formula and those measured by double theodolite, the average height of special cloud forms, and other statistics of cloud heights.

Thus the period from 1886 to 1894 saw the introduction and comparison of many new instruments and observational techniques. At the same time a constant practicing effort to improve local forecasting, even for periods of a week, continued. However, the most important scientific work stemmed from the cloud studies that began as a few hourly observations and culminated in remarkably accurate cloud height and motion measurements. These studies were among the most comprehensive in the world and the only ones of their kind in the western hemisphere. They were useful to forecasters and vital to those who followed in developing models of the global circulation. The results of these investigations at Uppsala, Sweden and Blue Hill were so impressive that in 1896 an international series of measures was undertaken by the principal countries of Europe and America.

This work, that required new cloud nomenclature, stimulated, through the efforts of Rotch, the organization of the International Cloud Committee, and, under Clayton's supervision, the first U.S. Cloud Atlas. That volume, *Atlas of Clouds,* was issued in 1897 by the U.S. Hygrographic Office.

By this time the Observatory had established itself as one of the best in observational meteorology both in the United States and abroad. Director Rotch supplied the newest instruments while keeping himself informed on all aspects of meteorology. The new instruments provided new observations, though not necessarily for use in a thought-out research program, but as a consequence the new data that became available were analyzed to form a program, a sequence quite normal for the period. Theories on the formation and maintenance of cyclones and anticyclones came mostly from researchers and observers living abroad attached to government or private organizations. But without trepidation, and with limited knowledge of thermodynamics, Clayton joined the theorists with his new upper air circulations that were revealed by the cloud observations.

References

Clayton, H. H. 1923: *World Weather.* New York: The MacMillan Co.

Davis, W. M. 1888. Local weather predictions. *Amer. Meteor. J.* 4: 409-412.

Espy, J.P. 1837 The Brunswick spout. *Amer. Phil Soc. Trans.* n.s. 5: 421-426.

GAN 1888. Local weather predictions. *Amer. Meteor. J.* 4: 373-376.

Hazen, H. A. 1888. Movement of upper air currents. *J. Franklin. Inst.* 126: 45-52; abstracted in *Amer. Meteor. J.*, 5: 275-276.

Koelsch, W. A. 1985. Ben Franklin's heir: Alexander McAdie and the experimental analysis and forecasting of New England storms, 1884-1892. *New Eng. Quart.*, 59:523-543.

Pickering, E. C. 1886. *40th Annual Report of the Director of the Astronomical Observatory of Harvard College.* Harvard College Observatory. Cambridge, MA.

Chapter VI

Atmospheric Sounding by Kites and Other Kite Experiments: 1894-1904

1894

In July William A. Eddy, a journalist from New York, came to Blue Hill for the purpose of lifting instruments into the lower atmosphere with his kites. Eddy had experimented with kites for some time and by flying kites in series he had succeeded in reaching considerable heights. Eddy (1891) and others had used kites to lift minimum thermometers aloft (see bibliography, Fergusson 1933 and McAdie, *Principles of Aerography, 1917*). The British meteorologist E. D. Archibald (1884) succeeded in measuring average wind speed over periods of about an hour with small Briam anemometers that were lofted by kites (item 18, appendix B) but continuous records had never been obtained. Rotch was aware of Eddy's experiments; in fact, he had tried to visit him in December 1893 to discuss the work. For the occasion of Eddy's visit to Blue Hill Fergusson constructed a lightweight thermograph that could be lifted by a kite. He used the basic Richard thermograph but substituted aluminum or hard rubber for cast iron or brass wherever possible. In this way he made an instrument that only weighed 1.1 kg (2.4 lbs.). A basket was placed over the instrument to screen the temperature sensor from radiation.

On 4 August 1894, using a series of five Malay tailless kites, the thermograph was carried aloft to an altitude of 427 m (1,400 ft.) above the hill as determined by angles measured at two stations 100 m (328 ft.) apart. An excellent record of temperature was obtained. The thermograph and type of kite used for the flight are shown in figure 29. The 4-h continuous recording of temperature obtained from this flight was the first of its kind in the world, marking the start of worldwide pressure, humidity, and sometimes wind soundings of the atmosphere.

The importance of having demonstrated a new method of obtaining quantitative sounding data in the lower atmosphere was recognized by Clayton (1894) in a letter dated 13 August 1894 to the Munn Company, a publishing house in New York. In it he stated that he and Fergusson believed it possible to measure, at relatively little cost, the pressure, temperature, wind, and possibly humidity in future flights. A description

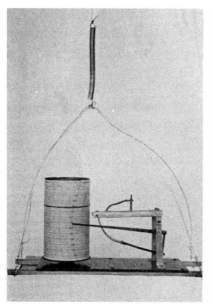

Figure 29. Type of Eddy kite and thermograph used to obtain the first temperature sounding of the atmosphere. Copyright President and Fellows of Harvard College. Gordon McKay Library—Blue Hill Collection.

of the thermograph and method of attaching it to the kite line was enclosed.

Rotch was in Europe at the time and occupied with the cloud commission and the death of his brother, Arthur, back in Massachusetts. For these reasons he apparently did not recognize the significance of the first flight, and made no mention of it. His enthusiasm was shown, however, after his return in September. In a letter to Samuel Cabot, in October Rotch (1894) told of the work in detail and asked Cabot if he would care to cooperate in the project.

Only one other flight was made through the remainder of the year as other events of interest took place. On 10 October 1894 Rotch attended the annual meeting of the MIT Corporation. He was made a member of the visiting committee to the Department of Architecture, and later a trustee of the Museum of Fine Arts; he was granted both memberships in place of his late brother, Arthur.

One afternoon in the fall, after Rotch had descended the hill by foot and reached Readville preparatory to taking the train to Boston, he noticed vivid, fire-like colors in the windows of the Observatory. He hurried back to the base thinking the Observatory was on fire; from there he phoned to learn that all was serene. Brilliant sunset colors reflecting off the Observatory windows caused the alarming sight. This realistic phenomena, often seen several times a year, has been cause for concern by others in the Readville area over the years.

Now that the Observatory staff took their principal meals at the base with Mr. and Mrs. Clayton, shelves were erected in the kitchen of the Observatory. These were used to store the increasing number of foreign meteorological reports. In addition to the seventeen journals received monthly or more frequently, about 200 books and pamphlets were added to the library in 1894.

The year commenced with W. H. Fergusson's departure; Clayton served as meteorologist, S. P. Fergusson was in charge of the instruments, and A. E. Sweetland was placed in charge of the observations. The observational program continued much as before except for the adoption of the International Cloud Classification.

Rotch's inexhaustible vitality continued as he pursued many diversified activities while adding the responsibilities of family life which, at that time, involved the search for a new town house. In March he delivered a lecture before the Boston Scientific Society titled "Methods Employed and Results Attained in the Studies of the Upper Air." He noted that it was received before a "large and appreciative audience." In this lecture, which was published in *Commonwealth* and summarized in the *American Meteorological Journal,* he not only told of mountain observatories, the use of cloud measurements and kites, but he said that the upper air held the key to forecasting, an uncommon concept at the time.

In April, while he was in Washington, D. C., at the National Academy, he had the opportunity to inspect the Weather Bureau's new meteorographs. Also in April he presented a history of meteorographs at the annual meeting of the Society of Arts in Boston. He showed slides of Mount Blanc and El Misti, Peru. Fergusson exhibited his newly constructed meteorograph, shown in figure 30, which Pickering was to set up on El Misti's 5,884 m (19,300 ft.) high summit. This meteorograph was capable of running four months by clockwork without attention. It recorded wind direction and speed, pressure, temperature and relative humidity on a strip chart.

Figure 30. Meteorograph made by S.P. Fergusson for use on El Misti. Copyright President and Fellows of Harvard College. Harvard Collection of historical scientific instruments.

With the taking of land by the Park Commission, Rotch became more concerned about his Observatory. He requested from the Commission a guarantee of its protection for the next five years. Meanwhile he asked for improvements in the road, which eventually took place in the spring. The instruments at the Observatory were appraised, for insurance purposes, at $3,000, and operational costs of the Observatory had now risen to about $4,000 annually. Publication costs totaled $5,000 for the five year period 1887-1891. Rotch paid $1,250 as his share of these costs. Although details were not given, Rotch drew up a codicil to his will on 14 August 1895 which included a $50,000 bequest to Harvard College. Although a graduate of MIT, Rotch probably chose Harvard because of his close association with the college observatory, his appointment as assistant in meteorology at Harvard, and to some degree the superior academic and endowed position of Harvard over MIT.

In the spring he made it a point to attend the Harvard and MIT graduations, renew acquaintances, and inspect new facilities. On 1 August John Clinton, who lived in the valley southwest of Blue Hill, was hired as janitor and Frank Brown's employment ended. Clinton hauled supplies to the Observatory well into the 1930s.

In addition to these activities Rotch spent much time editing and appraising the long reports of Fergusson, Clayton, and Sweetland that culminated in the studies discussed in the preceding chapter. He found time to drive Mrs. Rotch through the hills and to the towns and surrounding points of interest. Occasionally, one time in full moonlight, they climbed the hill from their rented house in nearby Milton. During the year he published two papers, one about the observatory on Mount Cimone, Italy. The other was titled "Mountain Observatories, the Happy Thought," obviously one of Rotch's favorite subjects (see bibliography). His private book collection was significantly enhanced by Benjamin Franklin's "Experiments and Observations on Electricity, made at Philadelphia in America," to which were added letters and papers on philosophical subjects, published in London, 1769.

By midsummer of 1895 experiments with kites were once again underway. In an effort to obtain greater lift while retaining stability, Clayton tried various tailless kites, but improvement was marginal. Other experiments were made with a so-called cellular, or box kite which was designed by Lawrence Hargrave of Sydney, Australia. Meanwhile the base station cloud theodolite had been moved to a rock about 91 m (298 ft.) northeast of the tower. Simultaneous observations from that point and the tower theodolite were used to determine meteorograph altitudes. Eddy returned to the Observatory on 17 August and on that day was successful in obtaining pictures of the ground from a camera that was carried aloft by a kite. The camera could operate from 80 min to 2 h after setting. According to Hart (1982) the first such pictures appear to have been made by Archibald in 1886. At that time small explosions were made to trip the camera shutter. Eddy (1898) presented a summary of early kite flights, photographs, and meteorological measurements from kites. His first aerial photograph from a kite was made near Bayonne, New Jersey, on 30 May 1895. One of these pictures, shown in figure 31, shows the Blue Hill Observatory.

Earlier, Fergusson had constructed a new meteorograph that contained an aneroid pressure sensor and thermometer, thus enabling rough altitude determinations from the pressure recording. This meteorograph was first flown on 19 August 1895. With the aid of seven kites in series along the line and a 16 m sec^{-1} (36 mi. hr^{-1}) wind it ascended about 400 m (1,360 ft.). Pull on the cord reached 45 kg (20 lb.).

In the fall Fergusson constructed another meteorograph in which he included a small cup anemometer (figure 32) whose geared-down revolutions could be recorded on the chart. It flew for the first time on 16 November and was used regularly thereafter.

For the year 1895 a series of twenty-eight ascents were made, with an

Figure 31. "Bird's eye view" of the Observatory taken by W.A. Eddy, 1895. Copyright President and Fellows of Harvard College. Gordon McKay Library—Blue Hill Collection.

average maximum altitude of 510 m (1,675 ft.) above sea level. The highest ascent, 28 August, reached 567 m (1,860 ft.) above the summit. The temperature and wind data, though sparse and obtained through a thin layer of the atmosphere, proved new and interesting, spurring the entire Observatory staff to seek ways of sounding to greater heights.

During the fall of 1895 the Observatory tower roof was completely replaced because the timbers had become rotten and leaks were severe.

In order to place the orifice of the recording rain gage level with the standard gage, a pit or channel was blasted in the rocks in the enclosure in November. The recording gage was then located in this depression, where it remained for many years.

Interesting measurements from an odometer wheel traveling up the summit road from its start were noted in October. The spring was located 1.0 km (0.62 mi.) from the start, "timberline" was noted at a distance of 1.2 km (0.75 mi.) or about 170 m (560 ft.), above sea level, and the Observatory gate at 1.5 km (0.93 mi.). This places the spring near a point above the mid-distance turn, near the top of a steep slope where water frequently runs across the road causing icy conditions in winter. In a letter to the park

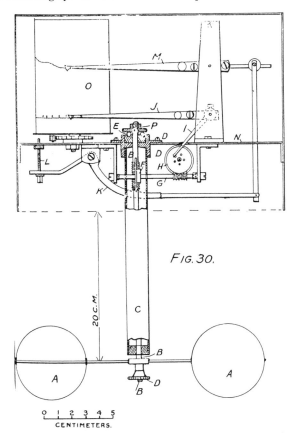

Figure 32. Plan of the Fergusson cup anemometer, which hung below the
meteorograph. Copyright President and Fellows of Harvard College. Gordon McKay Library—Blue Hill Collection.

commissioner dated 1 April Rotch (1895) positioned the spring "200 feet
from the summit, along the summit road." These positions do not agree.
At the time he noted that the spring did not dry up until midsummer and
proposed, to no avail, that it be cleaned out, as he had done several years
earlier, covered with a trap door and a stone arch for the benefit of the
public.

At year's end, Clayton's weekly weather bulletin, which contained
forecasts based on periodicities, was discontinued with the following announcement:

> With this issue, which ends the second volume, it is thought best to
> suspend the publication of the Weather Bulletin. The careful verification
> of the forecasts made by a number of capable and disinterested men seems
> sufficient proof that in the average when a sufficient length of time is taken
> the forecasts are better than chance and are hence based on facts; but at

times the succession of occurrences on which the forecasts are based reverse after the forecasts are made, so that the succeeding weather comes exactly opposite to that expected. This materially interfered with the utility of the forecasts, and might at such times prove injurious. It is hence thought better to suspend publication until this difficulty can be overcome.

This is a fine description of the unchanged problems of cyclic analysis that range from natural phenomena to the stock market.

The year 1895 marked the end of ten complete years of meteorological observations; these were summarized in detail along with the usual data for the year that were published in the annals.

1896

During the year 1896 much progress was made in the kite sounding program. Trials encountered and physical stamina required were severe for work at the hilltop site. Frequent breaks in the kite restraining cord led to the adoption of piano wire in its place. On one occasion winding of the wire under tension built up forces sufficient to break the ends of the reel, necessitating the design and construction of a stronger reel. Most of the flights were made with the Hargrave cellular, or box kites. Some of these kites had their cloth covered with varnish or paraffin for ascents through rain or snow. A combination baro-thermo-hygrograph constructed of aluminum by Richard of Paris was first used on 13 April (figure 33). The old original windlass, which merely consisted of a hand-cranked reel on a box, was replaced by a new windlass mounted on a wheelbarrow (figure 34). Devices were attached to measure the inclination of the wire and the amount reeled out. A dynamometer measured the pull. All of these devices and instruments, with the exception of the Richard meteorograph, were developed by Observatory staff and constructed by Fergusson, truly a remarkable achievement for so few individuals in such a short time.

In spite of these improvements difficulties were often encountered, as exemplified by the account of the 11 March 1896 flight. The original account of the flight is reproduced in figures 35a and 35b.

During the year ascents reached higher and higher and on 8 October an instrument was lifted 2,651 m (8,695 ft.) above the hill to establish a new world record. The plot of this sounding, shown in figure 36, illustrates the detail that could be obtained. Present day radiosonde or aircraft flights would be hard pressed to obtain this degree of detail. From this sounding

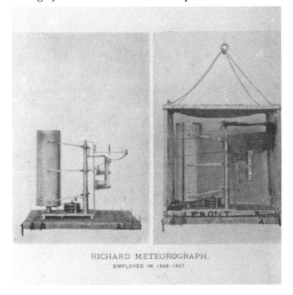

Figure 33. Richard meteorograph used 1896-1897. Copyright President and Fellows of Harvard College. Gordon McKay Library—Blue Hill Collection.

Figure 34. Kite windlass mounted on a wheelbarrow. Copyright President and Fellows of Harvard College. Harvard Collection of historical scientific instruments.

Figures 35a and 35b. Copy of the original account of a kite flight made on March 1896. Copyright President and Fellows of Harvard College. Gordon McKay Library—Blue Hill Collection.

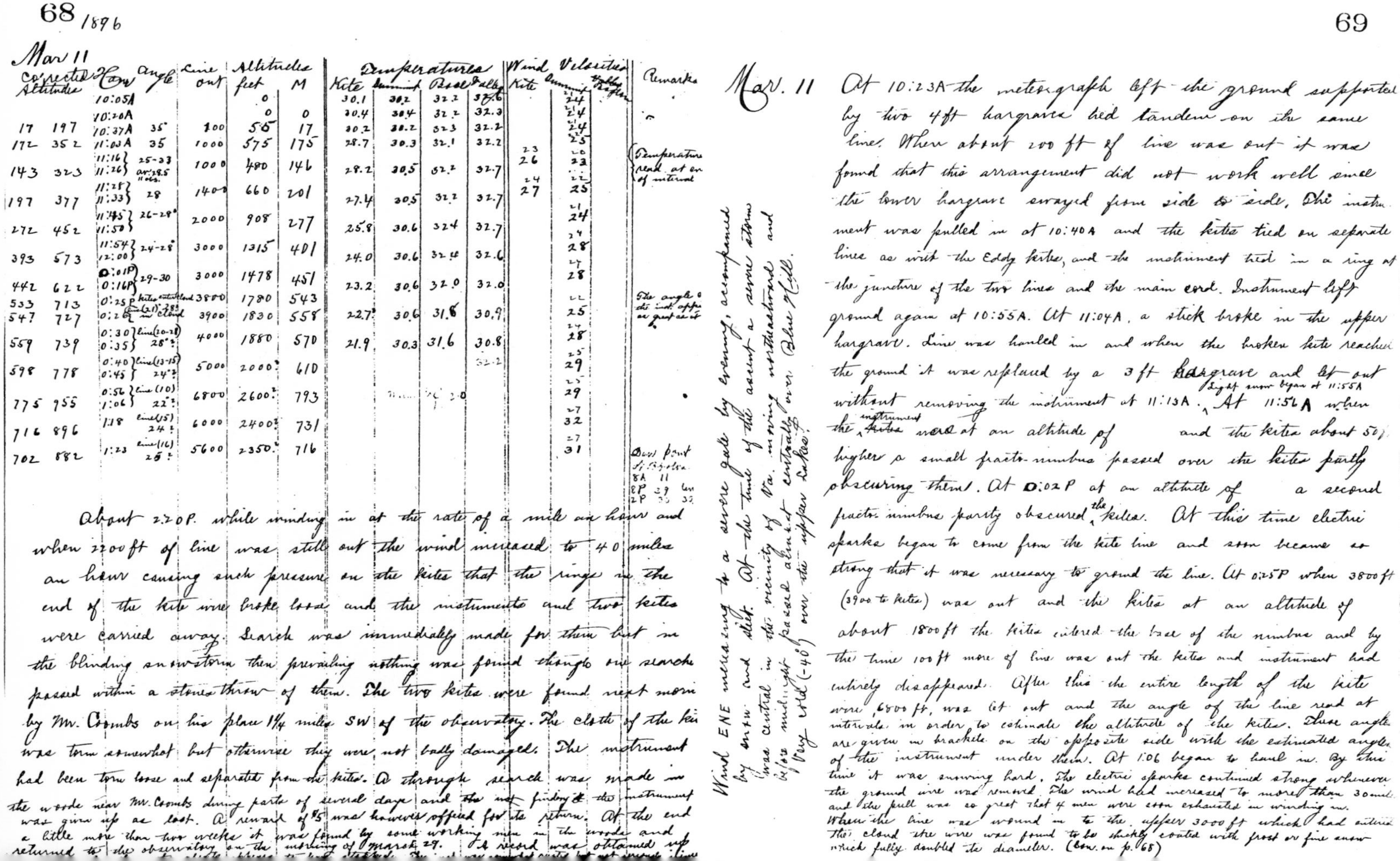

68 1896

Mar 11

Corrected Altitude		Time	Angle	Line out	Altitudes feet	M	Temperatures Kite	Summit	Base	Valley	Wind Velocities Kite	Summit	Valley	Remarks
		10:05A			0		30.1	30.2	32.2	32.6		24		
		10:20A			0	0	30.4	30.4	32.2	32.3		24		
17	197	10:37A	35°	100	55	17	30.2	30.2	32.3	32.2		24		
172	352	11:03A	35	1000	575	175	28.7	30.3	32.1	32.2	23	25		
143	323	11:16 11:26	25-33 avr 28.5	1000	480	146	28.2	30.5	32.2	32.7	26	23		Temperature read at end of interval
197	377	11:28 11:33	28	1400	660	201	27.4	30.5	32.2	32.7	24 27	25		
272	452	11:45 11:50	26-28°	2000	908	277	25.8	30.6	32.4	32.7		24		
393	573	11:54 12:00	24-28	3000	1315	401	24.0	30.6	32.4	32.6		28		
442	622	0:01P 0:16P	29-30	3000	1478	451	23.2	30.6	32.0	32.0		28		The angle of the inst. appear as great as it
533	713	0:25P	kites entered cloud	3800	1780	543								
547	727	0:2[illegible]	line (21) 28.3 in cloud	3900	1830	558	22.7	30.6	31.8	30.9		25		
559	739	0:30 0:35	line (20-28) 28°	4000	1880	570	21.9	30.3	31.6	30.8		28		
598	778	0:40 0:45	line (13-15) 24°	5000	2000	610				32.2		29		
775	955	0:56 1:06	line (10) 22°	6800	2600	793						29		
716	896	1:18	line (15) 24	6000	2400	731						32		
702	882	1:23	line (16) 25	5600	2350	716						31		Dew point ... 8A 11, 8P 39, 2P 33

About 2.20P. while winding in at the rate of a mile an hour and when 2200 ft of line was still out the wind increased to 40 miles an hour causing such pressure on the kites that the rings on the end of the kite wire broke loose and the instruments and two kites were carried away. Search was immediately made for them but in the blinding snowstorm then prevailing nothing was found though our searchers passed within a stones throw of them. The two kites were found next morning by Mr. Coombs on his place 1¼ miles SW of the observatory. The cloth of the kite was torn somewhat but otherwise they were not badly damaged. The instrument had been torn loose and separated from the kites. A through search was made in the woods near Mr. Coombs during parts of several days and the not finding it the instrument was given up as lost. A reward of $5 was however offered for its return. At the end a little more than two weeks it was found by some working men in the woods and returned to the observatory on the morning of March 29. A record was obtained up [illegible]

Wind ENE increasing to a severe gale by evening, accompanied by snow and sleet. At the time of the ascent a severe storm was central in the vicinity of Va. moving northeastward and before midnight passed almost centrally over Blue Hill. Very cold (-40) over the upper lakes.

69

Mar. 11 At 10:23A the meteorgraph left the ground supported by two 4ft hargrave tied tandem on the same line. When about 200 ft of line was out it was found that this arrangement did not work well since the lower hargrave swayed from side to side. The instrument was pulled in at 10:40A and the kites tied on separate lines as with the Eddy kites, and the instrument tied in a ring at the juncture of the two lines and the main cord. Instrument left ground again at 10:55A. At 11:04A, a stick broke in the upper hargrave. Line was hauled in and when the broken kite reached the ground it was replaced by a 3ft hargrave and let out without removing the instrument at 11:13A. (Light snow began at 11:55A) At 11:56A when the instrument were at an altitude of and the kites about 50 higher a small fracto-nimbus passed over the kites partly obscuring them. At 0:02P at an altitude of a second fracto-nimbus partly obscured the kites. At this time electric sparks began to come from the kite line and soon became so strong that it was necessary to ground the line. At 0:25P when 3800ft (3900 to kites) was out and the kites at an altitude of about 1800ft the kites entered the base of the nimbus and by the time 100ft more of line was out the kites and instrument had entirely disappeared. After this the entire length of the kite wire, 6800 ft, was let out and the angle of the line read at intervals in order to estimate the altitude of the kites. These angles are given in brackets on the opposite side with the estimated angles of the instrument under them. At 1:06 began to haul in. By this time it was snowing hard, the electric sparks continued strong whenever the ground wire was removed. The wind had increased to more than 30 miles and the pull was so great that 4 men were soon exhausted in winding in. When the line was wound in to the upper 3000 ft which had entered the cloud the wire was found to be thickly coated with frost or fine snow which fully doubled its diameter. (Con. on p. 68)

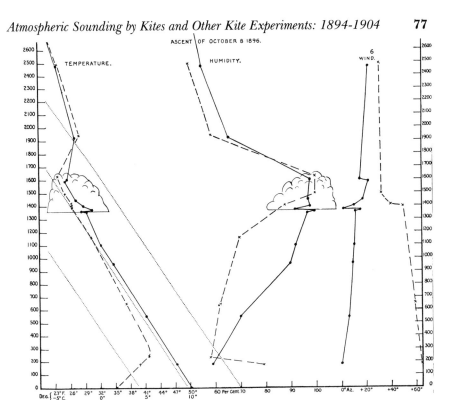

Figure 36. Plot of the 8 October 1896 sounding. Solid line represents ascent, dashed, descent; dotted lines are dry adiabats. Copyright President and Fellows of Harvard College. Gordon McKay Library—Blue Hill Collection.

and others, diurnal temperature curves were developed for different altitudes in the lower atmosphere and characteristic soundings were derived for different quadrants of cyclones and anticyclones. Details of this monumental work, part of which was supported by the Hodgkins Fund of the Smithsonian Institution, were set down by Fergusson and Clayton (see bibliography for 1897).

Also in 1886 Rotch sent a cloud theodolite to the Weather Bureau in Washington so it could be copied for use during the coming "cloud year." He then traveled to Washington to advise the bureau on its kite flying program, which was in the planning stage.

Rotch notes in his diary for 10 January 1896, "Attended meeting of visiting committee of Lawrence Scientific School at 50 State Street, R. S. Peabody, Chairman," and continued with the audacious remark, "advocated making school of higher grade than MIT."

Preparatory to the International Cloud Year, starting on 1 May, in which detailed cloud observations were to be made, a new station for cloud triangulation measurements was set up at the valley station. A new cloud theodolite was placed in service and a private telephone line connected to the Observatory. The new base line was 2,590 m (8,495 ft.) in length, enabling more accurate determination of high cloud heights. Nephoscope observations in conjunction with the cloud height determinations by triangulation provided good wind direction and speed measurements at various levels on many days throughout the following months. Rotch felt this program of such importance that he volunteered to publish similar data from Weather Bureau stations if the bureau could not do so.

A ninety year lease between the Park Commission and Harvard College was finally negotiated in June 1896. The area leased comprised about 1 1/4 acres of the summit on which the Observatory was located. A plot plan, prepared by the commission, is shown in figure 37. Contour elevations above sea level, outcrops of rock, and the positions of the Observatory, thermometer shelter, and rain gages are shown. Provisions of the lease were as follows:

> First. The said College shall have the exclusive use of the building now upon said parcel, subject only to inspection from time to time by said Commission, or other officers or agents of said Commonwealth. Second. Said College shall at all times have the use of all the land herein described outside said building in common with said Commonwealth and said Commission, and the public in general; it being the intention hereof not to limit or interfere with the rights of the public under said chapter four hundred and seven, section four, but so far as not inconsistent with said rights, to enable said College to use all said land, and the building hereby leased, for purposes of scientific study, observation, and research, in connection with said Observatory. Third. The surface of the ground included in said parcel, and all natural objects now thereupon, shall be left in their present state. Fourth. Said College and those connected with its work shall use in going to and from public highways the paths or roads now laid out over the adjoining lands of said Commonwealth, unless and until other paths and roads are laid out and built by said Commission (the right so to do being hereby expressly reserved), such use to be always at their own risk; and they shall not at any time require the establishment or maintenance of any other roads or paths for purposes of ingress and egress, nor any improvement of the existing roads or paths. Fifth. Said College shall never sublease this parcel nor said building, nor any rights hereunder; and should any of the terms, provisions, or agreements of this lease be broken, or the premises be used by said College or any person under it for purposes or in a manner other than those herein prescribed, said Commonwealth by said Commission may forthwith terminate this agreement without notice.

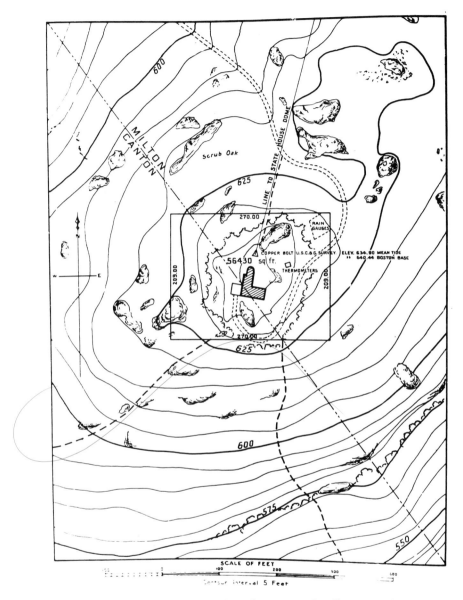

Figure 37. Plot plan showing land leased to Harvard College atop Great Blue Hill, 1896. Copyright President and Fellows of Harvard College. Gordon McKay Library—Blue Hill Collection.

The lease did not leave Rotch without concern. Prior to its execution on 29 May he wrote, "am doubtful whether reservation is sufficient to protect Observatory from encroachment of buildings in the future and so wrote Pickering." (Pickering was then director of the Harvard College Observatory.)

On 14 July Mr. and Mrs. Rotch, with their month-old baby and a nurse, sailed for Europe. In France and Germany he met his old colleagues de Bort, Hildebrandsson, Assmann, Vallot, and others. The weather was unusually wet and Rotch's attempt to climb Mount Blanc with Vallot failed due to wind and snow. In Paris he met Jaubert and discussed, with de Tourville, the use of balloons for sounding the atmosphere. These discussions were continued with Assmann in Berlin.

The primary purpose of his European trip was to attend the International Meteorological Conference at Paris where he represented the Harvard College and Blue Hill Observatories. He submitted a question on how thunderstorms should be counted for climatological purposes (for example, individually, by the hour or day), and participated in further development of weather symbols for international use.

The family returned on 29 October after an unusually stormy ocean voyage.

In the fall Robert DeC. Ward, professor of climatology at Harvard, spent ten days on the hill studying the methods of observation and their reduction.

In October selectmen from Milton and Canton placed a copper bolt in the rock beneath the floor about 30 cm (1 ft.) southwest of the tower. This bolt marked the boundary between the two towns. A brass pin placed in the floor about 46 cm (18 in.) southwest of the center of the tower was noted to have the position: 42° 12' 43.65" N, 71° 06' 52.78" W according to the State Trigonometrical Survey.

The idea of illuminating a spot on a cloud with a vertically directed beam of light and then determining the cloud height trigonometrically after sighting the spot from a known distance was considered at Blue Hill at least as early as 11 November 1896. Clayton (1896) inquired of the New York Calcium Light Company if they could provide or rent a spotlight capable of illuminating a cloud "3-4 miles high." Their response must have been unsatisfactory since such measurements did not follow.

In December Rotch (1896) asked August Reidinger of Germany if he would like to compare his method of lifting instruments into the atmosphere by means of a Parseval kite-balloon with the Blue Hill technique. The comparison, however, was never made at Blue Hill.

During the year both Fergusson and Clayton were called to the Dedham Courthouse to testify in weather-related cases.

1897

Beginning in 1897 the printing and distribution of monthly and annual meteorological data by the Observatory were terminated. Some reports in addition to the Harvard Annals continued to be published in the *Monthly Weather Review* and in the New England Section of the Weather Bureau's *Climate and Crop* series.

At the start of the year it was estimated that over 3,000 people had visited the Observatory in the preceding seven years.

Professor Todd of Amherst, a friend of Rotch with whom he hoped to write a book on meteorology and astronomy, visited the hill in January and promised a searchlight for cloud height measurements at night. However, no record of such measurements has been found.

It is interesting to note that at this early date President Eliot of Harvard and Professor Pickering of the College Observatory became concerned with the dangers of errant kites and the great lengths of kite wire that might be down across the countryside, a problem which eventually did play a role in termination of the work. At the time Rotch (1897) succeeded in dispelling fears of liability.

Much of Rotch's success in business and scientific affairs could be traced to his association with individuals important in these fields. As an example, he notes in his diary for 28 February 1897, "met Prof. Langley and O. Chanute of Chicago, S. Cabot and F. H. Peabody, talked of aeronautics and Chanute showed pictures of his soaring machines." Langley,[1] Secretary of the Smithsonian Institution, administered the Hodgkin's Fund, which helped finance the kite work. Chanute was a pioneer in flying; Cabot and Peabody were Boston financiers.

On 6 March 1897 a third generation kite wire windlass was put into use. This windlass was driven by a two-horsepower kerosene-fired steam engine. It was located just across the summit road south of the enclosure. The apparatus, a Hargrave kite, theodolite, and meteorograph are shown in figure 38. The windlass was modeled after one made by Sir William Thompson that was used for deep sea sounding. The wire was first wound around a strain pulley, thus avoiding the tremendous compressional forces that built up on the reels, and then onto the storage reel. Oil was automat-

---

[1] The solar radiation unit "langley" was named in honor of Professor Langley for his contributions to the knowledge of solar radiation. Later, just after the Wright brothers made their first flight, one of Langley's powered airplanes also made a successful flight.

Figure 38. Steam-driven windlass used in kite soundings beginning in March, 1897. S. P. Fergusson stands beside the shed. Copyright President and Fellows of Harvard College. Gordon McKay Library—Blue Hill Collection.

ically dropped on the wire to prevent rust. When not in use the entire apparatus could be covered by a shed that slid over the platform. Attached doors then closed the open end of the shed.

Later, in March, Rotch went to Boston Light at the entrance to the harbor to observe experiments with bells and whistles as fog signals. He also took the opportunity to make a kite flight there that carried a meteorograph 457 m (1,500 ft.) above the sea. This was the first sounding made away from the Observatory. Shortly afterward he, along with Clayton and Fergusson, visited the cable-laying ship *Mina* to study the cable-laying apparatus. He remarked that it was "just what we want." The significance of this is not clear because the new windlass on the summit had just proved satisfactory; perhaps he was already thinking of kite-flying from a ship at sea, a project which he eventually undertook.

The expediency in raising money for scientific work is beautifully illustrated, at least once, during this early work. On 9 January a proposal was sent to the Smithsonian Institute for constructing twenty kites and instruments to sound the atmosphere to "10,000 feet or higher." (Rotch 1897) The list of materials was brief, and the cost given as $500. Only ten days later a letter from Rotch (1897) acknowledging receipt of the funds was in the mail.

While in Washington in April, Rotch called on Langley and they discussed funds for balloonsonde work. At the time Langley showed Rotch his disassembled "aerodromes," presumably power-driven airplanes.

In reference to a circular received in which a monumental bust of the late Buys-Ballot was proposed Rotch (1897) wrote on 26 April the following to Dr. Snellen, director of the Royal Dutch Meteorological Institute: "I am very glad to be included in the International Committees to carry out the project and will contribute to honor the memory of this amiable and distinguished man, whom I first had the pleasure of meeting in Utrecht in the year 1885. I had the honor of subscribing toward the testimonial rendered him two years later in recognition of which a commemorative medal was sent me."

In a later letter, Rotch (1898) wrote that he "regrettably had received only one subscription, that from Alexander McAdie at the San Francisco Weather Bureau office."

Nevertheless, there were a total of 455 contributors; a bust of Buys-Ballot was cast in bronze by sculptor Pier Pandor and was unveiled on 9 July 1900. The bust now resides in the hall of the Royal Netherlands Meteorological Institute.

On 30 April the International Cloud Year was completed. The theodolite at the valley station and phone line connecting it with the Observatory were removed. A detailed report of the year's cloud data, combined with that obtained in 1890 and 1891, was prepared by Fergusson and Clayton (see bibliography for 1897.) In this excellent summary, data showing measured average and extreme heights for the different cloud genera were given. Other tables showed seasonal and diurnal height variations, maximum cloud speeds, and relationships between genera, height, and direction of motion, and a new comparison of measured cumulus cloud heights compared with those calculated by the dew point formula. These data were basic to cloud climatology for the area.

Through the late spring and summer of 1897 numerous items of interest were recorded. A *New York Herald* reporter came to the hill to observe a kite ascent; Ward brought his female students in climatology to the Observatory; barbed wire was placed around the thermometer shelter to prevent spectator interference; a trip was made to Falmouth, Maine, to inspect a huge kite capable of lifting a boy off the ground, and the Rotches once again rented a home near the north side of the Hill.

Friction continued to surface between the Central Weather Bureau and the Observatory, this time in the form of a letter to Fergusson in which Marvin abused Rotch and the Observatory staff. In an anonymous comment that appeared in the bureau's official journal, *Monthly Weather Review* (1897), it was claimed that the staff magnified their work at the expense of the national service.

In August Rotch attended the (American Association for the Advancement of Science) AAAS meetings at Detroit and the (British Association for the Advancement of Science) BAAS meetings in Toronto.

Experiments in attaching a lantern to a kite at night for sighting purposes were made in September. During an all-night flight the lantern could be seen to a height of about 2 km (6,600 ft.).

On 15 October 1897 a meteorograph was raised 3,571 m (11,713 ft.) above sea level, the highest for the year. The meteorograph was lifted by two kites while other kites supported the wire at distances of 2 km (6,500 ft.) and 3.5 km (11,480 ft.) from the top of the line. A total of 6.3 km (20,660 ft.) of wire was out when the kite reached its highest level; pull ranged from 45 to 68 kg (100 to 150 lb.). Tracings of the temperature, humidity and pressure records are shown in figure 39.

On the following day there were a host of accidents: first, the kite wire fouled a Gurley theodolite knocking it off its stand; then, while reeling in, a splice broke and 3.2 km (2 mi.) of wire and the meteorograph went trailing across the hills. The men went in search of the instrument while Rotch stayed behind to take the observation. In the course of doing so, he dislocated his shoulder, a frequent occurrence, but this time he could not get it back in place. He ran the distance to his summer home and Mrs. Rotch set his shoulder. Rotch apparently worried from time to time about his health and in December he was plagued by palpitations of the heart which his doctors said were not serious.

In an article published by Rotch in which he described the exploration of the free air by kites from Blue Hill (see bibliography) he presented the lapse rates and clouds preceding "warm waves." The information, which was used for forecasting, was also a good description of lapse rates below, through, and above a warm front as described many years later by the Norwegian school of meteorology. Apparently the concept of the warm layer gradually working down to the surface to create a "warm wave" was not understood because it was not discussed.

By year's end Rotch purchased the Denny estate which he had rented for the summer. This estate was located at 1632 Canton Avenue in Milton, just north of Blue Hill.

Lastly, for the year, it is interesting to note that Rotch attended an MIT alumni meeting and afterward discussed a proposed consolidation of

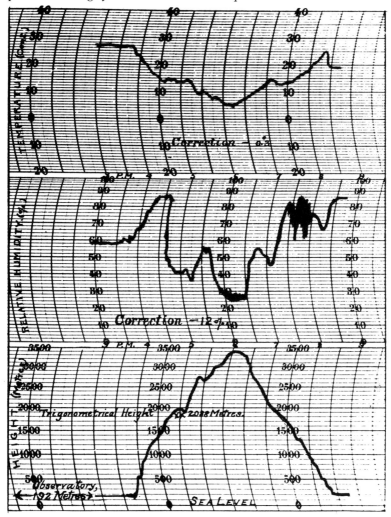

Figure 39. Tracings of kite meteorograph record made on 15 October 1897. Copyright President and Fellows of Harvard College. Gordon McKay Library—Blue Hill Collection.

Harvard and MIT that was under consideration by a Committee.

1898

Beginning on 1 January a new standard of wind measurement was put in use at the Observatory. Based on Fergusson's comparisons of anemometers, it was decided to reduce all previous speeds by 18 percent. Weather Bureau records continued on the old standard for many years.

Also beginning on this date the nephoscope observations were set up to give cloud speeds in m sec^{-1}. This was done by placing the eyepiece 16

2/3 cm above the mirror and following the cloud reflection for 60 seconds. The distance traveled on the mirror in centimeters times the cloud height in kilometers yielded cloud speed in meters per second.

Rotch noted on 9 March, "attended corporation meeting at MIT ... Negotiations with H.U. explained, which came to nothing." This was in reference to the previous year's proposal that MIT become a part of Harvard.

On 14 March Mr. and Mrs. Rotch departed for Strasbourg, France, where Rotch presented a paper at the International Aeronautical Committee meeting. At the meeting a date for international kite flights was discussed. By this time kite flights had been introduced in Europe and the U.S. Weather Bureau had commenced an ambitious program of flights called the "daily atmospheric survey." In an effort to improve their forecasts they attempted daily flights from sixteen locations, striving for a height of about 1,600 m (5,200 ft.) during each flight. Some of these soundings continued until the advent of the airplane sounding.

Rotch visited de Bort again at Trappes and witnessed an impressive balloonsonde which, after recovery some 100 km (62 mi.) to the east–southeast, showed that it had ascended to the 10 km (32,800 ft.) level. With a similar program in mind Rotch made a special visit to Riedinger, in Germany, to determine balloon costs.

While visiting at Utrecht, Rotch proposed the use of kites in Batavia and observed a captive balloonsonde that rose to a height of 2 km (6,560 ft.).

Mr. and Mrs. Rotch returned to Boston on 26 May and they prepared to spend the summer and fall in their new home in Milton. From there he could easily walk to the Observatory.

In August 1898 meetings of the AAAS were held in Cambridge, Rotch presented a paper on progress in the exploration of the air with kites at Blue Hill. About fifty attendees of the meetings came to the hill to observe a kite flight and later Rotch led a group to the White Mountains for climbing and a general outing.

Over the year the kite sounding work at Blue Hill had surged ahead. Thirty-five successful flights had been made to an average height of 2,240 m (7,347 ft.). On 26 August a meteorograph was raised 3,679 m (12,867 ft.) above sea level, the highest for the year. Throughout several days and nights kites were kept in the air, except for the brief periods required to bring them in to change the chart on the recording drum. These data were used to show, for the first time, the extent of diurnal temperature and humidity changes with altitude.

On one occasion, during a northeaster, a flight was made to 1,800 m (5,940 ft.) but when reeling in at night nearly a kilometer of wire, the kites and meteorograph broke loose. The men searched as far away as 8 km (5 mi.) covering an estimated distance of 325 km (200 mi.) on bicycles, without finding the equipment. Two days later the wire was found 5 km (3.1 mi.) to the east and the meteorograph, uninjured, on the Broken Hills, a part of the Blue Hills.

With higher and higher flights the amount of wire reeled out also increased, and on 31 October of that year a record-breaking 7,622 m (4.1 mi.) of wire were reeled out.

In the fall Samuel Cabot proposed to Rotch the use of hot air balloons to lift free meteorographs and by year's end models had been fabricated in his laboratory.

In November Rotch (1898) responded to a letter from V. Bjerknes by saying he was "flattered" by comments regarding the kite flying program. Bjerknes' paper, "Theory of Cyclones," had been received and Bjerknes made reference to Blue Hill reports on the sounding work with kites. This important work by Bjerknes proposed that baroclinicity was the basic source of his "circulation development" or cyclone development. In commenting on this, Kutzbach (1979) notes that Bjerknes had been unable to test his theorem with the use of observational data until the kite sounding observations from Blue Hill appeared in a report by Clayton in 1899 (see bibliography). Bjerknes then assigned J. W. Sandstrom, a student, to analyze the published data which consisted of a series of soundings to about 3 km (9,800 ft.), as a cyclone passed from 21 to 24 September 1898. Baroclinic zones were shown to be present ahead and behind the storm center corresponding to what were later known as frontal zones, to substantiate this landmark work.

Late in December Rotch delivered a series of six lectures at the Lowell Institute. The titles were:

1. "Ancient and Modern Knowledge of the Atmosphere"
2. "Cloud Formation and Classification, Measurements at Blue Hill, the International Observations"
3. "Notable Balloon Ascents and Results Obtained"
4. "Captive Balloons, Balloonsondes for Great Altitudes the International Ascents"
5. "Kites, History and Application to Meteorological Purposes at Blue Hill"

6. "Results of Kite Flights at Blue Hill, Future Work?"

These illustrated lectures were attended by audiences of 100 to 250 persons. Rotch was paid $900 for the lectures, which he delivered without a break during an almost continuous severe toothache.

1899

Cabot's work in developing a hot air balloon capable of lifting a meteorograph continued until the spring of 1899. Attempts were made to lift a balloon having a volume of about 57 m^3 (2,000 ft^3) by heating the air with a "gasoline furnace," a "kerosene furnace", and even burning kerosene and excelsior, but all failed to lift the balloon, let alone a payload, and the project was abandoned.

A boy, Arthur, was born to Mr. and Mrs. Rotch on 1 February 1899. In later years Arthur made many financial commitments to the Observatory and upon several occasions helped to pay the salary of this writer.

In the spring Langley consulted with Fergusson and Rotch regarding the use of recording instruments. He hoped to use these on power-driven model airplanes which he had successfully flown. No further mention of this proposal indicates that these instruments were not developed.

At that time Langley also proposed to Rotch that he make wireless telegraphy experiments in which kites would be used to lift an antenna. About the same time, 2 May, Alexander Graham Bell and Professor Cross of MIT met with Rotch to discuss wireless telegraphy. Bell was anxious to have Rotch try one of his kites, which he described as his "radial kite," to lift the antenna wire. Langley soon made $500 available for tests and Greenleaf Whittier Pickard was placed in charge of the experiments. This was the beginning of a long-lasting relationship between Pickard and the Observatory, which culminated in many significant experiments in radio.

Pickard soon made a spark transmitter which was tested over a 2 km (6,500 ft.) path to the Observatory. The transmitter was set up on the scaffolding for a new tower, which was to be called a "pagoda," on Rotch's estate in Milton. The first test was unsuccessful, but on 16 June signals were received at the Observatory. On 3 July the antenna wire for the transmitter was lifted about 25 m (80 ft.) above the hill by a kite and signals were received on Chickatawbet Hill, 4.8 km (3 mi.) to the east. Signals were then sent between Memorial Hall, at Harvard University and Blue Hill. By year's end signals were sent to Winthrop, a distance of about 18 km (11 mi.) and it was decided that elevating the antenna above the Observatory tower did not significantly improve the signal, in fact the atmospheric potential interfered and presented a danger. Rotch noted that the "wireless telegraphy experiments rivaled Marconi's." Only $100 had been spent on the project.

In August Clayton (1899) wrote to Melville Dewey, founder of the Dewey decimal system for cataloging literature. In this, Clayton presented an extension of the 551.5 category, assigned to meteorology, to include 89 divisions, some of which were subdivided into nine more categories. He had extended Dewey's scheme to facilitate cataloging the growing wealth of material in the Observatory library and asked for Dewey's comments, adding that if a satisfactory extension could be arranged he would advocate its use in other meteorological libraries. Comparison of Clayton's classification with the present day system shows wide differences, so it obviously was not accepted.

Rotch sailed alone for Europe on 7 September for a brief visit to present papers at the British and French Associations for the Advancement of Science. He noted, "reluctant to leave home" and "very nervous." He thought that a pain which annoyed him considerably might be appendicitis and he consulted the ship's physician. Ironically, a ruptured appendix caused his death thirteen years later.

Clayton published a paper in which he related the kite sounding data to their position in cyclones and anticyclones. He also showed time–height charts of changes in temperature and humidity as an anticyclone passed and a cyclone approached. He related the lower and midlevel observations to past high level winds derived from the cloud observations and concluded that low pressure sloped westward with height. He also concluded that "cyclones and anticyclones are but secondary phenomena in great waves of alternately warm and cold air which sweep across the United States." This was a remarkably close description of modern-day "Rossby waves" and their associated cyclones and anticyclones.

Within the year Fergusson published an article summarizing progress with kites at Blue Hill. It is interesting to note that Clayton was aware of basic thermodynamics, but the hypsometric formula for obtaining heights was not used. Instead, when the kites were visible their heights were determined from the length of wire reeled out and the elevation angle of the kites. From earlier triangulation it was found that the length of wire in the calculation should be reduced by 2%–3% due to sagging and other factors, such as turning of the wind with height. When the kites were obscured, pressure and a mean temperature of 0°C for the air column were used to estimate height.

Sweetland published accounts of two severe snowstorms, those of 26–27 November 1898 and 13–14 February 1899. In each of these he showed and discussed the surface synoptic charts.

1900

In the spring Rotch, Clayton, and Fergusson went to Georgia to observe the 28 May 1900 eclipse. Instruments, including a new wind vane and anemometer made by Fergusson, were set up at Washington and Wardsboro, both along the track of totality. Using data from this and other eclipse expeditions, Clayton incorrectly concluded he could explain the usual diurnal oscillation of atmospheric pressure. The eclipse cyclone, shown by the observational data, he called a "cold cyclone" and likened it to the normal nocturnal fall in pressure. The afternoon dip in pressure he conversely attributed to a "warm cyclone." Both of these "cyclones" he believed traveled around the earth in synchronization with the daily cooling and heating cycle to produce the well-known diurnal pressure oscillations.

After experimentation at the Observatory an acetylene generator was installed on 12 July 1900, along with pipes to carry the gas to the shop, office, and each of two bedrooms. One charge of the generator provided enough gas for light throughout the evening.

On 28 July Rotch noted that his pagoda was completed. This structure was a tower at the rear of his property, which he now called "Tower Hills," in Milton. From the tower he could obtain an unobstructed view above the trees. Later Rotch placed recording instruments on the tower. These data were used to represent the 119 m (390 ft.) level when extending the kite soundings down to the valley station level of 18 m (60 ft.) above sea level. It stood until shortly after the 1938 hurricane, which caused damage necessitating its demolition.

Mr. and Mrs. Rotch departed for Europe on 18 August 1900. They cruised the Mediterranean then journeyed to France where Rotch met his old friend, Vallot, for some climbing in the vicinity of Mount Blanc. Rotch presented an up-to-date paper on Blue Hill kite work at the Paris Congress on Meteorology. He visited de Bort's observatory at Trappes where twice-weekly balloonsondes were underway. Successful soundings to about 10 km (33,000 ft.) altitude were becoming routine. Balloon racing and navigation were popular in France at that time; on one occasion he witnessed the launch of seventeen balloons that took off for a designated landing spot. Before leaving Paris he attended a "spectacular" dinner for fifteen given by de Bort, and the evening's conversation was filled with "aeronautics." At a banquet held for Congress attendees Rotch gave an elegant speech on behalf of the foreign visitors. It was received with much applause. He then traveled to the Berlin Aeronautical Observatory where Assmann and others showed their newly completed kite installation. He noted, "the kites were copies from ours."

Soon after the Rotches returned home in October, the Observatory was visited by K. Nakamnoa, director of the Japanese Meteorological Service, and J. W. Gregory, chief of the scientific staff of the English National Antarctic Expedition. Both were interested in the use of kites to obtain sounding data.

During the year Clayton prepared a "second memoir" to his paper on cyclonic and anticyclonic characteristics shown by kites. Tabulated details of temperature, relative humidity, and wind were given for many soundings, and from periods of frequent flights mean values were determined for 100 m (328 ft.) increments of height. A comparison of the observations with the theories of Ferrel and Hann were made with brief reference to the work of Professor Bjerknes in Stockholm.

During the year Rotch combined and expanded the lectures that he had given in 1898 at the Lowell Institute to form a book titled *Sounding the Ocean of Air.* This popular book was published in 1900 (see bibliography).

For the year 1900, the mean height of twenty-five soundings was 2,576 m (8,450 ft.), a slight increase over that of 1899. A new record height of 4,815 m (15,793 ft.) above sea level was attained on 19 July. It is interesting to note that by this time meteorographs could be reeled in and out rapidly, sometimes at rates exceeding the equivalent of 100 m (328 ft.) min^{-1} of vertical height.

1901

At the start of 1901 two new observational programs went into effect. The audibility of a whistle in Hyde Park was noted. This whistle was located 4.3 km (2.7 mi.) northwest of the Observatory and the sound was observed at 0700 and 1700 EST. When steam from the whistle could be seen, the time required for the sound to reach the observer was noted by stopwatch. An estimate of optical refraction was also made by measuring the elevation angle to the base of a building atop Mt. Wachusett, 70 km (44 mi.) to the west–northwest. This observation of the building was later changed to that of Boston Light, enabling nighttime as well as daytime observations.

The unusually active life of Rotch, who was now 40 years old, appears to have caught up with him in the spring. For years he frequently attended lectures on all facets of science and some on other subjects. He administered the Observatory, studied and edited reports by the staff and prepared his own papers. He often spent mornings at his downtown office where he attended to his financial affairs, making stock transactions and arranging rentals and mortgages. Traveling to the Observatory from his Boston home involved a train ride and 3 km (1.9 mi.) walk to Blue Hill; when living in Milton he walked directly up the hill from his home. He attended virtually every scientific meeting of importance both home and abroad and maintained a personal relationship with scores of scientists, mostly in Europe. With Mrs. Rotch he shopped, regularly attended Emmanuel Church in Boston, visited relatives, attended dinners, dances, the theater, and opera. He endeavored to go riding every week. He served as member, director, or trustee to an ever-increasing number of organizations and during this spring period he was overseeing extensive renovation of his Milton home. In February he was first approached to travel to Sumatra to make meteorological observations during the 18 May total eclipse of the sun. He was undecided about the trip for months and finally gave it up after making a gift to the expedition. A trip was then planned, with Fergusson, to Europe, more in the nature of a vacation to recoup his strength than for business. Rotch introduced Fergusson to Teisserenc de Bort at Trappes and they discussed the kite and balloonsonde work. Rotch became fascinated with the idea of flying kites from ships at sea, where the ships' progress could be used to generate adequate wind for launch. Furthermore, no sounding data existed over oceanic areas. He wrote to the Prince of Monaco asking for the use of his yacht to make flights in the tropics but failed to obtain the charter. In Strasbourg, on 4 July, he made an ascent with Hergesell in a balloon filled with illuminating gas (figure 40). In two and one-half hours they ascended to 4.3 km (14,100 ft.) as Rotch recorded temperatures and pressures. An approaching thunderstorm forced a rapid descent; afterward Rotch noted his pulse had risen to 130 beats per minute and he "felt badly." Further discussions with Koppen were held on the exploration of the air with kites from ships and Koppen showed his kites that were to be used by an expedition to Antarctica.

Meanwhile at the Observatory in July a lightning stroke hit the metal roof on the southeast corner of the library. This was the first hit in the

Observatory's history and no damage occurred.

Immediately upon return from Europe, Rotch chartered a tugboat at five dollars per day and sent Fergusson and Sweetland into Massachusetts Bay to attempt trial kite flights. On 22 August 1901, with the aid of a surface wind augmented by the moving tug, the first kite meteorograph soundings over water were made. These were made at a time when the winds alone at sea or on Blue Hill were too light to have lofted kites.

Hurriedly, arrangements were made to attempt flights from a trans-atlantic steamer. Rotch and Sweetland departed from Charlestown, Massachusetts, on 28 August with the old hand windlass, kites and meteorographs. With the 8 m sec^{-1} speed of the vessel, it was possible to fly kites on five of the eight days required for passage. This experiment supplied the first observations of their kind as high as 700 m (2,300 ft.) above the sea, and demonstrated the feasibility of kite meteorograph soundings from ships at sea. Upon landing in England, Rotch, in his enthusiasm about the work, prepared a dispatch for the *Boston Herald* in which he described his experiments.

While in England Rotch presented to the BAAS a paper on the subject of kite-flying at sea. He then went to the Royal Society and looked up future Arctic voyages where vessels might be used as platforms to further kite

Figure 40. Professor Hergesell and Rotch in the basket of a free balloon at Strasbourg, 4 July 1901. American Meteorological Society—Blue Hill Collection.

soundings.

Upon return to America he tried, without success, to interest the Navy in supplying a ship from which kites could be flown in the tropics.

By year's end the library was rapidly filling so Rotch asked for architectural plans which would include a new library, storage space and other changes. Plans were submitted by architects, Tilden and Fox, and permission was obtained from the Park Commission to build, using stone from the hilltop.

During the year ten kite flights were made, attaining an average height of 2,440 m (8,003 ft.) and maximum height of 3,825 m (12,546 ft.) on 7 March. One flight was made with much difficulty, at a time of very low pressure, on 28 January.

1902

On 1 January the audibility observations were terminated. Rotch analyzed the data to compare theory with observation (see bibliography). Other than cases when the wind caused extraneous sound, he found, to no surprise, that the whistle was most easily heard when the wind blew from the sound source. Rotch also showed, using the summit and valley temperature and humidity data, that the sound was better heard during temperature inversions with drier air at the summit.

A severe snowstorm occurred on 17 February, and on the morning of the 18th the observer awoke to find the front door covered by a snowdrift.

During the months of January and February Otto Knopp, formerly of the Aeronautical Observatory of the Royal Prussian Meteorology Institute, assisted in kite construction at the Observatory. Knoff was the first of a long series of foreign students and colleagues that would follow in various positions at the Observatory.

Preparatory to construction of the new library wing and bedrooms, razing the kite shed on the southwest side of the Observatory started 31 March.

In late April Rotch traveled to St. Louis to discuss meteorological programs at the coming Louisianna Purchase Exposition. He suggested balloon races and kite "contests," for which $25,000 was appropriated. On his route home he stopped at Pittsburgh and met Santos Dumont, a Brazilian aeronaut, who had constructed and flown a "dirigible balloon." The man impressed Rotch and was invited to Boston, but he never visited the Observatory.

On 21 May 1902 the valley station was removed from the land of the Metropolitan Park and relocated at the new home of Mrs. Dean, about 0.4 km (1/4 mi.) to the west and about a meter higher. This three-story brick house with a glass-enclosed observatory on its roof still stands at 8 Colchester Street in Readville. At the time of the move open fields surrounded the area. Records continued as before and for irregular periods an anemometer was in operation.

Also in late May Rotch sailed for Europe to attend the International Aeronautical Congress at Berlin. By this time Assman had made numerous balloonsondes to heights of 15-20 km (49,000-66,000 ft.). He was aware of the decrease in lapse rate above 10 km (33,000 ft.); de Bort had made similar flights and he too noted the base of the generally isothermal layer at 13-14 km (43,000-46,000 ft.) in summer and lower in winter. He also found the base of the layer to be lower over surface low pressure systems than over areas of high pressure. No mention was made of the words "tropopause" and "stratosphere," but de Bort is credited with these terms at a later time. A photo of de Bort and Rotch at Trappes is shown in figure 41. One can almost read in the expression of the Frenchman his often repeated quizzical comment: "My wife (i.e., his observatory) takes all my money, but I do not care." (McAdie 1934).

In the summer Pickard returned to the Observatory to make measurements of atmospheric electricity and carbon dioxide.

Rotch continued his efforts to gain support for kite flying at sea. He applied for a $10,000 grant from the Carnegie Foundation for flights in the trade winds. This was backed up by a lecture at the Carnegie Institute on "the new meteorology." He later lowered the application amount to $5,000, but this was eventually refused. In an article in the *Quarterly Journal of the Royal Meteorological Society* (see bibliography) Rotch described the flights from a tugboat and steamer and outlined the costs of flights to 3 km (9,800 ft.). He estimated 160 pounds would cover the costs of two meteorographs, 18 km (59,000 ft.) of wire, and five kites. He gave up independently underwriting a trip to the tropics since an 800-ton steamer would cost $3,500 per month plus the cost of coal. He tried to interest the Hydrographic Office in studies of sea fog, in which kites would be used, off the coast of Maine and Nova Scotia, but without success. His only satisfaction came, in the form of an approval of the trade wind project, from the International Aeronautical Congress at Berlin.

In November 1902 alterations and additions to the Observatory were complete. The new library addition and storage room below measured 4.6

Figure 41. M. Leon Teisserenc de Bort and Professor A. Lawrence Rotch
at Trappes, c. 1902. A. McAdie. Bull. Amer. Meteor. Soc., 15, p. 174, 1934.

m x 8.5 m (15 x 28 ft.). The rooms on each level were fireproof. The lower
room was capable of storing six kites. Two new bedrooms extended to the
south; one former bedroom was made into a study for the director and the

other bedroom was made into a new hall. The tin roof was replaced by copper. A hot water heating system was installed throughout the building and a much-needed bathroom provided. An interior view of the tower first floor room with Rotch seated at the observer's desk is shown in figure 42. The instruments, their cases, and furniture were those that continued in use until about 1959.

Rotch's appointment as assistant in meteorology at Harvard was renewed in 1902 and it continued until his honorary professorship in 1906.

The routine observations yielded continuous records of atmospheric pressure, air temperature, relative humidity, wind direction and speed, precipitation, bright sunshine, and cloudiness at night in the vicinity of the pole star.

Over the year thirteen kite flights were made, mostly on "international days." Average height attained was 2,420 m (7,938 ft.) and on 5-6 February an overnight record was obtained at the 3-km (9,840-ft.) level. This flight, which reached 4,286 m (14,058 ft.) was the highest for the year. Flights on some of the international days could not be made due to wind conditions. A wind of about 6 m sec^{-1} (13 mi. hr^{-1}) but not over 16 m sec^{-1} (36 mi. hr^{-1}) was required for launch. Breakaways continued in spite of the use of piano wire having tensile strengths of 110-230 kg (240-506 lb.). One meteoro-

Figure 42. Interior view of first floor room of the tower, with Rotch seated at the desk. Draper weighing barograph at left, weekly barograph center, and standard barometers right. December, 1907. Copyright President and Fellows of Harvard College. Harvard collection of historical scientific instruments.

graph and kite were found 21 km (13 mi.) away and two kites and instruments were lost at sea in spite of a search by launch in Boston Harbor. These flights may have reached the 5-6 km (16,000-20,000 ft.) level, judging from the length of wire out and its inclination angle. This compares with a record height of 5-6 km (16,000-20,000 ft.) made by one of Dr. Assman's kites in December of this year.

During the year Sweetland assembled the fifty-year temperature record made at nearby Milton Center. From a four-year overlap with Blue Hill he adjusted the data to Blue Hill to form a record extending back to 1849. This adjustment was later refined by the present author (see bibliography). During the year brilliant red sunsets were frequently described by the observers. Clayton attributed these to a volcanic eruption in Martinique (see bibliography).

At year's end Rotch traveled to Washington to speak at the AAAS meetings. The new year was seen in at a gala party put on by Alexander Graham Bell for twenty distinguished guests.

1903

Red sunsets and whitish haze around the sun continued into 1903. This dust probably contributed to the darkness of the afternoon of 6 January, when with smoke, rain and snow at 1500 EST it was as dark as at 1800 EST. Again on 21 June, the longest day of the year, it was dark at 1700 EST, with heavy rain.

In April and May the base thermometer shelter was moved 30 m (98 ft.) and the rain gage 20 m (66 ft.) both northward, but with no change in exposure.

Also in the spring Rotch met Oliver Fassig, who had developed a "trace rain gage." Later the instrument was modified by Fergusson and tested on Blue Hill. In 1904 the instrument, called an ombroscope, was put in regular use. A clock-driven drum, on which a chart was placed, rotated beneath a small hole in a metal cover. Cross hatching lines and a time scale were printed on the chart in printer's ink which ran when wet. In this way a record of the duration of scattered or small precipitation droplets was obtained. The instrument is shown in figure 43 and a sample rain record in figure 44.

On 8 May 1903 Sweetland died from tuberculosis. He was thirty years old and had been associated with the Observatory since 1896. Rotch wrote that he was "a conscientious and accurate observer, always devoted to the

Figure 43. Ombroscope with cover open. Photo c. 1950. Courtesy J.H. Conover.

interests of the Observatory." Lewis H. Wells, who had worked part-time as an assistant over the past nine years, was engaged to take Sweetland's place at $75 per month.

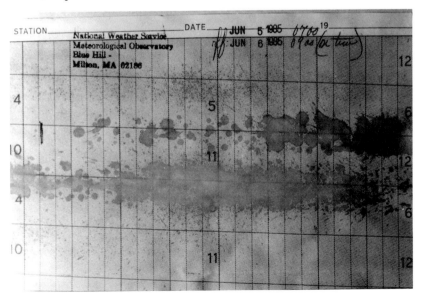

Figure 44. Rain record from the ombroscope. Chart drum rotates four times every 24 hours. Each vertical line represents 5 minutes. Courtesy J. H. Conover.

By midsummer the library had been furnished with steel shelving capable of holding about 5,000 volumes. Thin tile vaulted the room. This type of masonry vaulting originated with the Guastavino family in Spain. According to Frank Creedon and Bob Ralston, architectural historians who have studied the Observatory, the family came to America near the turn of the century and obtained exclusive patents for their method of construction. It consisted of thin tiles, usually in two layers, that were mortared together to form an unusually strong, arched and fireproof ceiling. The Observatory ceiling was most likely laid by subcontractors who were expert in that particular type of construction. Oak-topped tables with shelves below provided space to spread out maps and store oversized volumes. In the corners of the interior frieze Rotch had placed copies of eight bas-reliefs representing the allegorical figures of the winds that were on the first century B.C. Tower of the Winds in Athens. The library is shown in figure 45. The director's roll-top desk was used there by two succeeding directors. It now resides in the offices of the American Meteorological Society in Boston. At this time the library contained about 8,000 books and pamphlets. Pamphlets and climatic data were housed in the old, east wing library. Invitations (figure 46) were sent out for the official opening of the new library on 26 June 1903. Mrs. Rotch served tea at the reception, which was held in the new room. About sixty persons, including many neighbors from the surrounding area, came by foot and carriage to the opening.

In late July the Rotches once again sailed for Europe. Rotch visited the German Aeronautical Observatory and met Count Zeppelin, who asked for help in constructing an airship for the St. Louis Exposition. Rotch then went to Chamanieux and climbed to Vallot's Observatory near the peak of Mount Blanc. On the second attempt he succeeded in reaching the summit (see bibliography for 1904). The family then traveled to England where Rotch spoke at the BAAS meetings. They returned to Milton 1 October.

Meanwhile a trolley line was under construction from Mattapan, north of Blue Hill, to the base of the hill and points south. After its opening on an especially fine Sunday an estimated 8,000 people climbed Blue Hill. This prompted the Park Commission to construct a new path leading up the hill from what is now the large parking lot north of the Trailside Museum on Blue Hill Avenue, or State Highway Route 138. A drinking fountain was placed at the start of the wide path and the Blue Hill Street Railway Company constructed a platform, waiting room, and rest rooms at the base of the hill. A sample trolley schedule, the company's route, and connecting lines are shown in figure 47. On the summit of Great Blue Hill

Figure 45. New library in the west wing, c. 1903. Copyright President and Fellows of Harvard College. Gordon McKay Library—Blue Hill Collection.

the scrub oak which had come up over the past ten to twenty years was cut to improve the view and a circumferencial path was made around the top of the hill. On the northeast side a stone bridge was constructed over a small ravine and dedicated to Charles Eliot, son of President Charles Eliot and collaborator with Frederick Olmsted in the planning of the metropolitan park system.

On 29 December 1903 Rotch journeyed to the AAAS meetings in St. Louis. Plans for the coming exposition, including aero competitions and balloon soundings, were discussed.

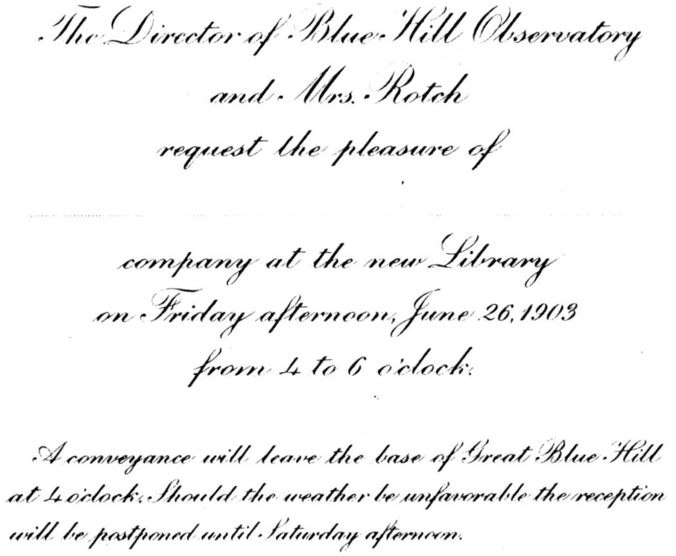

The Director of Blue-Hill Observatory

and Mrs. Rotch

request the pleasure of

company at the new Library
on Friday afternoon, June 26, 1903
from 4 to 6 o'clock.

A conveyance will leave the base of Great Blue Hill
at 4 o'clock. Should the weather be unfavorable the reception
will be postponed until Saturday afternoon.

Figure 46. Invitation sent to friends announcing the opening of the new library. Courtesy Rotch family.

1904

January was snowy and cold. By the ninth a snow cover of 64 cm (25 in.) was recorded. Rotch used snowshoes regularly for his trips up the hill and he notes first use of the "snow shoe" path which led from the summit

Figure 47. Trolley schedule and line routes, 1904. Courtesy Rotch family.

northwestward to the base near the present Trailside Museum. A temperature of -35.6°C (-32°F), the lowest ever recorded at the valley station, occurred on 5 January.

In late January Rotch lectured on one of his favorite subjects to an audience of 200 at the Appalachian Mountain Club. His illustrated talk was titled "Five Ascents to the Observatories on Mount Blanc" (see bibliography).

Crowds visiting the hill increased now that the trolley line made access to the base easy. On Sunday, 22 May, the number of visitors to the summit was estimated at 1800. These crowds prompted Rotch to consider surrounding the Observatory with a wall or fence.

About this time Rotch had the Casella Instrument Company of London make a device of his design for determining the true wind from a moving vessel. The instrument consisted of three wooden rules which were graduated in units of speed. In use, two of the rules, which were attached at one end, were adjusted as vectors to represent the ship's direction and speed and the other the observed wind; the connecting rule then indicated the true wind direction and speed.

Rotch showed the instrument to the Hydrographic Office with the hope that it would be used on naval and merchant vessels to obtain better observations, but little interest was shown.

Later, in a letter to McAdie, Rotch (1909) told how the man to whom he had shown the instrument, now employed by the Weather Bureau, "pirated" the device. According to Rotch it was described in the 1907 annual report of the Bureau's as their own invention. Since the instrument was not patented Rotch let the matter pass.

Other notes mention auto races that were now being held on the old trotting track in Readville.

On 11 August 1904, Fergusson put a new pole star recorder into operation. Use of the old Pickering records, which required a new glass plate every night, was terminated in favor of the new model which would accommodate seven nights on a single roll of film. The star's trace on film was magnified to a circle 7.5 cm (2.9 in.) in diameter, which made reduction of the data much easier.

In the summer the Rotch family once again went to Europe. Rotch traveled on alone to St. Petersburg, where the International Committee for Scientific Aeronautics met. While there he delivered a paper in French titled "La Temperature de l'Air dans les Cyclones et Anticyclones d'Apres les Observations a l'Aide de Cerfs-volants a Blue Hill Observatoire."

Rotch arrived home on 17 September and immediately departed for the exposition in St. Louis.

It was noted at the end of 1904 that publications of the Observatory were sent to 320 institutions or individuals interested in meteorology.

For this history, the interval of kite-flying experimentation at Blue Hill, which started in 1894, has been arbitrarily terminated in 1904 because the instrumental development and flying technology was perfected and a new method of sounding the atmosphere was begun at the Observatory. However, soundings by means of kites continued for many years, mostly on designated "international days," and publication of the data and their analysis continued. Thorough documentation was made by Clayton in three reports dated 1904, 1909, and 1911 (see bibliography).

In the 1904 report on diurnal and annual periods of temperature, humidity, and wind speed up to 4 km (13,000 ft.), Clayton first presented an excellent error analysis of the data. In the course of this, he showed that the summit shelter temperature was very close to that of the free air during the day but at night it was often lower. The nighttime difference was attributed to the adiabatic lift of air over the hill from a surrounding stable layer, rather than directly from radiation. From the ascensions of long duration, chiefly at Blue Hill, came the discoveries that the diurnal change of temperature, found at all heights on mountain stations, disappears at a height of about 1 km in the free air and that the temperature is generally lower and the wind speed higher on mountains than in the free air. Tables and graphs gave vertical gradients and actual values of temperature, relative humidity, and wind speed at different times of the day. Monthly values at different heights illustrated the annual periods of the meteorological elements.

In Clayton's 1909 paper he presented the distribution of temperature, relative humidity, pressure, and wind to the 3 km (9,800 ft.) level around cyclones and anticyclones. Graphical composites showed these distributions as departures from normal.

The 1911 paper was a study of clouds in the light of the kite-sounding data. Cloud levels and form were shown on individual soundings and a few time–height cross sections showed cloud changes and associated temperature–humidity changes. These studies were amazingly detailed, and were comparable to recent work except for their limited vertical extent of about 3.5 km (11,500 ft.).

This period of atmospheric sounding by means of kites provided the first routine measurements anywhere in the world, of temperature, humidity and wind in the lower atmosphere. The period was undoubtedly one of the most inventive and productive in the history of the Observatory. Soon after the first successful soundings, the technique was deemed important enough to warrant its initiation at stations abroad and in the United States.

Kite-flying activity through the years is summarized in table 1. The large increase of flights in 1896 occurred with the introduction of piano wire to restrain the kites. The decrease in 1901 accompanied the introduction of scheduled flights on "international days."

TABLE 1

Year	Number of Flights	Average Height (m MSL)*	Maximum Height (m MSL)*	Date of Maximum Height
1894	2	567	619	4 August
1895	28	510	759	28 August
1896	86	845	2843	8 October
1897	38	1,450	3,571	15 October
1898	35	2,240	3,679	26 August
1899	25	2,256	3,793	28 February
1900	25	2,576	4,815	19 July
1901	10	2,440	3,825	7 March
1902	13	2,420	4,286	6 February
1903	15	2,214	4,258	1 October
1904	14	2,300	4,468	4 March
1905	16	2,120	3,372	30 August
1906	12	2,214	3,618	5 April
1907	17	1,710	3,676	6 September
1908	22	1,930	3,570	3 September
1909	20	2,074	3,408	30 June
1910	8	2,217	2,611	3 February
1911	12		3,332	
1912	15		3,585	
1913	3		3,210	
1914	9		1,207	

*Launch Level was 192 m above sea level.

References

Anonymous. 1897. Further comment is unnecessary. *Mon. Wea. Rev.* 25: 166.

Archibald, E. D. 1884. An account of some preliminary experiments with Biram's anemometer attached to kite strings or wires. *Nature*. 31: 66-68.

Clayton, H.H. 1894 Author index and copies of letters. 1893-95. p. 205-206. Blue Hill Collection, Gordon McKay Lib., Harvard University.

——1896: Author index and copies of letters. 1893-95. p. 211. Blue Hill Collection, Gordon McKay Lib., Harvard University.

——1899: Author index and copies of letters. 1898-01. p. 185 and 353-358. Blue Hill Collection, Gordon McKay Lib., Harvard University.

Eddy, W. A. 1891. Meteorological kite flying. *Amer. Meteor. J.* 8: 122-125.

——1898. A record of some kite experiments. *Mon. Wea. Rev.* 26, 450-452.

Hart, C. 1982. Kites, an historical survey. *Paul P. Appel,* Mount Vernon, N. Y. P 171.

Kutzbach, G. 1979. The thermal theory of cyclones. A history of meteorological thought in the nineteenth century. *Amer. Meteor Soc.* 159-171.

McAdie, A. 1934. The discovery of the stratosphere. *Bull. Amer. Meteor. Soc.,* 15, 174-177.

Rotch, A.L. 1894: Author Index and copies of letters. 1893-95. p. 240. Blue Hill Collection, Gordon McKay Lib., Harvard University.

——1895: Author Index and copies of letters. 1893-95. p. 340. Blue Hill Collection, Gordon McKay Lib., Harvard University.

——1896: Author Index and copies of letters. 1896-98. p. 234. Blue Hill Collection, Gordon McKay Lib., Harvard University.

——1897: Author Index and copies of letters. 1896-98. p. 256. Blue Hill Collection, Gordon McKay Lib., Harvard University.

——1897: Author Index and copies of letters. 1896-98. p. 248 and p. 265. Blue Hill Collection, Gordon McKay Lib., Harvard University.

——1897: Author Index and copies of letters. 1896-98. p. 326. Blue Hill Collection, Gordon McKay Lib., Harvard University.

——1898: Author Index and copies of letters. 1896-98. p. 465. Blue Hill Collection, Gordon McKay Lib., Harvard University.

——1898: Author Index and copies of letters. 1898-01. p. 69. Blue Hill Collection, Gordon McKay Lib., Harvard University.

——1909: Author Index and copies of letters. 1905-09. p. 386. Blue Hill Collection, Gordon McKay Lib., Harvard University.

Chapter VII

Balloonsondes: 1904–1913

1904

The use of balloons to sound the atmosphere dates back about 200 years. At the suggestion of Cleveland Abbe of the U. S. Signal Service, S. A. King made, between 1870 and 1885, what are probably the first manned ascents with instruments in the United States (see bibliography for Fergusson 1933). These were followed with ascents in 1885 by W. H. Hammon and with others by H. A. Hazen in 1887, both under the auspices of the Signal Service. Hazen appears to have been the first to use a sling psychrometer during the ascents, and the height of 5,000 m (16,400 ft.) attained by him stood as a record for many years.

The use of balloons to carry registering apparatus for the study of the atmosphere was suggested by an unknown writer as early as 1809 according to Clayton and Fergusson (see bibliography 1909). On 21 March 1893, for the first time a special instrument, recording the time, pressure, and temperature, was successfully flown and retrieved by M. M. Hermite. de Bort followed with successful soundings in 1898; he devised a paper balloon that was enclosed in a net, thereby restraining the balloon from the expanding forces of the gas as it rose to lower ambient pressures. On 18 December 1893, at St. Louis, Hazen filled a large balloon with coal gas and lifted a barograph and thermograph. The balloon was found 150 km (93 mi.) from St. Louis, but according to Fergusson no further knowledge of the experiment was ever discovered. Later, in 1901, Assmann, of the Royal Prussian Astronautical Observatory, succeeded in making expandable rubber balloons that he filled with hydrogen to lift de Bort's recording instruments.

Over this period in which the use of unmanned balloons was perfected, the work at Blue Hill was primarily concerned with furthering the use of kites to sound the atmosphere. However, Rotch was well aware of the work in Europe and as a member of the International Aeronautical Committee he wrote to Assmann on 9 January 1899 saying that he was considering the use of balloonsondes (Rotch 1897). At that time he asked about the cost of balloons, an aspirated meteorograph, and the use of coal gas for filling the balloons. He also wrote to Hazen at the U.S. Weather Bureau noting that kites were "destined to come into general use in meteorological observations," but at the same time he asked if balloons could be successful in the eastern United States (Rotch 1897). In a letter to Hann, dated 27 March 1897 (Rotch 1897), he revealed that "I have been writing that some balloonsondes might be attempted in America." In that letter he asked Hahn if he thought it worth the trouble and expense to attempt such soundings, adding that the balloons would have to be "liberated several hundred miles to the west of Blue Hill," to avoid their falling into the ocean.

Rotch seized upon the opportunity to inaugurate balloonsondes at the time of the St. Louis Louisiana Purchase Exposition. The location was ideal—far from the ocean and surrounded by open farmland where the instruments would likely be found soon after descent. The exposition appropriated $2,500, which was used by Rotch to purchase balloons and instruments. The expandable rubber balloons, which had been developed by Assmann, were manufactured in Germany; de Bort furnished the instruments at cost.

On 15 September 1904 S. P. Fergusson, working with very little assistance, prepared for the first ascent at St. Louis. The balloon was filled and a parachute dropped over its top, but the hydrogen gas was so inferior that the instrument could not be lifted. In frustration he removed the parachute, tied the instrument to the neck of the balloon, and let it go at 1633 local time. The launch site, balloon, and instrument are shown in figure 48. These instruments were surrounded by an open wicker frame for protection upon impact. They contained a notification of their purpose and a small reward for immediate return (figure 49). In this case the instrument was found intact 80 km (50 mi.) from St. Louis in spite of its fall with no parachute and only the ruptured balloon acting as a drag chute. In one hour and twenty-six minutes a perfect record of pressure and temperature had been obtained from the ground to the 17,045 m (55,908 ft.) level and returned.

This marked the first successful high-altitude balloonsonde in America.

With the entire Observatory staff taking turns at the work, two additional series of soundings were made in 1904. Three more ascensions were made in September; one failed in October and five succeeded in November and December. The last set of balloons and instruments were paid for by Rotch. Clayton holding one of these balloons and the meteorograph is shown in figure 50.

Figure 48. First Blue Hill Observatory balloonsonde launch at St. Louis, 15 September 1904. Copyright President and Fellows of Harvard College. Harvard collection of historical scientific instruments.

1905

After the success of these balloonsondes, Rotch, in February 1905, applied for and received $1,000 from the Smithsonian Hodgkins Fund for the purchase of additional sondes.

In January, February, and March 1905, Rotch made climbing trips to the White Mountains as leader of Appalachian Mountain Club outings. Wells accompanied him on some of these trips. On the March trip a party climbed to the halfway house on Mount Washington where they stayed overnight. The following day Rotch estimated the wind on the summit from cloud measurements and decided not to attempt the climb, although three men and one woman, with difficulty, succeeded in reaching the summit. He made temperature observations on these trips, using a small barometer to establish the height, and attempted to relate these to kite flight data at Blue Hill, 228 km (142 mi.) to the south.

N O T I C E !

Hydrogen Gas!   **Keep Away From Fire!**

If the finder will carefully wrap up this Balloon, Cover and Basket, WITHOUT OPENING BASKET OR DISTURBING CONTENTS in any Manner, and Return by Express, Collect, to

J. A. OCKERSON,

Chief, Department of Liberal Arts,

World's Fair, St. Louis, Mo.

A Reward of Five (5) Free Admission Tickets will be paid for the service. Please fill out the card inside this envelope.

No.................   Sent up from St. Louis, Mo.,...

.. for the study of the upper air.

Found at ....

..

..

Date and Time Found ....

Name of Finder ....

Address ....

..

If Found Before NOVEMBER 4th, return ....card.

If Not Found until After NOVEMBER 4th, return...................card.

Figure 49. Notice attached to balloonsonde and postcard to be filled out by finder. Copyright President and Fellows of Harvard College. Harvard collection of historical scientific instruments.

Crowds on the hill had continued to interfere with Observatory activities and equipment, so, on 17 April 1905, the thermometer shelter was moved 19 m (62 ft.) to the northeast and about 0.5 m (1.6 ft.) lower. The fenced area or enclosure was then enlarged to include the shelter as well as the precipitation gages.

In the spring, to protect the building from crowds who sometimes sat in the windowsills, hammered on the walls, and generally disturbed the staff, a concrete wall and iron fence was built around the Observatory. The work was completed 16 May 1905 at a cost of about $2,000. These additions are shown in figures 51 and 52.

Early in June 1905 Clayton sailed for Europe on the Romanic to make the much-sought-after kite soundings in the trade winds. While crossing the Atlantic he made several successful soundings to the 1 km (3,300 ft.) level. After a pleasant stop in the Azores he sailed to Gibraltar to meet de Bort's private yacht, the *Otario*. All on board were French; M. Maurice, a colleague of de Bort at Trappes, assisted Clayton in the observational program. Stops were made at Madeira, the Canary Islands, and Cape

Figure 50. Mr. Clayton holding balloon and meteorgraph at St. Louis, c. 1905. Copyright President and Fellows of Harvard College. Gordon McKay Library—Blue Hill Collection.

Verde, where double theodolite observations of balloons were undertaken to measure the winds aloft. Nineteen kite soundings were made from the ship, which sailed as far south as 9° north. On the return trip they made a flight off the northwest coast of Spain during the solar eclipse of 30 August. At the same time Rotch observed the eclipse from Burgos, Spain. There he measured small pressure changes with a statoscope that was loaned to him by Richard of Paris. Measurements from captive balloons were made by other observers. Clayton, Rotch, and de Bort met in Paris to discuss the preliminary results of the expedition; then Rotch went on to Innsbruck and Berlin to participate in meteorological conferences while Clayton returned home.

Figure 51. Iron fence in front of the Observatory. Survey marker stands at right, c. 1906. Copyright President and Fellows of Harvard College, by permission of Houghton Library.

As president of the Boston Scientific Society Clayton (undated) gave an account of the expedition. He gave marvelous details of his impressions of people and places that he visited. In speaking of Granada he said, "Further on we were in a rich agricultural region and fair signoritas with lustrous brown Spanish eyes and beautiful complexions lighted the stations with their smiles." He also noted, "For the first time I saw the fig tree and the olive in full fruit, and this too in a land where one looks up to the summits of the Sierra Nevadas capped with snow."

Meanwhile a bronze tablet explaining the purpose of the Observatory had been placed on the outside of the northeast wall of the tower. It read,

Figure 52. Concrete wall surrounding the Observatory. Looking east–northeast, c. 1906. Copyright President and Fellows of Harvard College. Gordon McKay Library—Blue Hill Collection.

"Blue Hill Observatory, founded and maintained by Abbott Lawrence Rotch for research in meteorology, 1885."

In November Rotch applied once again to Carnegie for $5,000 to continue tropospheric investigations over the trades, but funds were never received from the Institute.

Rotch had served the American Academy of Arts and Sciences as librarian for some time and he had worked diligently toward finding a new home for the academy. In December 1905 he was proud to note "new headquarters at 28 Newbury Street, Boston."

Over the years Rotch attended automobile shows, often with his family, and watched the annual races in Readville. It is surprising that with this interest he did not purchase a car and continued to rely on his own horse-drawn vehicles or public transportation.

The Park Commission added a refectory to the group of buildings at the base on the west side of the hill during the year. This well-constructed wooden building matched the architecture of other buildings erected by the commission there and elsewhere within the Park. Also in this year the commission made a path down the east side of the hill and planted 30,000 pine seedlings. Some of these pines may have been set out in the area east of the upper half of the Blue Hill summit road where a fine red pine forest stood until its partial destruction by fire in the 1970s.

1906

Early in January Bjerknes, often considered the father of modern meteorology, visited the Observatory, where he witnessed a kite flight and discussed at some length his studies in hydrodynamics and atmospheric pressure distributions in the vertical. Undoubtedly these studies were a part of Bjerknes' work that led to the well-known term "dynamic height," attributed to him. While in Boston he lectured at MIT and was entertained by Rotch.

Also in January an Aero Club exhibit in New York featured the Observatory's work; kites, meteorographs, and balloonsondes were shown.

Rotch's close friendship with Kaiser Wilhem and the Prussian Meteorological Service was manifested on 27 February 1906 when he cabled the German Emperor, wishing him and his wife on the occasion of their 25th wedding anniversary "health, happiness and continuation of his glorious reign."

On 5 April, with four kites flying, the meteorograph at the 3.7 km (12,136 ft.) level, and 7.0 km (22,960 ft.) of wire out, the restraining wire broke. The next day the instrument, which fell in Weymouth about 14 km (45,920 ft.) to the east, was recovered, but the wire had crossed electric wires, and for the first time a complaint was made.

Clayton desired to resume his forecasting studies and in February 1906 the Weather Bureau in Washington promised him a professorship and salary of $2,500-$3,000. In June he entertained Canton friends at the Observatory and departed for Washington. However, he lost the appointment, so returned to Blue Hill for halftime employment. Rotch encouraged him to take his time while finishing reports of Blue Hill work and promised to assist in his forecast studies while he searched for a replacement. Meanwhile, M. E. MacGregor was hired as an assistant.

In late summer Fergusson traveled to the White Mountains for vacation and privately for kite work in the vicinity of Mount Washington. Fergusson installed a meteorograph on the mountain summit and flew kites from Twin Mountain, a few kilometers to the west, to compare free air measurements with those on the summit. In view of Rotch's interest in such comparisons, and even his making a trip up the mountain to help Fergusson pack up the meteorograph, it is surprising that he did not sponsor the work.

Rotch was unhappy with the continual problem of leaks in the tower. The outside walls had been painted repeatedly and coated with varnish or other materials nearly every year; the wooden roof had been replaced once. He therefore called in Tilden, his architect, to draw up plans for a new concrete tower.

An estimate of $3,050 was obtained from the Aberthaw Company for the work, but Rotch decided to postpone reconstruction unless the price of materials increased.

In the fall Rotch departed for Europe, reluctantly leaving the family at home. He spent time climbing in the Alps, then attended the Fifth International Aeronautical Congress at Milan, where he presented a paper on balloonsondes in America. He went on a balloon flight, climbing through clouds to 1,450 m (4,756 ft.) where the mountains presented a beautiful panorama. At this congress de Bort suggested "grand international ascensions." These would be scheduled four times per year; upper air data would be obtained from balloonsondes, pilot balloon observations, kites, and cloud motions. de Bort suggested the data be collected from islands in the Atlantic Ocean, Europe and America. Rotch then traveled to

Berlin and to another aeronautical meeting at the Technische Hochshule in Charlottenburg. Much time was spent with de Bort at the meetings and in Paris. Rotch failed to raise funds for further exploration in the tropics; nevertheless he paid de Bort $2,300 to help him continue the tropical observations.

At Blue Hill a new building with a concrete floor was constructed to house the kite windlass. A gasoline engine was tried in place of the steam engine, but it gave considerable trouble, and eventually the steam engine was returned to service.

The increased use of kites or balloons to lift recording instruments into the atmosphere prompted Fergusson to study the errors of the hair hygrometer, which was used in these instruments to measure relative humidity. He reported on this work in 1906 (see bibliography) in a well-written paper which included a summary and bibliography, an unusual feature in papers at that time. He showed calibrations and the sensitivity of hair hygrometers, but was unable to show quantitatively the effects of cold on these parameters.

In the course of the year, Rotch delivered numerous lectures, was elected a director of the Young Men's Christian Association, declined for the second time nomination for president of the Appalachian Mountain Club, and was elected as professor of meteorology at Harvard University, effective 1 September.

1907

In January Professor Rotch moved his office from the small room behind the tower to the new library; and his old office was put to use for the display of current periodicals. Meanwhile a new desk telephone was installed.

Rotch continued his snowshoeing activities, traveling to and from his Milton home on a new trail leading north from the Observatory. As a leader on Appalacian Mountain Club outings, he made other trips across the Blue Hills range and to the White Mountains. On 4 February he and Wells climbed Mount Monadnock, in southern New Hampshire, and exchanged signals by heliograph with Blue Hill, 110 km (68 mi) to the southeast. The experiment was a success even though the hill could not be seen through the haze.

Preparatory to reconstruction of the tower Fergusson commenced work in mounting anemometers on steel poles that were strapped to the chimney. From these instruments comparisons could be made with the instruments on the old and new towers. They would also furnish a record during the tower reconstruction.

Over the January to April period Rotch gave at least six lectures on a range of subjects including his tropical explorations, kite and balloon soundings, climbing in the Alps, and Benjamin Franklin. He had found from Franklin's letters that lightning rods had been installed on buildings in Philadelphia before his famous kite experiment. As an international expert on kite flying Rotch was called upon to illustrate a kite for use in Webster's dictionary.

In the summer of 1907 Fergusson returned to Twin Mountain, New Hampshire, to resume his kite flights for comparison of free air temperature, humidity, and wind with those elements on Mount Washington. This time Rotch sent Clayton to help him with the work. Five flights were made in a variety of weather conditions.

On 29 September 1907 Fergusson went to St. Louis for an extensive series of balloonsondes. This work was partially supported by a $500 grant from the Hodgkins Fund.

Preparatory to the Gordon–Bennett free balloon races from St. Louis, the team of Captain Hildebrandt, O. Erbslock, and H. Hiedimenn came to Blue Hill and asked Rotch to accompany Erbslock in the race. On the day before the race Rotch asked Clayton to take his place, which he did, and on 21 October 1907 nine balloons took off from St. Louis at 1600 CST at five minute intervals to race for the most distant point. Clayton immediately made use of a balloonsonde that had previously been released to guide Erbslock to a level of good westerly wind. They drifted until 0900 EST of the 23rd when it became necessary to land at Asbury Park, on the coast of New Jersey, or float over the sea. They covered 1,410 km (876 mi) in forty hours to win the race and set an American record for distance traveled.

Meanwhile Rotch witnessed two dirigible-balloon races at St. Louis and on his way home visited the new Weather Bureau research station set up on Mount Weather, Virginia. At New York he attended the Aeronautical Congress and commented that the army and navy were making significant advances in ballooning. This also seemed to be the case in Europe.

On 1 November Rotch attended a victory dinner with Clayton which was hosted by the German aeronautics in New York.

1908

In 1906 and 1907 many kite sounding flights had failed due to breakaways or problems with the windlass. This was in large part due to Fergusson's absence during his involvement with the balloonsonde work at St. Louis. Rotch therefore placed Fergusson back in charge of the program in January 1908 with the hope of continuing soundings on the international days.

Professor Jagger of MIT made inquiries about the possibility of constructing a small seismographic station on the leased land atop Great Blue Hill. It was his desire to locate instruments on "a geological stratum different from that of the Cambridge site." Further correspondence on the subject was not found and such a station was never erected.

In the spring President Eliot suggested that Rotch go to the University of Berlin as an exchange professor for the following winter. Rotch (1908) replied that he was flattered but felt that he could not perform to the credit of the University among such prominent men as those at Berlin. The subject came up the following year and that time Rotch was much relieved to learn that Dr. Schmidt of the University of Berlin did not feel able to recommend an appointment. He said that as welcome as Rotch was, the University did not consider meteorology important enough to invite a colleague to come to teach a small number of students (Rotch 1909).

In February Rotch authorized the Aberthaw Company to reconstruct the tower. It had become porous and during very high winds vibrations affected the traces on the recording instruments.

In March the sledding of sand to the summit began and soon afterward hauling began in earnest. The wind vane and anemometer were moved to the pipe supports that had been attached to the chimney. Prior to this time a comparison of wind speeds between locations atop the temporary masts and the tower was made. The barometers were moved to the old library. Demolition started on 25 March and by 2 April forms were in place for the initial pouring of concrete. The tower was completed at a cost of about $5,000 by 4 June. Reconstruction must have been a monumental task, since the concrete appears to have been mixed by hand and hauled up in buckets by rope and pulley. By 10 June instruments were placed in their new locations. The anemometers that were placed 5.3 m (17 ft.) above the flat roof of the new tower and 15.3 m (50 ft.) above the ground or 5 m (16 ft.) above their former level. However, the comparative simultaneous measurements of the wind showed no constant difference between the old and new exposures. It was suggested at the time that the increased height of the anemometers might be advantageous because the vegetation had been increasing in height since 1893 when the Park Commission took over the hills. This increase was attributed to their control of forest fires.

Inside diameter at the first floor was 5.3 m (17 ft.) and the outside diameter 6.3 m (20.6 ft.). The new height was 10 m (32.8 ft.). The walls were of reinforced concrete and of double thickness, 10 cm (4 in.) each, separated by a 25 cm (10 in.) air space. The reinforcement consisted of square, twisted steel bars, 23 cm (9 in) on center, running at right angles to each other, embedded in the concrete near the surface and extending into the wall. The outer surfaces were described as "picked concrete."

The floors and roof were 15 cm (6 in.) thick, also reinforced with twisted bars that projected into the walls. The roof pitched to two drain holes on the southeast and southwest. Construction is shown in figure 53 with forms for the top floor and roof parapet in place. Reinforced concrete webs separated the inner and outer walls at 0.9 m (3 ft.) intervals. The tower walls were poured in sections corresponding to each floor's height, as shown by seams at these levels. The stairs were of wood with pipe railings, and the windows were double-sashed. A sliding hatchway provided access to the roof. The parapet of the tower roof contained eight notches that

Figure 53. Tower reconstruction, May, 1908. Note the anemometer and wind vane on temporary masts attached to the chimney. Copyright President and Fellows of Harvard College, by permission of the Houghton Library.

were set at the cardinal and intervening points of the compass. Provision was made in the parapet near the north notch for a mast or flagpole roughly 20 cm (8 in.) in diameter, from which the weather flags could be flown. The bronze plaque that had been set in the old tower in 1905 was reset in the wall facing northeast. The upper fireplace flue that was built into the wall was soon sealed because of the excessive rain that entered the building. The fireplace on the second floor met a similar fate. However, the one on the first floor, made of bricks with a concrete mantle complete with grate, remained intact, but with a sealed flue until the 1960 renovations.

Figure 54 shows the tower in cross section and the arrangements made for the wind instruments. Sections of threaded 5 cm (2 in.) pipe were placed at strategic positions leading through the floors. From these short pipe stubs 5 cm (2 in.) pipe masts could be erected and guyed with rods. Vane direction or motion generated by the anemometers was then transmitted by lightweight aluminum rods, inside the pipes, to recorders below on the appropriate floor. The Draper wind recorder was positioned on the third floor near the northeast wall and its cup anemometer directly above. A spare mast was placed near the southeast wall with provision for transmission to the second floor. The wind vane and windmill anemometer were mounted on two masts near the south–southwest, wall with a steel ladder leading to the mastheads. Their shafting extended to the first floor. Two additional floor pipes near the southeast and northwest walls served as spares.

The new tower (figure 55) never shook, but the author can recall low frequency sounds that were transmitted throughout the tower from the numerous pipe masts and guy rods above. These strange sounds were not noticeable until winds reached about 35 m sec^{-1} (78 mi. hr^{-1}). Over the course of time water apparently leaked into the hollow wall chambers, causing the wooden forms which had to be left in during construction to rot. This decay showed in the form of rivulets of a dark brown, tarry substance which frequently stained the interior walls. Within six months leaks developed over the stairs where the old building joined the tower. Ironically, in spite of many expensive repairs, these leaks have never been completely controlled and the problem, although intermittent, remains.

In May the idea of advertising with kites and balloons was tried at Pittsfield, Massachusetts. Balloons with advertising notices were released by Rotch and Clayton. In one instance, a balloon was found at Randolph,

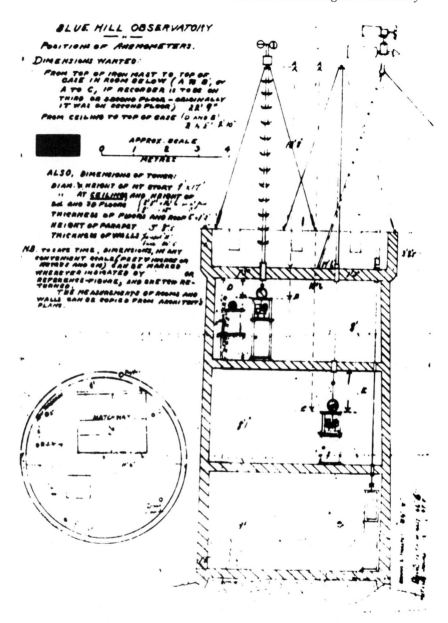

Figure 54. Cross section of the new tower facing east; shows arrangement of the wind instruments. Copyright President and Fellows of Harvard College, by permission of the Houghton Library.

Vermont, forty-five minutes after release and 160 km (99 mi.) to the north–northeast.

Clayton then supervised more balloonsondes from Pittsfield. An attempt was made to have the meteorograph fall before reaching the sea by

Figure 55. The Observatory and new tower viewed from the northeast. Copyright President and Fellows of Harvard College, by permission of the Houghton Library.

coupling two balloons to give a rapid ascent and detaching the meteorograph and parachute within one hour by an automatic release. Notwithstanding the unfavorable terrain for recovery, three of the four instruments released were found. A flight that was made on 7 May was found ten months later, the intact record showed that it had reached 17,695 m (58,040 ft.), the highest ascent in America until that time. The meteorograph sent up on 8 May was also found intact more than seven months later.

In July Clayton made a balloon flight to an altitude of 1.6 km (5,248 ft.) near North Adams, Massachusetts, and in September he traveled from Springfield to Bridgewater, Massachusetts, by balloon, a distance of 134 km (83 mi.).

In the fall of 1908 Rotch attended meetings of the BAAS in Dublin and the German Meteorological Society. While in England he had a long talk with Wilbur Wright concerning the flight of birds and later, in France, witnessed one of Wright's aeroplane flights over a 15 km (9 mi.) course. From near London Rotch made a uniquely British balloon flight. He ascended through clouds to a height of 1.8 km (5,900 ft.), then descended, had tea, and went off again to a height of 1 km (3,280 ft.) to cover a total distance of 60 km (37 mi.).

de Bort had taken air samples from a "great height" and shown the presence of helium in the sample. He gave Rotch vacuum bottles to obtain samples, for comparison, at sea level and in unpolluted air, during his shipboard return to America.

In November 1908 Clayton left for a month's tour of lecturing in Ohio. Upon his return he offered his resignation, effective April 1909, which Rotch accepted. Clayton had finished the balloonsonde analysis, grown tired of traveling, and wanted to get back into forecasting.

During the year Ward had encouraged Rotch to bring advanced students to the Observatory. Quite possibly, as a result, a young Harvard man by the name of Charles F. Brooks visited the hill on 25 October 1908. After this trip, Brooks wrote in his observation book and later transcribed in the Observatory Journal the following: "As I stood on Blue Hill early in the afternoon there was a shower in the SE and a heavy one W. The one SE diminished and passed over yielding only a few drops. The day was sultry and cumulus domes rose here and there. The sun would shoot long beams of light through the clouds making very beautiful scenery." Twenty-three years later Dr. Brooks became the third director of the Observatory.

On 6 November 1908, A. H. Palmer was hired to assist in the observations. Palmer was a graduate of the University of Minnesota and was doing postgraduate work at Harvard. Thus began the employment at the Observatory of a long series of Harvard students over many years.

1909

Rotch took his professorship seriously and with guidance from Ward he encouraged Palmer, in 1909, to take an A.M. degree. Palmer's thesis was titled "Change of Wind at Different Heights in the Free Air" (see bibliography). For this he used Observatory data that had been obtained from kites, balloonsondes, and clouds. In addition to average speed profiles with height, he presented tables of speeds from each direction for heights to 3.1 km (10,168 ft.). Other tables showed the extreme speeds that were encountered at various levels up to the highest cirrus clouds. In June Rotch proudly noted, "My student was awarded his degree."

In January Rotch went on a snowshoeing trip up Mount Greylock in northwestern Massachusetts. It is interesting that the superintendent of the reservation proposed that a weather observatory be constructed on the summit. This writer, many years later, through the winter of 1950-51, operated a monthly thermograph on the summit to determine if temperatures from the 1,067 m (3,500 ft.) high mountain could help in forecasting precipitation type in eastern Massachusetts.

In February a member of the Argentinian Weather Service came to the Observatory to study the methods of observation. It is quite likely that, later, Clayton took a position as meteorologist with the Oficina Meteorologica Argentina as a result of this visit. Clayton left the Observatory on April 1909 after twenty-three years of almost continuous service. He became dean of the School of the Aeronautics Association Institute in Boston for one year. Rotch had noted that numerous press articles had referred to Clayton as "Professor" and "Director of the Observatory." Finally in September Rotch (1909) reminded Clayton of these improper references. An exchange of letters took place in which the blame was placed on the media, but not to the complete satisfaction of Rotch. Nevertheless, after Clayton's departure Rotch noted, "His researches published in the annals and elsewhere are among the most notable recent contributions to American Meteorology," and in his final letter he wished Clayton "fame and fortune in the pursuit of aeronautics."

Rotch represented the United States at the sixth meeting of the International Committee for Scientific Aeronautics in Monaco in April 1909. There he proposed a standard form for the international exchange of data obtained from kites and balloonsondes and he helped in arranging a system of telegraphic stations over the globe. It is interesting to note that at this meeting Bjerknes proposed that kite flights or balloonsondes be made at noon, and pilot balloon soundings be made in the morning and afternoon in an effort to gather synoptic upper air data. The name "pilot balloon" was given to balloons that were released preceding manned balloon ascents since, in a sense, their indicated drift could be used subsequently to pilot the manned balloon. Later the balloons were used for any winds aloft measurement and the term shortened to "pibal."

Hergesell proposed the use of single station pilot balloons, since he had determined from double theodolite runs that deviations of the balloon's height from its assumed height seldom exceeded four percent. Also during this congress, de Bort made use of the terms "troposphere" and "stratosphere," probably for the first time.

On 6 May, also for the first time, a meteorograph was kept aloft at Blue Hill with kites through a thunderstorm with no serious accident.

Lightning struck the Observatory for the second time on 18 July 1909; this time the flagpole on the tower was hit, but sustained no serious damage.

In July double theodolites were set up at the base to follow pilot balloons. After some unproductive starts a balloon was followed successfully by theodolites for the first time in United States on 7 July 1909. Soon afterward theodolites were set up on Hemenway Hill, 1,180 m (3/4 mi.) to the east, and at the Observatory. However, this base line proved inconvenient and was soon abandoned.

After these successful pibals, Rotch ordered two new theodolites and additional rubber balloons for the sondes.

In November the tower was waterproofed with a dressing of hot paraffin at a cost of two hundred and eighty dollars.

In December, for a "grand international series" which was now in effect, three balloonsondes were launched from Pittsfield on the 10th, 11th and 12th and two pibals were launched from Blue Hill.

During the year Rotch was made an honorary member of the Austrian Meteorological Society and the Harvard Chapter of Phi Beta Kappa.

1910

In 1910 Rotch published *Conquest of the Air* (see bibliography), an up-to-date, as well as historical, book on the subjects of dirigible balloons and flying machines. The last chapter, on the future of aerial navigation, made special note of the importance to the military of the coming age of flying. It was also conceived that an airplane might carry mail and two or three passengers at 45 m sec^{-1} (100 mi. hr^{-1}) with "tolerable regularity." A letter in which Wilbur Wright commends the book is reproduced as figure 56.

In spite of Rotch's successful publication of this book, he did not abandon the subject in favor of a new field, as is so often the case. His primary interest continued in aviation and he spoke whenever possible on the meteorological needs of aviation in the future. At the time "aeronauts," as they were called, simply did not fly when the weather was not good or surface winds exceeded about 7 m sec^{-1} (15 mi. hr^{-1}). Rotch was thinking beyond the problem of surface weather; he was concerned with charting winds aloft, and laying out favorable courses for long distance flights. He attended aviation shows at Mineola on Long Island, New York, and Atlantic, Massachusetts, on Boston Harbor, which kept him informed concerning the machines and their development. In January 1910 he attended a dinner at the Tavern Club in Boston that was given to honor O. Chanute and the Wright brothers. Wilbur Wright's speech on the future of flying made a lasting impression on Rotch.

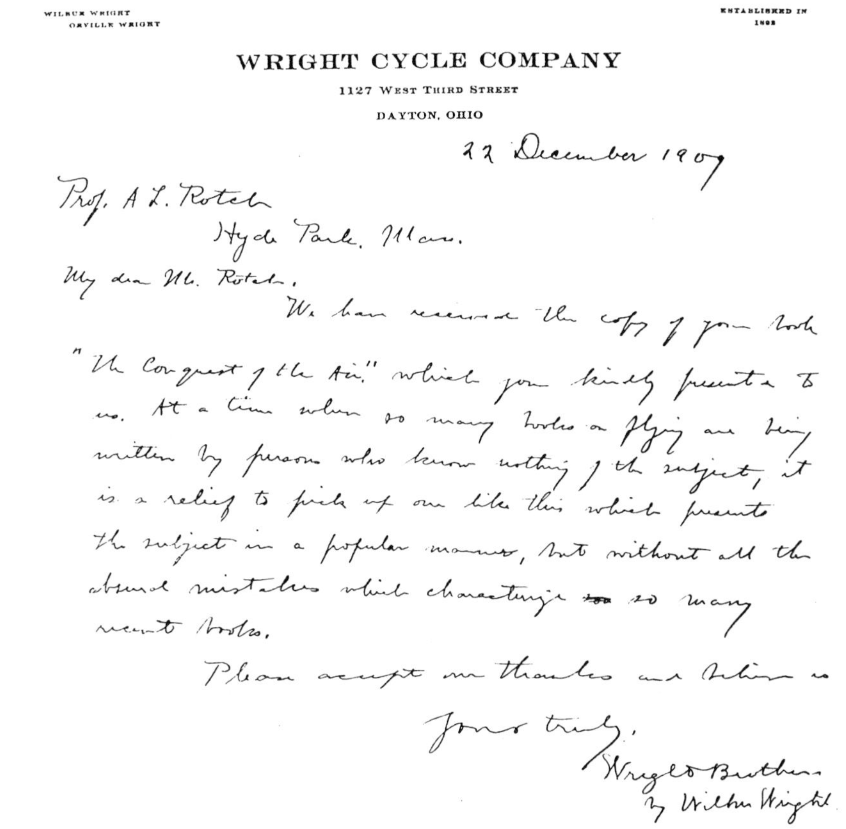

WILBUR WRIGHT
ORVILLE WRIGHT

ESTABLISHED IN
1892

WRIGHT CYCLE COMPANY

1127 WEST THIRD STREET

DAYTON, OHIO

22 December 1907

Prof. A L. Rotch
Hyde Park. Mass.

My dear Mr. Rotch,

We have received the copy of your book "The Conquest of the Air" which you kindly presented to us. At a time when so many books on flying are being written by persons who know nothing of the subject, it is a relief to pick up one like this which presents the subject in a popular manner, but without all the absurd mistakes which characterize ~~so~~ so many recent books.

Please accept our thanks and believe us

Yours truly,

Wright Brothers
by Wilbur Wright

Figure 56. Letter from Mr. Wilbur Wright. Copyright President and Fellows of Harvard College, by permission of the Houghton Library.

In a letter to the Aero Club of America dated 30 May, Fergusson (1910) statistics of record sounding heights up to that time were given. The highest balloonsonde in the United States was made from Indianapolis, Indiana, on 6 October 1909. It reached 19,433 m (63,740 ft.) and it was sent up by the Weather Observatory, a facility of the Weather Bureau. The Blue Hill record was 17,700 m (58,056 ft.) in a sounding from Pittsfield, Massachusetts, on 7 May 1908. The highest kite sounding was from Mount Weather, Virginia, on 6 May 1910. It reached 7,205 m (23,632 ft.). According to Fergusson a pilot balloon from Blue Hill reached 18,600 m (61,000 ft.) on 8 October 1909. Since this was probably tracked by a single theodolite, the height must be based on a fixed rate of ascent assuming no loss of gas. A leaky balloon, not at all uncommon in those days, could have made the balloon seem much higher than its true height.

Fergusson left for a year's absence on 22 July 1910. He had been offered a professorship at the University of Nevada in Reno, Nevada. There he planned to work on the problem of how to make snow-pack measurements in the Sierras and to construct a meteorograph that would run for long periods, unattended, atop Mount Rose. His leaving significantly hindered the sounding balloonsonde program, which attempted to continue flights on the "international days." After Fergusson's departure an engineer from Boston came to the Hill to operate the windlass on flight days and at one point a meteorograph was shipped to Fergusson for repairs. Due to these difficulties, soundings were obtained on only four days in 1910. Pilot balloon runs, which were less complicated, were made on eleven days.

Other events of the year included the removal of the trigonometric survey tower, which had stood in front of the Observatory since before its founding in 1885. In the summer, band concerts were regularly held at the base of the Hill on the west side. The pole star recorder was moved from the enclosure to a position next to the west side of the tower.

Rotch once again declined to teach during the coming winter at Berlin University; this time the invitation was made by President Lowell.

In July members of the National Education Association met in Boston and a tour of the Observatory was arranged. Over 200 delegates arrived atop the hill to be relayed through the Observatory in groups.

In August and September Rotch went to Pasadena with a stopover at Flagstaff, Arizona, to visit Percival Lowell's astronomical observatory. A conference, mostly of astronomers, met at the Mount Wilson Observatory. Delegates were quartered at the Mount Wilson Hotel, close to the observatory. Rotch's old friend, McAdie, came from San Francisco, but most of the time Rotch was uncomfortable and ill with a throat ailment. The highlight of his stay was Abbot's talk on the solar constant.

At this time the International Solar Commission made use of the homogeneous climatological record at Blue Hill. They presented the data in a study of secular changes of climate.

In October Professor and Mrs. Rotch attended a dinner given for President and Mrs. Taft at the home of W. Endicott. About twenty guests were present and Rotch noted that he had an "interesting talk with the President of Panama Canal."

In December Rotch spoke on "The Air and its Navigation" before 300 students at the Dartmouth College Scientific Society.

About this time Rotch wrote that the Observatory library "has probably the best collection of meteorological books, pamphlets and journals in the United States outside the City of Washington." The pamphlets had been classified, like the books, according to the Dewey decimal system, amplified by Clayton for meteorology and related subjects. About 250 volumes or pamphlets were added each year.

Palmer was sent, in June, to Dwight, Massachusetts, to investigate a claim that a balloonsonde had killed a farmer's cow. He concluded that the cow, indeed, had died but apparently not from the falling instrument.

In a letter July (1910) Rotch claimed no liability for the cow but offered $35 for the safe return of the balloonsonde. In later correspondence (Rotch 1911) the farmer apparently claimed that the cow had eaten the parachute and died. Once again Rotch repeated his offer but no record of a payment exists.

1911

Fergusson's absence was felt again in 1911 when instrumental problems developed. This time men from the Boston Elevated Railway were hired to repair the Draper anemometer.

On 1 June 1911 the base station thermometer shelter was moved about 320 m (1,050 ft.) south to Park Commission land and about 4 m (13 ft.) higher than its former location. This site, just south of the present-day Metropolitan District Commission rest rooms and on the edge of a recently made duck pond, was kept until 1943.

Fergusson elected in March not to renew his employment at the Hill but he did return in June to repair instruments and build a new kite. He brought with him a new meteorograph that he had constructed which could be mounted inside the large Hargrave type kite. After assisting in a kite flight, he and Mrs. Fergusson left the area, but he avowed his continuing interest in the Observatory.

On 15 June 1911 an airplane that had taken off from Atlantic, or the site of the old Squantum Airfield, about 13 km (8 mi.) northeast of Blue Hill, circled the hill three times, once very close to the tower, for the first time in history. About this time flights were also made from the Readville race track, near the valley station.

Beginning on 9 July 1911 forecasts for the next thirty-six hours were displayed on a board attached to the iron fence outside the Observatory. These forecasts were based upon reports received by telephone from the

Boston office of the Weather Bureau and modified through use of the Blue Hill observations. Twenty-four-hour forecasts communicated by flags continued as before. At this time it was estimated that under favorable conditions, the flags could be seen by about 35,000 people living within or passing through a circle having a radius of 6.5 km (4 mi.) centered at the Observatory. Wells was largely responsible for these forecasts.

In midsummer Rotch and Palmer published *Charts of the Atmosphere for Aeronauts and Aviators* (see bibliography). In it they made the distinction between an aeronaut who piloted a balloon and an aviator who piloted a flying machine. The book contained much basic data, obtained by the Observatory, such as average pressure, temperature, and wind, including maximum speeds from sea level to 9.2 km (30,000 ft.). Other charts depicted monthly and even hourly temperature and wind from the surface to the 3.6 km (11,800 ft.) level. Hodographs were also included for Blue Hill and the northeast trade wind area in winter and summer. These charts represented Rotch's latest endeavor to apply what he called "engineering meteorology" to the developing field of aviation.

The book brought much publicity to Rotch. Aero clubs were springing up all over the country and many of them sought Rotch for popular lectures or articles on the subject of the weather and flying.

Also in the summer Percival Lowell proposed an astronomical museum in Boston and invited Rotch to participate with a section on meteorology. This never came to fruition.

The Rotch family spent August at Northeast Harbor in Maine. While there many temperature observations were made on hillsides during foggy conditions. Later he used these data at a Weather Bureau symposium in Washington during a discussion of fog formation.

On 3 October, Messrs. Linsly and Charles Brooks visited Rotch at the observatory to discuss enrollment in a course called "Geology 20" at Harvard and their work toward their A. M. degrees. Brooks elected to study ice storms of New England for his thesis work. The two students came to the Observatory frequently for the remainder of the year.

Also in October 1911 the MIT Corporation decided on the "Riverbank site" for the new location of the institute. Rotch had been involved in this decision, but his primary choice had been the Fenway part of Boston.

In December photographs of mountain observatories in Europe and United States were hung on the walls of the library above the stacks. These pictures were of much interest to many visitors until their removal and disappearance in the 1970s.

At year's end Rotch reiterated his concern for the need of new and extensive upper air data in an address titled "Aerial Engineering," which he presented to a section of the AAAS in Washington.

He also noted that his private collection of historic books and pamphlets had now been completely cataloged and reached 814 in number.

1912

In January Rotch made further study of "winds that would affect an aeroplane voyage across the Atlantic."

Cabinets were constructed for housing the charts from the automatic instruments. These were set up on the second floor of the tower and the records moved from the safe to the new cabinets in February.

Rotch's students were already having some effect on research activities. In March Linsly proposed the study of valley fog through the use of captive sounding balloons raised from the valley. From the interest in ice storms created by Brooks's thesis, a kite flight was made to about 0.8 km (2,600 ft.) through freezing rain.

Also in March Rotch accepted an invitation to deliver three lectures at the University of Michigan. Their titles were: "Methods of Studying the Upper Air," "Physics of the Atmosphere," and "The International Survey of the Air."

On 31 March 1912, without hint of sickness or discomfort, Professor Rotch made the last entry in his diary. The next day he was too ill to write for the first time since the diary's beginning in 1877. Five days later he was operated upon for appendicitis which had gone undiagnosed because of an abnormally positioned appendix. The next day, 7 April 1912, he died.

News of Rotch's death appeared in newspapers coast to coast; many carried long articles telling of his achievements. He was honored at his funeral by eight pallbearers, including the presidents of Harvard University, MIT, and the Museum of Fine Arts. The others were faculty members of Harvard and MIT.

Hundreds of letters and telegrams of condolence poured in from Europe and America along with a few from other parts of the world. Elaborate documents, signed by friends or officials, were sent from many organizations in which he held honorary membership.

A letter (figure 57) from Petersfield, England, to President Lowell of Harvard suggested that a Professorship in Dynamical Meteorology be founded in memory of Abbott Lawrence Rotch. It was signed by:

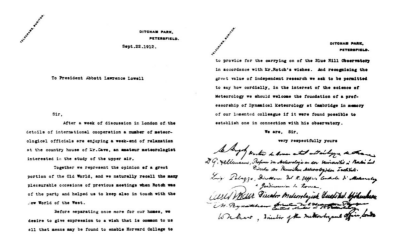

Figure 57. Letter to President Lowell in which a professorship in dynamical meteorology was suggested in memory of Abbott Lawrence Rotch. Copyright President and Fellows of Harvard College. Division of Applied Sciences.

1. A. Angot, Directeur du Bureau Central Meteorologique de France
2. Dr. G. Hellmann, Professor der Meteorologie and der Universität Berlin und Direcktor des Preussischen Meteorologischen Instituts
3. Luigi Palazzo, Direttore del R. Ufficio Centrale di Meteorologiai Geodinamica in Roma
4. Carl Phail, Director, Meteorologisk Institut, Kjobenhaun
5. N. Rykacheu, Directeur de L'Institut de Physique Central St. Petersburg
6. W. N. Shaw, Director of the Meteorological Office, London

(Note that Shaw was not yet knighted; his first name was William.) A copy of this letter now hangs in Pierce Hall at Harvard, near the Blue Hill Collection.

According to Rotch's wishes, the Observatory was bequeathed to Harvard University with $50,000 which was to be set up in an endowment fund to operate the facility. However, this transfer did not take place until March 1913. The burden of maintaining the Observatory and supervising its activities for the intervening year was carried by Mrs. Rotch. Palmer and Wells continued with the kite flights on international days, the forecasts, and routine observations. From March 1913 until 30 September 1913 Robert DeC. Ward assumed general supervision of the Observatory.

Rotch was universally recognized for establishing and maintaining the now famous Observatory, which had developed a unique climatological record and, in the United States, the most complete meteorological library outside Washington. Rotch had the means to accomplish this and sponsor the cloud climatology of heights and motions, the kite sounding and balloonsonde programs, and the use of pilot balloons. Some of the work was initiated by Henry Clayton, who performed the meaningful analyses, while the ingenious Sterling Fergusson tested, developed, and maintained the instrumentation. Interpretations in the dynamic sense were almost totally lacking, although the data were suitable for work of this nature.

A conspicuous feature of this period in the Observatory's history was the youth of its staff, the energetic plans for investigation, and the early publication of results. Rotch was twenty-three years of age at the time he established the Observatory, Gerrish was nineteen, Clayton twenty-four, Fergusson nineteen, Wells twenty-seven and Sweetland twenty-three when they came to the Observatory. At the time of McAdie's first temporary work he was twenty-two years old.

Aside from maintaining the Observatory and its programs, Rotch's most important contributions to meteorology were his participation in international meetings, lecturing, over the last years of his life, on the coming need of meteorology to aviation and teaching, which he carried on without pay at Harvard. His lectures were much in demand and, within a few days after his death, an official letter from the French Ministry of public instruction arrived in which Rotch was named as exchange professor at the Sorbonne for the year 1912-13. He had just begun to effectively integrate his teaching with the Observatory, and only speculation can say how his students would have developed and what new research they might have initiated.

Rotch was not only a renowned meteorologist, but also a competent businessman; he managed a fortune so carefully that it enabled him to advance the science of meteorology, maintain his Observatory in topnotch condition, and amply provide for his family. He was a sensitive and kind man, making regular visits to older servants, relatives, and less fortunate friends, whom he helped financially. Rotch was never inactive; at times he became very nervous and suffered from severe headaches and hypochondria, but he remained a warm, personal gentleman who loved his family. Trips abroad, often with Mrs. Rotch, combined business with pleasure; dinner parties, the theatre, opera, and concerts made their social life most enjoyable. He was an outdoors man. He loved to climb, having made five ascents to the observatories on Mount Blanc in the French Alps. This love of the mountains was undoubtedly an underlying factor in his interest and advocacy of mountaintop observatories. He probably had visited more mountain observatories and knew more about them than any other person, as shown by his comprehensive documentation of their construction, operation and programs. In summer the entire family, which included two

Figure 58. Abbott Lawrence Rotch, 1861-1912. Copyright President and Fellows of Harvard College. Gordon McKay Library—Blue Hill Collection.

girls and a boy, vacationed at the seashore. In winter they made excursions to the mountains and in the meantime, they frequented the Blue Hills, traveling by carriage, foot or on snowshoes. A portrait of Rotch that was probably made within two years of his death is shown in figure 58.

Over his lifetime Rotch authored about 183 papers, reviews, and books, some of which were in French and German. Some of these were of the popular variety in which he announced the most recent meteorological accomplishments at home and abroad.

The following list, showing membership and positions held in various organizations, evidences how widely spread Professor Rotch's interests were:

1. Aero Club of America
2. Aero Club of New England (first president)
3. American Alpine Club
4. American Antiquarian Society
5. American Association for Advancement of Science (vice president Section D)
6. American Member of International Jury of Awards for Instruments of Precision
7. American Meteorological Journal (associate editor, ten yrs.)
8. American Philosophical Society
9. Appalachian Mountain Club
10. Astronomical and Astrophysical Society of America
11. Berliner Verein für Luftschiffahrt
12. Boston Society Of Natural History (Trustee)
13. British Association for the Advancement of Science
14. Chevalier Legion d'Honneur (1889)
15. Club Alpin Français
16. Deutsche Meteorologische Gesellschaft
17. Harvard Aeronautical Society (president)
18. Harvard Travellers Club
19. International Commission for Scientific Aeronautics (American member)
20. International Meteorological Committee
21. International Solar Commission
22. Massachusetts Institute of Technology (Corporation Member eighteen yrs.)
23. Massachusetts Institute of Technology (president, Alumni Association 1902)

24. Museum of Fine Arts (on board of trustees)
25. National Geographic Society
26. Oesterreichische Gesellschaft für Meteorologie
27. Paris Exposition (1889)
28. Physical Society of London
29. Preussicher Kronen Orden (1902)
30. Preussicher Rother Adler Orden (1905)
31. Royal Meteorological Society
32. Sociedad Astronomica de Mexico
33. Société Astronomique de France
34. Société Belge d'Astronomie
35. Société Meteorologique de France
36. Société Royale de Medecine Publique de Belgique
37. The Aeronautical Society
38. The American Academy of Arts and Sciences (librarian)
39. The American Geographical Society
40. The Royal Aero Club (London)
41. Washington Academy of Sciences

Following Rotch's death, Mrs. Rotch gave $4,000 for the construction of a fountain for the public at the top of the hill. She had heard her husband remark many times about climbers' need of water during the warm months. A bubbler was set up by the Park Commission in the summer of 1913 at the junction of the summit and circumferential roads and a pipe laid from the Observatory water tank, a distance of about 610 m (2,000 ft.). Mrs. Rotch then employed Bela L. Pratt to design a memorial to house the bubbler. The memorial which depicts birds flying above hills, was erected 17 July 1914 (figure 59). The inscription reads:

> In memory of Abbott Lawrence Rotch, founder of the Blue Hill Observatory, pioneer in the study of the upper air, a life devoted to science for the good of mankind.

In later years a favorite prank was to jam the bubbler open and thereby drain the Observatory's water supply. Still later, vandals defaced and damaged the monument, which necessitated discontinuing the water supply. In 1968 the monument was cleaned and repositioned in front of the Observatory, inside the iron fence.

In September Palmer left the Observatory and Charles F. Brooks arrived as research assistant. His interest in clouds was immediately shown in the form of detailed time–height cross sections of the clouds, after a style used by Clayton.

On 2 December 1912 work started on wiring the Observatory for electricity and on 3 December a large pipe was laid which connected the Observatory to the Canton water supply. A pump at the base on the south side of the hill, on Hillside Street, pumped water into the Observatory tank. A float and switch at the tank provided automatic pumping of water when the tank was low. In winter the pump was drained but in later years water

Figure 59. Memorial and fountain erected on the summit of Blue Hill in memory of Abbott Lawrence Rotch. Courtesy Rotch family.

could be obtained upon request, which required the services of a park attendent who manually controlled the pump and saw that it was properly drained.

1913

On 7 February a kite flight that employed five kites broke away in the evening. The wire lay across New Haven train tracks in Braintree and with the passage of a train it became entangled in the wheels. It was estimated that 5 km (3 mi.) of wire wound around an axle, necessitating a halt in the train's movement.

Only a month later, on 6 March, two kites were out of sight in a cloud from which heavy rain was falling. There had been no lightning previously, but occasional sparks had been coming off the wire. Suddenly there was a blinding flash in the kite house. Brooks jumped out the door, saw a line of white-hot beads falling from a line where the kite wire had been, and heard an even peal of thunder, without the usual crash, for the line of the flash had been straight along the kite wire. Two men were shocked, one severely, but no permanent injuries resulted. The engineer picked up his belongings and went down the hill, never to return.

June journal notes by Brooks stated that the first ripe blueberries were observed on the Hill. A few days later Wells used his prerogative as chief observer to correct the date for the first ripe berries inside the enclosure. Wells cherished his observership and interference with "his" observations was not appreciated. Unfortunately, Mrs. Rotch appears to have been turned against Brooks. This and the shortage of funds probably led to his resignation on 1 September 1913.

References

Clayton, H. H. Undated. Twelve thousand miles in pursuit of air currents. 13p Presented at the Boston Scientific Society.

Fergusson, S. P. 1910. Author index and copies of letters. 1910-13 p. 13. Blue Hill Collection, Gordon McKay Lib., Harvard University.

Rotch, A. L. 1897: Author index and copies of letters. 1896-98 p. 251. Blue Hill Collection, Gordon McKay Lib., Harvard University.

——1897. Author index and copies of letters. 1908-98 p. 278. Blue Hill Collection, Gordon McKay Lib., Harvard University.

——1908. Author index and copies of letters. 1905-09 p. 307. Blue Hill Collection, Gordon McKay Lib., Harvard University.

——1909. Author index and copies of letters. 1905-09 p. 399. Blue Hill Collection, Gordon McKay Lib., Harvard University.

——1909. Author index and copies of letters. 1905-09 p. 313, p. 414. p. 425. Blue Hill Collection, Gordon McKay Lib., Harvard University.

——1910. Author index and copies of letters p. 47. Blue Hill Collection, Gordon McKay Lib., Harvard University.

——1911. Author index and copies of letters p. 97. Blue Hill Collection, Gordon McKay Lib., Harvard University.

Chapter VIII

Directorship under Alexander McAdie:
1913–1931

1913 (continued)

On 1 October Alexander George McAdie (figure 60) was appointed professor of meteorology and director of the observatory.

Professor McAdie, born in 1861 in New York, attended the College of the City of New York, where he earned his A.B. at age nineteen and A.M.

Figure 60. Alexander George McAdie, 1863-1943. Director of the Blue Hill Observatory, 1913-1931. Copyright President and Fellows of Harvard College, Director of Applied Sciences.

at twenty-one. He earned two competitive gold medals in English composition. Ten years later, in international competition for the Smithsonian prizes, he received a medal for an essay titled "Equipment and Work of an Astrophysical Observatory." He joined the Signal Service in 1882 as a student in meteorology. Soon he was sent to Harvard to study atmospheric electricity under Professor John Trowbridge. While there he received a second A.M., in 1885. Rotch had discussed his Observatory project with McAdie and invited him to be the first observer, but McAdie declined on account of other plans. However, he did come later as relief observer and to conduct experiments in atmospheric electricity, as already noted.

Except for a short period in Indianapolis, Indiana, as a teacher, and about a year at Clark University as fellow in physics and lecturer in meteorology, he remained in the government weather service until 1913. While associated with Clark, he nearly succeeded in convincing the university to open a "great observatory and laboratory . . . that shall be the peer of any meteorological observatory in the world." He received approval of a thesis for a Ph.D. degree based on his work at Blue Hill and Washington only to learn that one hundred copies were required. This was financially impossible so he withdrew as a Ph.D. candidate and never published his work. A discouraged man, who had hoped to carry on his research as a professor associated with a great observatory at a well-known university, returned to government service in Washington. Subsequently he served briefly in New Orleans and in 1895 became forecast official at San Francisco.

In 1903 he was appointed professor of meteorology in charge of the Climate and Crop Service of California. For his fog studies he obtained upper air data from kites flown from Mount Tamalpias and from expeditions to the summits of Mounts Whitney, Shasta, and Ranier. A mountain near Mount Whitney, first climbed by J.E. Church, has been named by Church, "Mount McAdie, in recognition of Alexander McAdie's outstanding work in meteorology." McAdie forecasted freezes and patented four plant and fruit protectors. McAdie was honorary lecturer in meteorology at the University of California and received an honorary S.M. from Santa Clara College.

One of McAdie's most interesting papers was titled "Natural Rainmakers" (1895). In his description of a thunderstorm he wrote, "The thunder cloud is noteworthy in another respect; namely, that the water in it may be cooled below the freezing point and yet not frozen. A snowflake or ice crystal falling into it may suffice to start a sudden congelation, just as we may see ice needles dart in all directions when the chilled surface of a still pond is disturbed. We liken this monstrous cloud to a huge gun loaded and quiet, but with a trigger so delicately set that a falling snowflake would discharge it." Brooks commented later (M.R.B. 1949), "Here is a clear statement of the fundamental basis of the Bergeron–Findessen theory of rain, propounded forty years later, and of the artificial production of precipitation by 'seeding' an overgrown cumulus with dry ice, to produce ice crystals, ten years later still."

Numerous other articles and several books were published by McAdie before coming to Blue Hill. Of special note were his book, *Climatology of California* (1903) and pamphlet, *Protection from Lightning* (1894), which required four editions totaling 40,000 copies.

During the earthquake and fire of 1906 in San Francisco, he managed to keep the weather station open, and recorded details of the quake. Through the use of undamaged instruments supplied by a friend, he was able to resume meteorological observations almost immediately.

McAdie was always engaged in literary and educational affairs. In the absence of President John Muir of the Sierra Club, McAdie, as vice President, dedicated a monument erected by the Women's Clubs of Napa County on the spot near Mount St. Helens where Robert Louis Stevenson spent his honeymoon. He wrote many essays, two groups of which were published in book form (1909 - 1912).

Because of his interest in Blue Hill and outstanding work in California, he was called to Harvard to serve as the Observatory's second director. This pleased Mrs. Rotch and she made every effort to see that he and Mrs. McAdie were comfortable upon their arrival. She also realized that available funds from the endowment would barely keep the Observatory in repair and pay for the few necessary items such as fuel and instrument supplies. McAdie's salary was paid by Harvard but Mr. Wells' salary of $150 per month was paid by Mrs. Rotch until April 1915. Financially the Observatory was to see difficult times.

1914

On 16 January electric lights were placed in service. The cost of wiring and fixtures was borne by Mrs. Rotch. Power lines were strung between short poles which led up the south side of the hill.

Kite flights were continued on the international days whenever winds permitted and the routine observations were maintained almost entirely by Wells.

Under the Department of Geology and Geography McAdie offered two courses at the Observatory. One focused on meteorological instruments. The other was an opportunity to carry on investigations in advanced meteorology. Only one student attended regularly.

McAdie managed to have the expense of printing the Observatory's observations and reports in the Harvard College annals transferred to the Astronomical Observatory. This helped to lower the overall expenses to $3,415, a drastic reduction from former years.

1915

By 1915 Wells had settled into the routine observations that he would continue for many years. In addition to the check observations at 0800, 1400, and 2000 EST, hourly observations of the prevailing type and amount of cloudiness were made. No journal notes were made. The only notes made pertained to the advance of the seasons, such as the dates of freezing and thawing of nearby ponds, the first ripe blueberries, first cherry blossoms, last frost in spring, and first frost in autumn. In the absence of weather maps from Boston, the old method of obtaining reports from the railroads was resumed. Arrangements were made with the New York, New Haven and Hartford, Boston and Albany, and Boston and Maine Railroads, to receive weather reports each morning from locations throughout New England and parts of New York and Ohio. From these reports Wells skillfully prepared weather forecasts which were posted on the front fence and also supplied to those who telephoned. In return for the data, the railroads were given regional forecasts.

McAdie studied local nighttime temperature inversions and combined the forecasts by Wells with local observations to forecast frost conditions at the cranberry bogs in southeastern Massachusetts. Apparently, as a result of this work, McAdie's interest in plant physiology, which he acquired in California, was revived; For a part of August he made measurements of evaporation for each twenty-four-hour period.

McAdie also began a campaign about this time, which was to continue throughout his directorship, for the use of basic units of measurement in meteorology. He argued long and eloquently in favor of the "kilobar" as a unit of atmospheric pressure. T. W. Richards, in 1903, had suggested that the pressure of 1 dyne cm^{-2} be called a bar. Using this unit 106 bars approximated the pressure of a column of mercury 29.53 inches high, a value somewhat less than the standard atmosphere. Unfortunately, when this unit was first used in meteorology, 106 bars were referred to as 1,000 millibars. McAdie correctly argued that the term should have been 1,000 kilobars. He changed most of the units of pressure at the Observatory to kilobars but continued to publish the check readings of the mercurial barometer in millimeters, presumably because that was the scale engraved on its case. McAdie changed to publishing temperature in degrees absolute and continued publishing precipitation amounts in millimeters; wind speeds he continued to publish in m sec^{-1}, a practice started by Rotch.

Fergusson had been advising McAdie regarding the kite flying program and in 1914 McAdie had electrical conduits laid to the kite house and asked for estimates of the cost of a five horsepower 220 volt motor to haul in the kites. However, with the start of hostilities in World War I, all international cooperation ended in the exchange of sounding data and the kite sounding program quietly ended.

1916

Some clock stoppages in the recording instruments occurred at this time, marking the beginning of deterioration of the automatic records. The situation was aggravated by inadequate funds for repairs and the limited mechanical ability of Wells.

McAdie published a paper on the winds of Boston in 1916 (see bibliography). In this he presented statistical data from Blue Hill and the Boston Weather Bureau with some interesting interpretations of winds at the surface and aloft during sea breezes.

Contrary to his pursuit of the use of basic units McAdie conceived a new scale of temperature. He assigned absolute zero as 0 and the freezing point of water as 1000. This, he argued, avoided the use of negative signs, as was the case with the Celsius scale, and yielded a convenient unit which he called a "grad," as opposed to a degree. He said degrees should only be used to designate a portion of a circle. Later he named his new scale of temperature the "Kelvin Kilograd" scale. A "grad" corresponded to about one-half of one degree Fahrenheit.

The long weather records that had been made by Samuel Rodman and his son, Thomas, in New Bedford, Massachusetts (see chapter 1), were acquired by the Observatory at this time. They included temperature, rainfall, and weather records for the period 1812-1905. They were the most valuable series in existence for that period of time in southeastern New England.

At year's end the Library contained 7,920 bound volumes and 15,122 pamphlets. Forty-five periodicals were regularly received. The Blue Hill publications were sent to 340 institutions and individuals.

1917

McAdie showed new interest in aviation by noting the number of hours each day that he termed as "safe" for aviation. Using a gust recorder at Blue Hill he defined safe as when the wind averaged 10 m sec^{-1} (22 mi. hr^{-1}) or less and the variation was less than 50% of the five-minute maximum speed. This was to give some measure of the extent of "bumpiness," but no other meteorological elements that could be hazardous to aviation were considered. This statistic was computed for nine months then dropped with no particular analysis.

In telling of this work, McAdie indicated that a wind of 60 m sec^{-1} (134 mi. hr^{-1}), had been measured at Blue Hill over a 5-second period. He did not state what instrument was used to measure this extreme speed.

In this period McAdie converted an old weighing rain gage to an evaporation recorder with the help of a technician from Harvard's Jefferson Physical Laboratory. Evaporation measurements were made for a short period and these were used in McAdie's brief studies in horticulture. Also at this time he devised a "saturation deficit recorder" which graphed the dry and wet bulb temperature and relative humidity. He used this to obtain measurements in and around plants close to the ground.

Waldo E. Forbes died in 1917. He had been considered a staff member without pay and had helped obtain funds for the Observatory as well and made generous donations himself. One of his meteorological interests was the formation of sea fog between the Elizabethan Islands and Cape Ann of Massachusetts. Forbes also performed an interesting analysis of various long-period temperature records to determine whether an anomalous warming occurred on the tenth of May. Similar anomalous days in May were referred to as the "ice saints" in Europe. He concluded that no such phenomena occurred, at least in southeastern New England.

McAdie published "The Winds of Boston II" in 1917. This included very little new material and some erroneous ideas, one of which was an assumption that the winds above the gradient level determined the pressure distribution. It is difficult to understand such statements coming from a man of his background.

By midsummer of 1917 the Observatory endowment fund had reached $58,000 and income for the year was $9,400 from endowment income and gifts, the latter having come largely from Mrs. Rotch. Expenses were about $7,600.

Soon after the United States entered World War I, McAdie, with Harvard's permission, volunteered his services in meteorology. In 1917 the officer in charge of a group of naval students at MIT received word that training in meteorology was needed before they embarked for overseas duty. A young officer, Roswell F. Barratt, was sent to Blue Hill to investigate McAdie's standing offer of assistance. McAdie agreed to teach eight men, although he cautioned that six would be more reasonable. After trudging up the hill each day in winter for six weeks, they were ready to be shipped out. The officers had become so fond of McAdie and so stimulated by his teaching that they decided to make an effort to have him go abroad with them. Barratt went to Franklin D. Roosevelt, then assistant secretary of the navy, to enlist McAdie, but McAdie steadfastly refused. McAdie was then called to Washington, where Roosevelt left Barratt alone with him in his office for a last attempt at persuasion. McAdie finally agreed to go for one month and the group happily departed for London. There they separated, seven officers to French seaports and one to Ireland. While in London, Barratt and McAdie became acquainted with Sir Napier Shaw and they were elected fellows of the Royal Meteorological Society. McAdie visited and assisted these officers, but returned precisely at month's end, much to the disappointment of the men. A steady procession of officers followed for training at Blue Hill. During this period McAdie published *The Principles of Aerology* and *A Manual of Aerography* for the United States Navy (see bibliography 1917 and 1918), both of which were used in histruction of naval officers. McAdie invented the term "aerography" and proposed that "aerology" replace "meteorology." His influence is shown by present-day usage of these terms by the navy.

1918

McAdie was asked by the Navy Department to undertake the organization and supervision of an Aerograph Section of the navy in 1918. He was commissioned as lieutenant commander of the Naval Reserve Force and took a leave of absence for six months beginning in April. Fifty-four

officers had been trained at Blue Hill and placed on ships in the North
Atlantic or bases in Europe and at home. Among those trained was Francis
W. Reichelderfer, who later held the position of chief of the United States
Weather Bureau from 1938 until 1963. All were graduates of American
educational institutions, as shown by table 2.

Table 2

Educational Institutions Represented in the Aerographic Detachment Trained at
Blue Hill Observatory

Amherst	3	College City of New York	1
Brown	1	Univ. North Carolina	1
Univ. California	1	Northwestern	3
Univ. Chicago	1	Penn State	1
Columbia (Law)	1	Univ. Penn.	1
Columbia (Mines)	2	Princeton	8
Cornell	5	Univ. of Va.	2
Dartmouth	2	Univ. Washington	1
Denison	1	William and Marshall	1
Harvard	4	Williams	1
Univ. Kansas	1	Wooster	1
Lehigh	2	Worcester Polytechnic	1
Mass. Inst. Tech	5	Yale	2
Univ. Minnesota	1		

In recognition of his teaching, McAdie was appointed Abbott Lawrence Rotch Professor of Meteorology at Harvard University.

1919

McAdie remained on active duty until after the Navy-Curtis flying boat
trips across the Atlantic. He was sent to Halifax harbor aboard the *U.S.S.
Baltimore* to supervise flight forecasts from Long Island, New York, to
Halifax, Nova Scotia, and from Halifax to Trepassey Bay, Newfoundland.
Pilot balloon observations were made from the ship and two weather maps
covering a large area were drawn daily.

Letters from the Navy, (McAdie 1918-19), and from Franklin D.
Roosevelt, in which McAdie was commended for his work, follow:

Navy Department
Washington

June 15, 1919

MY DEAR COMMANDER MCADIE:

I am in informed by the Director of Naval Aviation that your request of April 19, 1919, for inactive duty was referred to the Bureau of Navigation, which has final cognizance of these matters, with a recommendation that it be approved and to become effective upon completion of your duty in connection with the trans-Atlantic flight. That duty has been completed, and your inactive duty orders were forwarded on May 29th.

I wish at this time to express my deep appreciation of your interest, efforts, and results of your services in this war. The fact that at the present time the aerographic work in the Navy in equipment and personnel is abreast that of the British Navy is a sufficient commentary on your efforts.

My best wishes for continued success in your work.

Sincerely yours,

Franklin D. Roosevelt,
Assistant Secretary
of the Navy.

Lieut. Comndr. Alexander McAdie
Blue Hill Observatory
Readville, Mass.

NAVY DEPARTMENT
U.S. Naval Observatory
Washington, D.C.

September 8, 1919

Enclosure

Sir:

I beg to thank you for your letter of August 30, 1919, in which you state that that we may consider ourselves at liberty to call upon you for your advice and information in connection with aerographic work.

The Naval Observatory appreciates very much what you have done personally for the Navy and for what has been done by the staff of the Blue

Hill Observatory, not only in connection with the training of aerographers for the flying corps, but in reference to aerography in general. In particular we are grateful for the benefits of your experience which you have so kindly given to the Naval Observatory from time to time both through personal visits and correspondence.

So far as the Naval Observatory is engaged in scientific instrument work, we will be glad to maintain the pleasant and helpful relations existing with Blue Hill, relations which are sure to be of benefit to the Navy, and we hope of mutual advantage.

I am glad to think that the Navy is coming to realize the importance of scientific study of weather conditions at sea and in the air. In bringing about this state of mind, your influence has been very great.

I hope you will visit the Naval Observatory whenever you feel so inclined.

The question of Navy instruments at Blue Hill will be taken up in separate correspondence.

Yours Truly,

J.A. Hoogewerff
Rear Admiral, U.S.N.
Superintendent.

Prof. Alexander McAdie
Director, Blue Hill Observatory
Readville, Mass.

NAVY DEPARTMENT
WASHINGTON

September 19, 1919

Sir:

I am in receipt of your letter of August 20, 1919, in which you request that certain instruments belonging to the Navy Department, which were installed in the Blue Hill Observatory during the war for the purpose of training aerologists for the Naval Air Service, be allowed to remain at Blue Hill.

I beg to inform you that the Navy Department has decided to grant your request pending further developments in the matter of future training of Naval personnel in Aerology.

The Department fully appreciates the assistance that has been given by the Blue Hill Observatory, not alone to the Navy, but to the scientific world generally. The removal of the instruments will be effected, if at all, only for the purpose of using them elsewhere when urgently needed.

The Department hopes that in case it becomes necessary to transfer the Navy material from Blue Hill, the pleasant relations existing between the Navy Department and yourself and your staff will continue as heretofore.

Sincerely yours,

Josephus Daniels.

Prof. Alexander McAdie
Director, Blue Hill Observatory
Readville, Mass.

In 1919 McAdie published a lengthy article on a quick method of measuring cloud heights and velocities. The method employed a somewhat modified nephoscope and the dew point formula, and therefore applied only to low clouds. The dew point depression was determined from dry and wet bulb thermometers; then differing values of height per unit of dew point depressions were used depending on the cloud type and general weather condition. The technique was not new and reference to similar work was reported in the Blue Hill annals.

1920

Detailed cloud observations, which included nephoscope readings when possible, in the morning, afternoon, and evening, ended. The other observations continued as before.

At about this time George P. Paine, on a fellowship of the National Research Council, began a series of investigations upon the aerodynamics of the psychrometer. This work he performed in Cambridge.

1921

In 1921 the director continued his efforts to increase the endowment but the Board of Observers postponed the endeavor because much of the university was already in the midst of a fund-raising drive. Income for the Observatory was small and if gifts from loyal friends, including Mrs. Henry Parkman, Jr., formerly Mrs. A. L. Rotch, had not continued, the Observatory almost certainly would have closed. Annual expenses at the Observatory now ran around $9,400. Only about $300 was allocated for equipment and supplies, resulting in more instrument failures. The check observations were maintained but the autographic records noticeably suffered.

During the year McAdie published tables of saturation weights and pressures and devised a hygrograph suitable for use in hospitals, where it was thought moisture content was important, especially in operating rooms and surgical wards.

This hygrograph (figure 61) employed ventilated wet and dry thermometers and a hair sensor for relative humidity. These were ingeniously connected through a series of levers to facilitate the graphing of temperature, dew point, and relative humidity.

A stationary psychrometer, also devised by McAdie and made about this time by Negretti and Zambra, is shown in figure 62. Note the thermometer scales in Kelvin kilograd.

Figure 61. Temperature, relative humidity, and dew point recorder invented by A. McAdie. In operation the vertical cylinder at right swung over the wet and dry bimetal sensors and air was pulled over the sensors and up the tube. A screened case covered the instrument. Copyright President and Fellows of Harvard College. Harvard collection of historical scientific instruments.

Figure 62. Psychrometer and nonogram devised by McAdie. Copyright President and Fellows of Harvard College. Harvard collection of historical scientific instruments.

Meanwhile, McAdie continued his preoccupation with units of measurement by publishing a paper titled "Symbols in Science" (see bibliography). This paper included about one hundred physical symbols, many of which were seldom used in meteorology at that time. However, inconsistencies appeared in the use of negative exponents. Pressure–height tables were prepared, presumably for a standard atmosphere; no virtual temperatures were specified.

1922

In 1922 Willard J. Fisher published a paper on low sun phenomena. In this he tabulated the duration of sunsets and sunrises. McAdie suggested that the structure of the atmosphere could be derived from the data, but no results were published. McAdie published papers on CGS units and the monsoon as a rain-maker. Wells continued his forecasts for the railroads and for varied business interests. The navy also used forecasts of fog in fleet maneuvers on the coast and of weather in general during speed trials.

David G. Byrnes served as a clerical assistant for a period beginning in 1922.

A note of interest by McAdie was that twenty or more institutions in the country gave courses in aeronautical engineering while only one course, in aerology, taught at Harvard, was available.

On 1 July the valley station was closed; it had operated 34 years.

1923

Friends of the Observatory raised the endowment to $121,000 in 1923. Nevertheless, the university asked McAdie to keep his expenses below $10,000 per year. The budget included the salaries of McAdie, Wells, and a janitor; supplies, nothing for instruments, small amounts for books, and experiments, and almost nothing for maintenance.

In a brief article titled "The Truth About Rain-Making," which appeared in the *Harvard Alumni Bulletin*, McAdie correctly explained the formation of a contrail and postulated that the rain gush in a thunderstorm originated with the coalescence of droplets which in turn caused a subsequent lightning discharge. Mention was also made of the new experiments by Bancroft and Warren in which they planned to drop electrified sand into clouds to cause rain.

McAdie's penchant for writing was at its best when he accounted for the great flood as follows:

> My own idea of the Flood is that, following a prolonged dry spell, there came a surge of cold air southward from the Caucasus. Then the hot breath of the Persian Gulf went streaming north and, as the air rose higher, and higher, the rain began. About this time Captain Noah shoved off. But it was criminal negligence on his part not to take a rain gage, or a pair of them, aboard; for now we shall never know how later floods compare with THE FLOOD.

McAdie addressed the legal problem of rainmaking:

> And then, granted that we do succeed in the gentle art of making rain, who shall say when and where to begin and when and where to stop? Ah! There's the rub! A mighty big rub too! Of course we would begin by Congress creating in its wisdom an Interstate Rain Regulation Commission. Most appropriately lame ducks could be appointed, for we have all heard of pouring water on a duck's back; going in one ear and coming out the other, as the freshman thought. The Commission would not render a decision in any case until five years or more after the rain in

question has ended. Then an appeal could be taken to the Courts, and who knows but that ultimately the case might go to an International Tribunal. We fear that the English might make it rain every day in London as a token of kindly interest and affection; and that the Germans would make it rain *immer und ewig* on the French from sheer *schrecklichkeit*; and the Turks, if they could raise money enough to pay the rain-makers, would make it pour all over the Greeks. Finally, to come nearer home. The chaps who failed to get their degrees might surreptitiously but cheerfully subscribe for a first class, crackerjack thunder-storm on Commencement Day.

During the year McAdie published two small popular books. *A Cloud Atlas* contained many pictures that he had taken in California; the other, *Making the Weather*, combined the contents of his shorter articles on the subject.

1924

The winter of 1923-24 was unusually stormy. In all, there were thirteen storms in which the wind exceeded 36 m sec^{-1} (80 mi. hr^{-1}). In one of these on 11 January a section of the copper roof on the south side was torn off by a gust of 37 m sec^{-1} (83 mi. hr^{-1}). The copper sheets were carried some 5 meters (15 feet) and wrapped around the chimney while a part dropped over the Observatory entrance on the north side of the building. A temporary roof was put on during the next two days but much of this was torn off five days later when the wind reached 35 m sec^{-1} (78 mi. hr^{-1}), also from the south. A third roofing was, in turn, demolished and final repairs were not made until April at a total cost of approximately $2,800.

"How big is an acre?" appeared in the *Atlantic Monthly* in July of 1924. In this article McAdie eloquently pursued his favorite subject of units. He solved the question through the conversion of inches to a link, square links to a pole, poles to a square chain and square chains to an acre. Then he compared the difficulty of calculating the area with that of giving the number of cents in $435.60. These and other simple examples were used to illustrate the importance of metric and absolute units.

At the time, Wells' salary was $2,000, and $150 was allocated for instrument repairs each year. No money was available for new instruments, books, research, or improvements.

1925

The Observatory Visiting Committee, with sanction of the university, began a fund drive in 1925. Six members of the committee contributed a total of $66,000 to start off the drive. Meanwhile, McAdie reminded the president of Harvard that the director's name appeared on official lists as a professor in the Department of Geology and as a member of at least two faculties, and as such rendered services, yet the Observatory had never received a dollar from the University to lighten the burden on its endowment. McAdie went on to say that the time was ripe for the establishment of a Department of Aerology in Harvard University.

New work during the year pertained to the determination of the dust content of air at Blue Hill using an Owens instrument, the study of thunderstorm development, and the results of Paine's work on the aerodynamics of the psychrometer. The latter work, which was performed in a laboratory in Cambridge, consisted of four parts in which theories were developed, and a final part where theories were verified using an Assmann psychrometer (see bibliography).

1926

The drive to raise funds netted $85,000 for the eighteen months ending June 1926. Although this increased the endowment to $176,000, the Visiting Committee and McAdie had set the "desirable" amount at $500,000. In addition, Mrs. Henry Parkman, the former Mrs. Rotch, gave the house at the base to the Observatory. With this relative affluence, the Harvard Astronomical Observatory ceased to pay for publication of the Blue Hill observations and papers. Further effort to obtain a department of aerology is evidenced by McAdie's comment to the president of Harvard that "the graduate of 1930 who has no knowledge of air structure and the many problems connected with it will be insufficiently trained, culturally as well as scientifically."

Wells continued his forecasts, which were now used daily by the railroads, public utilities, and educational institutions. Special forecasts of wind and fog were provided to crews working on salvage of the sunken S-51 submarine, and for tests of destroyers and cruisers off New York and Hampton Roads, Virginia. University policy did not permit the Observatory to charge for these services; therefore these services did nothing to supplement income.

During the year McAdie lectured to a class in Naval Aviation at M.I.T. The Observatory instrumentation was improved by the gift of a new 3-cup anemometer.

1927

In 1927 summary tables of Blue Hill Observatory observations were published and distributed to all leading observatories and a number of selected libraries.

The remainder of the copper roof that had not been replaced in 1924 was now re-roofed with new copper.

Friends continued to come forward with donations and at this time the endowment fund reached approximately $200,000.

In another *Atlantic Monthly* article McAdie wrote the following in regard to the dissipation of fog (see bibliography):

> Latent in every gramme atom of hydrogen and of oxygen are tremendous stores of energy. There may be in a single fog droplet energy enough, if let loose, to run a hundred automobiles for a year. Ah—to be able to unlock that energy! The newer physics teaches that mass is interchangeable with energy; and so even the edge of a fog droplet might, if converted, furnish a vast supply of heat. The physicists of today are hiking down a trail which leads ultimately to the liberation of atomic energy. The quest is on; the finish may not come for years; but the steady increase in knowledge will undoubtedly lead to methods that can be used advantageously in dissipating fog.

He continued by describing the accidental dissipation of ground fog from the propeller wash of an airplane as a cotton field was being sprayed by an entomologist. Forty-two years later similar tactics were used to dissipate ground fog when helicopters were employed on a grand scale by Plank and Spatola (1969).

1928

McAdie expanded his teaching activities to average fourteen hours a week for eight Naval Officers.

In the light of increasing use of aircraft McAdie sought and obtained a grant from the Daniel Guggenheim Fund for the Protection of Aeronauts. This money was used to prepare papers on the hazards of lightning and on icing from super-cooled fog droplets (see bibliography).

McAdie proposed that a first-class seismograph be set up at the Observatory in connection with the Department of Geology. He estimated its cost at $200,000. The proposal was never seriously considered by the University.

McAdie was sole representative from the United States on the International Commission for the Study of Clouds, which was constituted in 1921. In 1928 he prepared a report for the commission in which he discussed previous work of the Observatory and added a section on "cloud formations as hazards in aviation." Thirty-six illustrations showed characteristic formations. Nearly 300 were sent to observatories and interested persons. Cost of the report, which was about $1,200, was paid by Frank H. Bigelow and Henry W. Cunningham.

1929

McAdie summarized his work for the Guggenheim Foundation in an article titled "Weather Hazards in Aviation," which appeared in *Scientific Monthly* (see bibliography). Hazards he discussed were thunderstorms and lightning, fog and ice formations and *anakatabats*, a Greek word that he used for up-and-down marches or what was commonly called "bumpiness." He described the phenomena accurately in most cases, and gave the current status of research on each of the hazards. He described the roll cloud preceding a thunderstorm and for the first time in Observatory literature mentioned the term "cold front," the ability to forecast its motion, and associated thunderstorms in summer. Line squalls, now called squall lines, he also described accurately from the visual standpoint, but their cause he incorrectly attributed only to the convergence of air of different origins.

1930

The routine observations at the Observatory were not published for the first time, undoubtedly because of the lack of funds. Expenses had risen to about $14,000 per year.

A preliminary edition of a new *International Cloud Atlas* was published in 1930. McAdie helped in the selection of photos, some of which were taken from the Blue Hill collection. At this time he also published a book called *Clouds* in memory of Professor Rotch. This twenty-two page book contained fifty-two plates and three figures.

Allotments from a new fund, the Milton Fund, were used by McAdie to devise instruments for use at airports which might improve aviation forecasts. Modifications of the dew point formula were used for the determination of low cloud bases, and cloud motions were used to position cyclone centers in reference to the observer. From these data he advocated the introduction of "ice warning semaphores" to be located along airways at about 150-km (100-mi.) intervals. Presumably these visual signals were to warn low-flying pilots of icing conditions before they were encountered.

1931

Professor McAdie, now in his sixty-eighth year, retired in July. He had served as director for eighteen years. At this time the position commanded a salary of $8,000 per annum plus the use of the base house.

In summary, McAdie was a kind, witty, and articulate man. His personal charm played an important role in raising $170,000 for endowment, an outstanding service to the Observatory. He received little financial support from the University; in fact, at times the Observatory ran into the red by as much as $800 in a year and he personally made up the deficit, including interest for which he was charged.

He not only instructed young naval officers in meteorology during World War I, but, at the request of the navy, launched a whole new branch of the U.S. Navy for the handling of its aerological needs.

While serving at Blue Hill, he wrote numerous papers and spoke frequently on the use of metric units. He fought a losing fight, correct as he was, for the use of "kilobar" in place of "millibar," a unit of pressure already adopted by the International Meteorological Organization.

Some of McAdie's writings were purely philosophical (see bibliography). His scientific work was liberally sprinkled with anecdotes, sometimes leading to inconsistencies, and incorrect assumptions. He frequently interjected Greek phrases and loved to introduce new words; such as "nephelolater," meaning one who watched clouds. Some brilliant reasoning in regard to cloud physics appeared in his writings from time to time, but serendipity was lacking. No single outstanding scientific program developed over the years of his directorship, quite possibly due to the difficulty in obtaining financial support. He dabbled in new instrumentation and developed useful charts and nonograms, as shown by the following list spanning his years at Blue Hill:

1. Cryoscope 1920 (a form of psychrometer)

2. Chart and apparatus for correcting altimeter readings (patent application filed 30 September 1930)
3. Altimeter corrector, 1920, 1921, and 1926
4. Nephoscope (patent No. 1446574, 17 February 1923)
5. Psychrometer (patent No. 1488994, 1 April 1924)
6. Altimeter adjuster, 1927
7. Method and means for computing attained height of aircraft (patent application filed 16 August 1927)
8. Chart and apparatus for taking altimeter readings, 1927
9. Multiscale thermometer, 1930
10. Thermo-dynamic thermometer with McAdie multiscale (ordered 20 May 1931)
11. Barometer dial corrector
12. Bia-nephoscope
13. A set of glass plates that could be spaced one above the other to represent layers of the atmosphere, 1907 (Isotherms and winds were shown on each plate)

The observational program at the Observatory was not monitored closely by McAdie. Wells was in charge throughout McAdie's directorship. He lived alone at the Observatory, except for occasional trips to the valley. He spent long hours recording hourly cloudiness and prevailing types from 0700 to 2000 EST along with other observations every day of the year. He became obsessed with the program and sought to make everything look all right. The check observations, which were made by eye, may have been faithfully executed, although a study by Mitchell (1961) raises doubt. In a comparison of the Blue Hill temperatures with eighteen surrounding stations, the scatter was found to be slightly higher, especially in summer, than for periods before and after the "McAdie period." The autographic records suffered from instrumental problems that McAdie did not attend to. In later years, if a chart was changed later than the usual time, the record was meticulously erased and transferred to the new chart to indicate that it was changed at the usual time. This practice was a totally unacceptable scientific procedure that should have been corrected.

With the flair for teaching and background in physics and mathematics which McAdie possessed, it is surprising that he did not become interested in the new work of the Bergen School before M.I.T. started its atmospheric sounding program and studies in dynamic meteorology. This was never mentioned and it seems inconceivable that the work passed without notice.

McAdie was a member or fellow not only of the American and British Meteorological Societies, but also of the American Physical Society, Association of American Geographers, Ecology Society of America, Seismological Society of America, American Association for the Advancement of Science, the American Academy of Arts and Sciences, and the American Antiquarian Society.

The affectionate regard in which Professor and Mrs. McAdie were held was expressed in the presentation of a silver bowl and plate to them by the Blue Hill Visiting Committee at a dinner in their honor in 1931, when they were leaving Milton. After retirement they moved to Hampton, Virginia, and in one of his last letters, still thinking of the future, he wrote, "You know, I think that atmospheric electricity will ultimately take the place of oil in producing power . . . . A thunderstorm is simply a protest to man. Why not use me? — and the atmosphere is the best original electric machine we know." McAdie died two years later on 1 November 1943.

References

Koelsch, W. A. 1985. Ben Franklin's heir: Alexander McAdie and the experimental analysis and forecasting of New England storms, 1884-1892. *New Eng. Quart.*, 59:523-543.

McAdie, A.G. 1894. *Protection form lightning*. U.S. Wea. Bur. Bull. 15, 21p. 2d Ed. 14 May, 1894, 21p. 3rd. Ed. 1985, 26p.

——1895. Natural rain-makers. *Pop. Sci. Mon.* 47: 642-648.

——1903. *Climatology of California*. Washington, D.C. U.S. Govt. Print. Off., U.S. Wea. Bur. Bull. L., 270 pp.

——1909. *Infra numbem*. San Francisco: A.M. Robertson, 42 pp.

——1912. *The ephebic oath and other essays*. San Francisco: A.N. Robertson, 62 pp.

——1918-19: Blue Hill Observatory, Rept. of the President of Harvard University: 183-184.

McAdie, M.R.B. 1949. *Alexander McAdie, Scientist and writer*. Wakefield, (MA): Murray Print. Co., 421 pp.

Mitchell, J.M. 1961. The measurement of secular temperature change in the eastern United States. Res. Paper no. 43. U.S. Dept. of Commerce, U.S. Weather Bureau. Washington, D.C., 80 pp.

Plank, V.G. and A.A. Spatola. 1969 Cloud modification by helicopter wakes. *J. App. Meteor.* 8: 566-578.

Chapter IX

Charles Franklin Brooks becomes Director; the Observing Program is Restored and Mount Washington Reoccupied: 1931–1934

Charles F. Brooks was born 2 May 1891 at St. Paul, Minnesota. As a youth, the severe thunderstorms that occur in that part of the country impressed him and served to spark his interest in the weather. He applied climatology early in life when he prepared snow-shoveling contracts based on the season or by the storm. At Harvard, he received an A.B. in 1911, an A.M., under Professor Rotch, in 1912, and a Ph.D. in 1914. His Ph.D. in meteorology was the second to be awarded in this country. He then went to Washington, D.C., to work for the Department of Agriculture. There he prepared the climatic section of the *Atlas of American Agriculture*. He then took a position as instructor in geography at Yale University.

Later he served as instructor in meteorology for the U.S. Army Signal Service at College Station, Texas. Following this work he returned to Washington to edit the Weather Bureau's *Monthly Weather Review* from 1918 to 1921. For the next decade he taught meteorology and climatology in the Graduate School of Geography at Clark University. There his family lived on a sixteen acre woodland belonging to the university, an excellent place to raise a large family. Brooks was already a keen observer and especially interested in the mesoscale distribution of temperature on clear, calm nights. Late, one very cold night, while making a temperature survey on foot, he was apprehended by the police in a dump as he stood, flashlight in hand, whirling a hand psychrometer. Lengthy persuasion was required to avoid a trip to the local psychiatric hospital.

The American Meteorological Society was organized in 1919 largely through the personal energy and enthusiasm of Brooks. He was elected secretary at the first meeting and served in that office until 1954. He also served as editor of the society's *Bulletin* during its first nineteen years.

In 1931 he took the post of professor of meteorology at Harvard and director of the Blue Hill Meteorological Observatory. In view of Brooks's education at Harvard and his close association with Rotch and Ward, he was well qualified for the directorship. He had already worked at the

Observatory, was happy to reestablish a relationship with Harvard, and looked forward to the new challenge.

1931

After Brooks arrived at the university's base house in midsummer, his first work went into putting the observational program back in order. He succeeded in "borrowing" S. P. Fergusson from the U.S. Weather Bureau where he had advanced to a position in charge of the Instrumental Testing Laboratories.

Fergusson was given the title of research associate for one year, in which time he was to restore the instrumentation. Thermometer calibrations were begun immediately and corrections applied. Meanwhile, Brooks personally checked many of Wells' observations to ensure accuracy. By 1 November Wells had resumed more detailed cloud observations and Brooks had started recording cloud positions in a small circle representing the celestial dome.

Robert de Courcy Ward, long-time professor of climatology at Harvard, died suddenly on 12 November. However, his courses were continued in Cambridge, almost without a break, by Gardner Emmons and Brooks.

Since Brooks served as secretary to the American Meteorological Society and editor of its *Bulletin*, headquarters for the society automatically moved with him to the Observatory.

The base station was reopened 31 November.

1932

The meteorological observations had been published up until 1930, but because of the expense Dr. Brooks elected not to resume publication.

However, monthly summaries, which included temperature and precipitation, were published each month as part of a New England Weather Bureau report.

By this time, limited space at the Observatory had forced some curtailment in the collection of climatological data from foreign countries. A. Hamilton Rice, who had recently founded the Institute of Geographical Exploration at Harvard, 2 Divinity Avenue, Cambridge, generously offered to make the lower stacks of the institute building available for the Observatory's voluminous climatological collection. With this collection were included the file of the U.S. daily weather maps of the Department of Geology and Geography, complete from 1871, and the maps and collateral material used for the courses in meteorology and climatology at the Geographic Institute.

At the same time the entire meteorological and climatological holdings of the Harvard College Observatory were given to the Blue Hill library. The climatological portion was merged with the Blue Hill collection in the Geographical Institute stacks.

In March, under the auspices of the Observatory, the weather station atop the Geographic Institute building on Divinity Avenue was started. Wendell Smith took most of the observations and changed the recording charts weekly.

On 10 May the Hazen shelter, which had been retired to the backyard of the Observatory, was returned to the enclosure and put in use. A new wire fence completed the enclosure.

Extensive repairs were made to the Observatory and the base house which was occupied by Brooks and his family. New coal furnaces were installed with attached domestic hot-water heating systems. At the Observatory an outside coal bin was added to the south side of the building to help accommodate the twenty tons of buckwheat coal used annually.

During July and August S. M. Serebreny, assistant in aeronautical meteorology at New York University, with Charles H. Pierce, student at MIT, made morning and afternoon pilot balloon runs. Serebreny was studying the Observatory's methods of meteorological observations and turning of the wind with height. Pierce also helped Wells with the observations, but Wells insisted upon entering the data in the books even when others did the actual observing. The only known picture of Wells shows him assisting in one of these pilot balloon runs (figure 63).

The financial situation improved with a gift from Henry S. Shaw, chairman of the board at General Radio, for shelving and books in the library. This marked the beginning of a long and friendly relationship during which Shaw made substantial donations to the Observatory. Another notable gift by a Harvard graduate established the Geophysical Research Fund, which yielded an annual income in excess of $1,500. This income was put to use immediately to appoint Oliver L. Fassig, Bernard Haurwitz, and H.H. Kimball as part time research associates and to employ several part time assistants. Kimball quickly moved to start the solar radiation observational program. Pyrheliometers to measure the total radiation on a horizontal surface and normal incidence radiation were purchased from the Eppley Laboratories. An equatorial mounting with

Figure 63. Mr. L.A. Wells, left, assisting with a pilot balloon run. American Meteorological Society—Blue Hill Collection.

electric motor drive was set up in the south notch of the parapet; a normal incidence pyrheliometer was mounted on it so it could track the sun from sunrise to sunset. Continuous recordings began on 10 December. Radiation through different path lengths of the solar beam in selected spectral bands was also measured. From these, calculations of the turbidity, or the dustiness of the atmosphere, could be made, leaving the way open for approximations of the total water vapor content of the atmosphere. Kimball and the recorders are shown in figure 64.

The International Polar and Cloud Year, a program to gather data, especially upper air data, over temperate latitudes, began on 1 August. More than $6,000 became available from the Geophysical Research Fund, the Milton Fund of Harvard, the Hodgkins Fund of the Smithsonian Institute, and a generous gift from J.J. Storrow, Jr., brother-in-law of Rotch's son, Arthur. At this time Brooks put entirely new forms for the instrumental and cloud observations into use. Samples of these forms which remained in use through December of 1964 are shown in figures 65a-d. New observations were the grass minimum temperature each night, regular temperature readings at the ground surface and 15 cm (6 in.) below

Figure 64. Dr. H.H. Kimball beside Englehard and Leeds and Northup "Micromax" solar radiation recorders, c. 1939. American Meteorological Society—Blue Hill Collection.

the surface and visibilities in all quadrants. The detailed cloud observations were entered in a separate book that provided space for each layer, its position in the celestial dome, name and international code, height, direction and speed as measured by nephoscope, and optical phenomena.

Meanwhile Fergusson, and an assistant, Philip A. Towle, had been making new thermo-hygrographs, anemoscopes and anemographs which were to be used during the coming eclipse on 31 August. Eleven parties, including Edward Brooks's oldest son, were stationed over the path of totality in New England and eighteen more stations were equipped by the Observatory. Data were obtained on both land and water and by pilot balloons. Much of the path of totality was cloudy but extensive observations were obtained. A comprehensive analysis of the data was published in 1941 (see bibliography). Significant temperature changes were found at the surface as with previous studies. The increase in winds aloft when released from convection was too small to lower the pressure more than an average 0.3 mb and the layer of air appreciably cooled was too shallow to effect an observable sinking of pressure surfaces aloft. Therefore, little redistribution of air could be expected, and little evidence of a pressure rise at the surface attributable to inflow aloft was found.

E. Monroe Harwood arrived to serve as assistant on 15 September. He had studied at Clark University and quickly adapted to the observing program.

METEOROLOGICAL OBSERVATIONS - PRESSURE, TEMPERATURE, HUMIDITY, W

STATION | LATITUDE | LONGITUDE | ALT. OF BAROM. CISTERN ABOVE M.S.L

DATE	MERCURIAL BAROMETER: TIME 75TH MERID.	ATT. THER.	BAROMETER AS READ (BAROM. NO. 2.72)	COR'N. FOR TEMP.	STATION PRESSURE	SEA-LEVEL PRESSURE	BAROGRAPH: READING	BAROM. TEN-DENCY (SYMBOL)	PSYCHROMETER: TIME 75TH MERID.	DRY BULB	WET BULB (MUSLIN LAST CHANGED ------)	DRY MINUS WET	DEW-POINT	REL. HUM.	VAPO PRES SURE
Feb. 1	802	14.8	750.70	-1.81	748.39	✓	747.25		8.05	29.0	28.4	0.6	26	88	3.(

STATION | LATITUDE | LONGITUDE | HEIGHT OF ANEMOMETER ABOVE GROUND

DATE	TIME	DIRECTION: BY EYE (16 PTS.)	BY ANE-MOSCOPE (32 PTS.)	ANEMO-SCOPE MIN. FAST OR SLOW	VELOCITY: ESTI-MATED (BEAU-FORT)	BY ANEMO-METER (LAST 5 MIN.)	ANEMO-METER TIME (MIN. FAST OR SLOW)	ANEMO-METER DIAL READING	WEATHER: GENERAL: SIN. LAST OBS. (0-9)	CURRENT WEATHER .00-99 (BY SYMBOLS)	≡ R̷ ∞ △ OR ▲: TYPE	BEGAN	ENDED	PRECIPITA: RAIN: TYPE	BEGAN	ENDED	DURA-TION (MIN)	AMT.	TYP
Feb. 1	800A	W	W	3f	7	4.7	Cor	3.57	0	00									

METEOROLOGICAL OBSERVATIONS - CLOUDS, VISIBILITY, GROUND TEMPERA

STATION | LATITUDE | LONGITUDE | ALTITUDE OF GROUND ABOV

DATE	TIME	CLOUDS: CODE: C_L	C_M	N_L	N	C_H	GENERAL MOTION: HIGH DIR.	VEL.	MIDDLE DIR.	VEL.	LOW DIR.	VEL.	CEIL-ING	VISIBILITY: N	E	S	W	GEN.	DIST. MTS.	REL. HUM.	SUSPEN-DED IM-PURITIES IN AIR	GROUND TEMPERATU: GRASS MIN.	GRASS SET MIN.	AIR TEMP. MINUS GRASS SET MIN.	GROU SURF
Feb. 1	800A	0	0	0	1	9							Unl	6	6	6	6	6	0			18.9	25.7		29

METEOROLOGICAL OBSERVATIONS - CLOUDS IN DET/

STATION | LATITUDE | LONGITUDE | ALTITUC

DATE	TIME	WIND DIR. & VELOC.	SKY COVER: TOTAL AND AVERAGE DENSITY	SKY ABOVE 10° AND AVERAGE DENSITY	KINDS AND CHARACTERISTICS: CODE	SYNOP-TIC SYM-BOL	NAMES	AMOUNT: VISI-BLE	SUR-MISED	POSITION	DEN-SITY 0 TO 4	HEIGHT: HOM-TO	ABOVE OBSER-VER FT.	ABOVE M.S.L. FT.	MOVEMENT: TIME	AZIMUTH FROM	ANGULAR VELOCITY Thousandths of radians per sec	ACTU VELOC M/

Figures 65a-d. Observation forms put in use in 1932. The first line of each page is shown. Blue Hill Meteorological Observatory.

With the approach of the International Polar Year, Joseph B. Dodge, Appalachian Mountain Club hutmaster at Pinkham Notch, New Hampshire, Robert S. Monahan, mountaineer, and other enthusiasts decided to press for the reestablishment of a meteorological station atop Mount Washington (Monahan 1933). The project was endorsed by the International Commission and plans for the mountain station advanced rapidly. In those Depression days funds were scarce, but enough money was raised along with material contributions to obtain instrumentation, fuel, food, and other meager furnishings for wintering over in the old stage office. A voluntary crew consisting of Monahan, Salvatore Pagluica, and Alexander McKenzie were assembled as observers and radio operator, while Dodge served as field director from Pinkham Notch.

Brooks gave meteorological training and the Observatory supplied observation forms, a mercurial barograph, thermohygrographs and wind measuring equipment from the eclipse expeditions. Fergusson hauled instruments to the summit and installed them. A theodolite and associated equipment to make pilot balloon runs, a nephoscope, and total solar and sky pyrheliometer and recorder furnished by the Eppley Laboratories were also sent to the mountain. Base stations, which also used surplus Blue Hill instrumentation, were set up at Pinkham Notch on the east, and at the base of the cog railway on the west.

Thus began, in October, the present Mount Washington Observatory, and a renewed association with the Blue Hill Observatory.

While traveling in Norway in 1931, Brooks was shown an experimental heated anemometer by Sverre Pettersson of the Norwegian Weather Service. In view of the icing problem plaguing anemometers at the mountain station Brooks decided to acquire a similar anemometer. From Brooks's recollection of the instrument and Fergusson's knowledge of anemometry, D. W. Mann, of the Mann Instrument Company, designed and constructed a new instrument (Pagluica et al. 1934). This instrument, known as number 1 heated anemometer, consumed 600 watts of electricity. It was tested on the mountain during November and successfully recorded super hurricane winds under rime icing conditions. Plans for another anemometer incorporating some design changes were made immediately and sent to Mann.

The number of observations at Blue Hill were increased again on 1 October when hourly maximum visibilities, expressed in the International Code, were added to the hourly cloud observations of prevailing type and amount for the hours 0700 through 2000 EST.

Fergusson enjoyed the congenial atmosphere of the Observatory once again and decided to stay on indefinitely under Brooks.

On 10 December Brooks noted in the journal, "Wells, Harwood, Brooks, H. Wexler, Pierce, Haurwitz, Fassig, Fergusson, Gray, Greer, Emerson and Gast working here today, only Kimball of staff absent." Gray, Greer, and Emerson were from the Eppley Laboratories, and Gast was from Harvard Forest. The Observatory was once again "off and running," reminiscent of the days of Rotch.

1933

Originally much of the detailed observational program at Blue Hill was to end with the International Polar Year; however, the data proved to be of such interest that all but the pilot balloon program was continued.

From normal incidence solar data screened by Schot red and yellow filters, Dr. Kimball prepared a paper on the turbidity of air masses having different sources. Haurwitz analyzed the 1933 solar radiation data in the first of a new series, the *Harvard Meteorological Studies* (see bibliography for 1934). Dr. Brooks continued his cooperation with the Woods Hole Oceanographic Institution in supervising the investigation of the sea surface temperature of the western Atlantic by E. M. Harwood and P. E. Church.

In March a meteorograph capable of operating a week unattended and recording all the basic weather elements, except precipitation, was assembled by Fergusson and set up by Phil Towle and his father, Frank, in the fire lookout atop the hotel on Mount Wachusett (figure 66). This station was located 615 m (2,017 ft.) above sea level and 70 km (94 mi.) west northwest of Blue Hill. It operated until June, when it was removed for modification.

The new heated anemometer for Mount Washington arrived on the summit 8 March and was known as the heated anemometer number 2 (see Pagliuca et al. 1934).

Figure 66. Edward Brooks and Phil Towle stand before the hotel and fire lookout on Mount Wachusett. American Meteorological Society—Blue Hill Collection.

The work was paid for by a contribution of several hundred dollars from Henry Shaw and a grant of $75 from the American Academy of Arts and Sciences.

Through the winter of 1932-33 experiments with radio transmissions on 5, 2 1/2 and 1 1/4 m wavelengths were underway, primarily between the summit of Mount Washington and Shaw's residence in Exeter, New Hampshire. Shaw then interested Al Sise of the Shepard Broadcasting Company of Boston in the work and he expanded the tests to more distant sites. In the course of his travels from hilltop to hilltop with his portable equipment Sise brought his friend Arthur Bent with him to Blue Hill, where Bent was introduced to Brooks. Bent, a radio enthusiast, had become devoted to the mountains while working in the Appalachian Mountain Club system as a youth and did much to help the Mount Washington Observatory, of which he later became director. Sise's car and attached portable antenna are shown on Blue Hill in figure 67. On 10 May, Sise and Bent returned to Blue Hill with new equipment and made a good contact with Mount Washington on 5 m radio. This historic event marked a world record distance of 229 km (142 mi.) for transmission at that wavelength. The path was slightly over the horizon which previously was thought to make reception impossible. According to McKenzie (1980) the next day's contact was very weak at Mount Washington and unreadable at Milton. Propagation "anomalies" had been encountered which demanded explanation. Shaw then arranged to support Sise, who was made a research assistant at Blue Hill. At this time Pickard, who had experimented with radio as early as 1899 on Blue Hill, became involved in the propagation studies. By this time, Pickard was an eminent radio engineer and inventor, he had founded the Institute of Radio Engineers, and was its second president. Signal strengths were measured over various paths by Shaw at Exeter, New Hampshire; by Pickard at Seabrook Beach, New Hampshire; the Mount Washington station, and by Sise in Brookline, Massachusetts.

A new transmitter, having a frequency of 60.0 MHz, or wavelength just off the 5 m band, was constructed using funds from an anonymous donor. It was set up on 7 November on the third floor of the Observatory on Blue Hill and given the call letters W1XW. With this new more powerful equipment daily contacts with Mount Washington became routine (figure 68).

Because of the fact that the Mount Washington observations seldom represented those of the free air at the same level, and interest was increasing in the application of the observations to synoptic problems, a

Figure 67. Al Sise's car and portable antennas in the fog on Blue Hill, December 1932. American Meteorological Society—Blue Hill Collection.

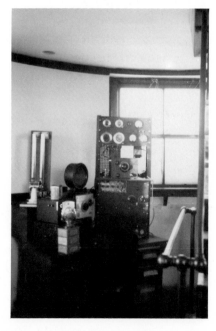

Figure 68. Transmitter on third floor of the Observatory, c. 1933. American Meteorological Society—Blue Hill Collection.

new comparative program was launched to evaluate these differences. In August Fergusson, Towle, Irving Schell, and Christos Harmantas went to Twin Mountain, New Hampshire, and Mount Washington to make further measurements between the free air, the summit air and air above the summit. An MIT meteorograph was mounted on an airplane (figure 69) and was flown from Twin Mountain while Fergusson made a few kite flights to about 400 m (1,300 ft.) above the mountain from the summit (figure 70). These data supplemented those of past years when Fergusson flew kites from Twin Mountain for similar investigations. Unfortunately they did not yield comparisons in bad weather and in seasons other than summer when the plane could not fly.

During the summer the Appalachian Mountain Club maintained numerous huts in the mountains and some of these were supplied with instruments from Blue Hill. Data from these stations were also used by Schell in the comparison of free air temperature with those on the mountains.

In the early fall additional funds were received. Of these $4,000 assured operation of the Mount Washington Observatory for another year. The new staff was made up of Pagliuca and McKenzie, both incumbent observers, Wendel Stephenson, and Robert G. Stone. Stephenson, a graduate of the University of Chicago, worked in a summer camp in New Hampshire where he fell in love with the mountains. He took work where available, and even helped the year before in rehabilitating the stage office

Figure 69. Airplane and meteorograph used to make free air measurements for comparison with those on Mount Washington. Some members of the Brooks family are present. American Meteorological Society—Blue Hill Collection.

Figure 70. S.P. Fergusson, back to the camera and wearing suspenders, flies a kite and meteorograph mounted inside the kite from the summit of Mount Washington. American Meteorological Society—Blue Hill Collection.

for winter use. Stone had been a graduate student under Brooks at Clark University. He was without work but especially interested in weather and climate, so he quickly agreed to join the team when asked by Brooks.

At Blue Hill, Harwood resigned on 24 November and Schell, along with Wells took over the observing duties.

Fassig went to San Juan where he was made visiting professor at the School of Tropical Medicine. He cooperated in planning an intensive study of public health in relation to climate and assisted in establishing a solar and sky radiation observing program.

At this time a "triple register" recorder, standard equipment at Weather Bureau stations, was sent from Blue Hill to Mount Washington. This recorder had pen arms that could be used to indicate wind direction every minute and another pen to indicate wind speed.

1934

In April and May R.F. Baker, a student who had arrived in January, made the first comprehensive measurements of ultraviolet light at Blue Hill (see bibliography for 1935).

On 12 April a super-hurricane wind was measured with the heated number 2 anemometer on Mount Washington, (Pagliuca et al. 1934 and McKenzie 1984). One-tenth-mile electrical contacts were timed with a stopwatch on the mountain and at 1321 EST three successive contacts were measured to give a corrected speed over a period of 1.17 seconds of 103.3 m sec^{-1} (231 mi. hr^{-1}), the highest wind ever measured at the earth's surface. Between 1237 and 1255 EST the contacts were broadcast, picked up at Blue Hill and timed by stopwatch or recorded on a high-speed-chronograph. During that period corrected speeds between 48 and 96 m sec^{-1} (107 and 214 mi. hr^{-1}) were measured over 5-second periods. Ultimately a duplicate of this anemometer was ordered by the Chinese weather service for use at its station on Mount Tai-Shan, where ice-forming conditions were similar to those on Mount Washington.

Pierce left Blue Hill on 25 May to take a position as forecaster with Transcontinental and Western Airlines, later Trans World Airways. Much later Pierce returned to the Boston Weather Bureau and served as deputy meteorologist-in-charge and chief forecaster.

Also in the summer Stone came to Blue Hill for full time work after spending most of the winter on Mount Washington. His initial work was to help with the observing but he soon began to assist Brooks with everyday matters associated with the American Meteorological Society. He also helped in editing material for the *Bulletin of the American Meteorological Society* and the new series, *Harvard University Meteorological Studies*. He became especially interested in the library and started restoration of its files. With the aid of new gifts, missing volumes were obtained, exchanges were negotiated, and binding reinstituted, a colossal job after neglect since 1912.

On 4 June a new survey party from the U.S. Geological Survey found it necessary to erect another tower over their benchmark on the rocks in front of the Observatory. This eyesore stood for several years before demolition.

Wells became ill with acute appendicitis on 24 June and after much persuasion was removed to Milton Hospital. The operation was successful but his wife of a short time and doctor convinced him that he should retire.

Wells had continued with much of the observing under Brooks but it was difficult for both him and the new young and energetic staff. He was fiercely loyal to the Observatory, the U.S. Navy and New Haven Railroad, and wanted no change. He treated some observations as strictly his own prerogative, while radio and the newer observations he shunned. He kept newspapers pasted over the German illustrations on the dining room wall to show his distaste for the Germans, who fought the United States some sixteen years earlier. Brooks, in his gentle way, tolerated these actions but the situation was strained until Wells's resignation. Shortly after Wells's retirement Towle prepared to move into his bedroom and in the course of cleaning he and Stone found hundreds of exchange publications that had been sent to the Observatory over the years and stuffed under the bed. In addition, recent publications that Brooks had addressed for mailing were also found. The two had quite a time sorting out the material, which had provided a haven for silver fish and beetles. Wells had served thirty-one years full time, preceded by nine years part time, by far the longest service of anyone up to that time.

H. S. Rice came to the Observatory 9 July as research assistant. Rice, a mathematician associated with Wentworth Institute, continued his assistance for many summers.

The newly rebuilt meteorograph was returned to Mount Wachusett in the fall. A large chart drum turned without attention for at least two weeks as pressure, wind direction and movement, temperature, and relative humidity were recorded. A two-week record could be contained on a single sheet of millimeter paper about 30 cm (12 in.) high and 1 m (39 in.) long.

In the autumn, a system for turning on the 5 m transmitter at Blue Hill for a few minutes every hour was set up for the purpose of studying signal propagation. Signal strengths were then recorded by Pickard at Seabrook Beach, New Hampshire, and starting in January 1935 by Ross A. Hull at Hartford, Connecticut. Hull, editor of *QST*, the official publication of the American Radio Relay League, was the foremost radio amateur of the day. He had noticed that he could frequently communicate with many amateurs in the Boston area from his Hartford location, presumably under certain weather conditions. Other Blue Hill signal strength measurements were made at Mount Washington and from boats traveling along the New England coast.

On 22 November a meeting of the Boston Section of the Institute of Radio Engineers was held. Sise and Pickard reported progress on the investigation of high-frequency radio transmissions over the past year. Average diurnal and seasonal variations in transmissions over several wavelengths, including 5 m, were presented. It was noted that strong signals often preceded bad weather. Brooks offered a hypothesis that this apparent connection with weather changes might be due to an increase in refraction, which is found at times when temperature inversions develop aloft during a weather change from good to bad. Later, Hull, through use of the Boston airplane soundings, showed this hypothesis to be correct for explaining most of the transmission variations at the wavelengths under discussion. Meanwhile, record transmissions on 2 1/2 and 1 1/4 m wavelengths over longer and longer distances were reported as more amateurs used the frequencies.

As the year ended Brooks, with the cooperation of A. J. Connor, climatologist of the Canadian Weather Service, finished maps of the climatic elements over North America and the West Indian region. The maps were published in the Koppen-Geiger *Handbuch der Klimatologie*.

The Mount Washington observations were now under study for use in forecasting local snowstorms, work that continued for many years.

Harwood returned 1 December to take the place of chief observer Wells.

A mailing consisting of 557 copies of *Harvard Meteorological Studies* (Number 2) was made at year's end. This report, by Jerome Namias, was the first detailed analysis of subsidence in the atmosphere. Several hundred copies were taken immediately by the Weather Bureau, army and navy meteorological services. To meet the large countrywide demand, "An Introduction to Air Mass Analysis" by Namais and others was edited and partially written at Blue Hill and published serially in the *Bulletin of the American Meteorological Society*.

Near the end of the year a modified Nipher shield was constructed and placed around the standard 20.3 cm (8 in.) diameter precipitation gage. The shield (figure 71) was 1.42 m (56 in.) in diameter with its top level with the precipitation collector. An opening at the base of the shield allowed the discharge of snow, and an outward flaring collar diverted the wind from that opening. Two of the old New England Meteorological Society's gages were also put into operation to provide comparative measurements.

On the evening of 26 December a northwest gale commenced. At 0050 EST on the 27th the cinemograph stuck at 33.5 m sec^{-1} (75 mi. hr^{-1}). Shortly afterward, the wooden mast from which the weather flags were flown in the past broke off and the radio antenna and three anemometer masts went down with it. Masts supporting the wind vane and windmill remained intact. By 0300 EST the wind began to subside and during the day Fergusson and Bent started temporary repairs.

Figure 71. General view of the enclosure looking west–northwest. Standard precipitation gage surrounded by the modified Nipher shield. Mosquito screening was added later to reduce splash. During the heavy snows of 1947-1948 the gage and shield were raised. The Fergussun weighing gage is shown just to the right of the Nipher shield. New England gages were positioned to the left and right, shown by arrows. c. 1936. American Meteorological Society, Blue Hill Collection.

1935

By mid-January repair of the tower masts was complete. A new wooden mast was installed, to support radio antennas, and the pipe masts were securely guyed with stainless steel rods.

Use of Blue Hill by skiers was first recorded in the journal on 27 January when "many skiers on hill" was noted. At that time all skiing was done on the summit road or bridle trails.

References

McKenzie, A.A. 1980. The good men do lives after them: Henry Southworth Shaw, 1884-1967. *Mount Washington Obs. News Bull.* 21: 55-59.

——1984. World record wind, measuring gusts of 231 miles per hour. *Mount Washington Observatory*, Gorham, N.H. 03581, 30 pp.

Monahan, R. S. 1933. *Mount Washington reoccupied. Stephen Daye Press*: Brattleboro.

Pagliuca, S., D. W. Mann, and C. F. Marvin. 1934. The great wind of April 11-12, 1934 on Mount Washington, N. H., and its measurement. *Mon. Wea. Rev.*, 62: 186-195.

Chapter X

50th Anniversary and the Radio-Meteorograph Program: 1935–41

1935 (continued)

On 1 February 1935 the Observatory celebrated its fiftieth anniversary. The event, as described by Brooks in the *Harvard Alumni Bulletin*, is reproduced here in full except for a few figures which are shown elsewhere.

Blue Hill Observatory—Fifty Years Old

Skyrockets shot from the top of Blue Hill on the night of 31 January announced to all who could see them the fiftieth anniversary of the first regular observation of the Blue Hill Observatory. On the following day, the celebration continued with an informal luncheon and an hour of short speeches which were heard not only by those on Blue Hill but also by the staff of the Mount Washington Observatory and by others.

The luncheon was served by the ten members of the Observatory staff present at this time to the twelve guests who had come to the party. Although the sky was clear, and the air crisp and pure, the deep and badly drifted snow made walking fairly arduous, especially for the few who were not equipped with snowshoes or skis; consequently appetites were good, and the luncheon of baked beans, sandwiches, cocoa, pie and cake disappeared rapidly. A photograph of the group was taken by Mr. Fergusson of the Observatory staff. Then, while the regular 2 p.m. observation was being made, the visitors inspected the Observatory.

Those present included: Mr. I. Tucker Burr, Dr. Edward Wigglesworth, Mr. J.J. Storrow, Jr., and Mr. Arthur Rotch, members of the Harvard Overseers' Committee to visit Blue Hill Observatory; Professor Harlow Shapley and Professor Willard P. Gerrish of the Harvard College Observatory; Mr. G. Harold Noyes, of the U.S. Weather Bureau; Professor Hurd C. Willett and Mr. C. Harmantas, of the Massachusetts Institution of Technology; Mr. Helm Clayton and Miss Frances Clayton; and Mr. John Clinton, 81 years old, who retired three years ago after 38 years' service as caretaker; he was the oldest one who climbed the hill for the celebration (figure 73). The members of the Blue Hill Staff present were: Mr. S. P.

Fergusson and Dr. B. Haurwitz, research associates; Messrs. I.I. Schell, R.G. Stone, and A.E. Bent, research assistants; Mr. E.M. Harwood, Jr., Chief Observer; Mr. P.A. Towle, general assistant; Miss L.H. Block and Mr. E.M. Brooks, voluntary assistants; and Professor C.F. Brooks, Director.

At 2:30 the group gathered in the radio room on the top floor of the tower, heard the hourly automatic tone signal, the strength of which is noted on Mount Washington, N.H., at Seabrook Beach, N.H., and at West Hartford, Conn., and then listened to a conversation between Blue Hill, Mount Washington, Pinkham Notch, and Exeter, N.H. Mr. A.E. Bent, in charge of the radio station W1XW, at Blue Hill, established contact with Mr. A.A. McKenzie, in charge of the Observatory's station on Mount Washington, and through him talked with Mr. Joseph B. Dodge, founder and manager of the Mount Washington Observatory and observer at the Pinkham Notch Station. Mr. Dodge sent his greetings, and Professor Brooks replied, with compliments to Mr. Dodge for his successful management of the Mount Washington Observatory, now well along in its third year. Next, also through Mount Washington, contact was made with Mr. Henry S. Shaw, a member of the Overseers' Committee for Blue Hill, at his home in Exeter, N.H. Mr. Shaw voiced his cordial greetings, praised the attainments of the Observatory, spoke of his interest in the present investigations, and expressed confidence in the future. He remarked that certain recent developments were probably not even dreamed of at the time the Observatory was founded, for example, radio; and asked Professor Gerrish if he had any idea that there would be any such development as radio. Professor Gerrish replied to Mr. Shaw's greetings, and admitted that radio was not thought of even as the remotest possibility. Mr. Bent then summarized briefly the nature of the Blue Hill radio equipment, including both 5-meter and 2 1-2 meter sending and receiving sets, used in experimenting with transmission over distances which approach the limits possible with such equipment.

At about 2:50 Professor Gerrish was called on for his description of the opening of the Observatory, and Mr. Clayton and Mr. Fergusson followed. Professor Brooks then read a communication from Professor McAdie, now in Virginia, and Mr. Harwood presented greetings from Mr. L.A. Wells. The remainder of the program included statements by Dr. O.L. Fassig and Dr. H.H. Kimball, research associates, concerning their special fields; a summary of meteorology in relation to the Astronomical Observatory by Professor Shapley; greetings by mail from Dr. S.S. Drury, chairman of the Harvard Overseers' Committee on Blue Hill, and from Mr. John Woodbury, a member of that committee; a tribute from Mr. G.H. Noyes, representing the Weather Bureau; and an outline of the work of the Observatory today and a look into the future by the present director. They are all presented on the following pages.

Professor Willard P. Gerrish

It is impossible for me to realize that 50 years have elapsed since a winter evening, January 31, 1885, when Mr. Rotch and I ascended Blue Hill to open the Blue Hill Observatory. Mr. Rotch had engaged me to take charge of the station, as observer, and this was the final trip of several which we made during the construction of the building. The work of construction had been delayed, and though the building was closed in, the only floor which was finished was that in the circular room at the base of the tower. In this room there were a stove and telephone; nothing else. There was no furniture of any kind. We slept for the first few nights in blankets, on the floor, and subsisted on camp fare. Our instrumental equipment was of the simplest, consisting of a barometer and a set of thermometers.

Observations were to be made at 7 a.m., 3 p.m., and 11 p.m. daily, and I made the first observation at 11 p.m. that night, in order to cover the full period of 24 hours on the following day. Mr. Rotch had brought up a few skyrockets to celebrate the event. These he set off at midnight from the ledge near the building, and the Observatory formally and officially opened.

The interior of the building was gradually finished, and furniture and equipment were supplied as fast as they could be accommodated. A steward, a colored man whom I had engaged, finally appeared and assumed the duties of cook, chambermaid, and general caretaker. With the exception of Mr. Rotch, who made frequent visits to the Observatory, and an occasional guest, this man was my sole companion during the year of my stay.

The honor of being the first observer has grown remarkably with the years, as the Blue Hill Observatory has developed from a small private enterprise into an institution of international fame.

H. Helm Clayton

In the years 1884-1885 I was acting as student assistant at the astronomical observatory in Ann Arbor, Mich., when word came to us that A. Lawrence Rotch of Milton, Mass., was building an observatory on Blue Hill to be devoted to meteorological research and that he was looking for an assistant. I was already actively engaged in meteorological research and it seemed an opportunity to join with others in this work, so I applied for the position. On February 1, 1886, I began work as observer and assistant at the observatory. I entered the work with all the enthusiasm of youth and we planned an intensive study of cloud forms and cloud movements which in a few years brought forth a large volume giving the results. In this volume was first formulated the law of the equality of mass movement at different heights in the atmosphere, which is now generally recognized by meteorologists.

In the meantime, Mr. S. P. Fergusson and later Mr. Arthur Sweetland were added to the staff of the Observatory. In 1894, stimulated by some preliminary work of William A. Eddy, we entered into the active work of exploring the air with instruments lifted into the air by kites. This was pioneer work, the success of which was greatly aided by the genius of Mr. Fergusson in inventing and adapting instruments for the purpose. Later, sounding balloons with recording instruments were sent up from St. Louis, Mo., and Pittsfield, Mass., reaching heights of about 16 kilometers, or ten miles into the atmosphere. These records demonstrated the existence of a stratosphere over the United States such as previously has been observed in Europe. Mr. Rotch's unflagging interest made the work of the Observatory widely known all over the world and led to similar work in other countries. While this work was in progress, the Observatory became a feature of the Boston newspapers for many years.

In 1900 I carried kites and balloons to Argentina to initiate exploring the upper atmosphere in that country, and later was put in charge of the forecasting service of the Argentine weather service.

For nearly a quarter of a century I was attached to the Blue Hill Meteorological Observatory. To slightly misquote Tennyson:

> Here about the hills I wandered
> Nourishing a youth sublime
> With the fairy tales of science
> And the long results of time

The Observatory is now manned by a new staff of young enthusiasts, to each of whom I wish a long and fruitful career.

S. P. Fergusson

My connection with the Observatory began in September, 1887, when I was employed by Mr. Clayton to relieve him of some of the routine work and provide more time for research. At this time the work of the Observatory and interest in meteorology were growing rapidly, and, beginning with 1890, I became a regular member of the staff; the original period of temporary service, intended to continue three months, extended to twenty four years. Then, after six years at the University of Nevada and fifteen in the Weather Bureau of Washington, I was invited by Dr. Brooks, in 1931, to rejoin the staff at Blue Hill as research associate and share the work of a new period of expansion.

The Observatory was founded during a period of rapid growth of interest in meteorology, when the Signal Service, then carrying on the work of the present Weather Bureau, was dominated by an able scientific staff headed by Abbe, Ferrel, Mendenhall, and others.

The conditions of work at Blue Hill have been most favorable for maintaining a high quality of service; the staff has been encouraged to serve the best interests of meteorology without the personal or administrative restrictions apparently inescapable in our governmental or other large institutions; European meteorologists have often expressed surprise at the relatively large productivity of so small a staff with such modest resources. Other conspicuous features of the history of the past fifty years are the youth of the staff, definite plans for all investigations, and early publication of results. Professor Rotch was 23 years of age at the time he established the Observatory, Professor Gerrish was 19, Mr. Clayton 25, I was 19, and Mr. Wells was 27, at the time of our full connection with the Observatory began, while Professor McAdie was 22 at the time of his first temporary work in 1885.

On the occasion of the fiftieth anniversary of the Observatory there is evidence of another period of advance in meteorology, indicated by the introduction of studies of air-mass analysis, the extension of aerological explorations, and the large number of young men beginning work in our universities and the national Weather Bureau. Under the able direction of Dr. Brooks, the Observatory is taking a most important part in this expansion, and the second half-century will undoubtedly record greater advances than have been possible during the first. We of the older generation offer our hearty congratulations on the work already accomplished and assurance of our continued interest and support.

From Professor McAdie's letter

In 1882, I went to the Signal office in Boston, and in some way soon met Lawrence Rotch, then an undergraduate at the Massachusetts Institute of Technology. A friendship began which lasted without interruption until his death. In 1882, 1883, and 1884, several times in his company, indeed as his guest, I went to the top of Blue Hill. He had a fixed intention to build an observatory. He offered me the position of observer, but I could not accept, being tied up at Cambridge. Then he got Gerrish (about as young now as ever). I relieved Gerrish occasionally. Clayton came after the first year, and then Fergusson, and later others. The first kites were flown in 1885 to get a measurement of atmospheric potential. Again in 1890 and 1891 measurements were made; and on one occasion during the approach of a thunder shower, we had the kite-wire attached to the new, large electrometer. The sparks were coming rather large and frequently, and the problem was to disconnect. We finally got the instrument disconnected, and the danger to the Observatory was over, but it was "lively."

In 1913, I was surprised by a letter from Ward, speaking for the President of Harvard, offering me the position of director. A week was needed to think over the matter. I was a professor in the Weather Bureau, practically in charge of the Pacific Coast. The salary was $500 less, and my friends did not wish me to accept. The President had said that he hoped "I had enough of the Harvard spirit to accept the offer." It was a challenge, and, because of my friendship for the founder, I accepted.

Figure 72. Tablet on the tower of the Observatory. American Meteorological Society—Blue Hill Collection.

During the War, as you know, some 58 men were trained at the Observatory, all but one being commissioned. The director went overseas and for six months was away.

Time and space will not permit my naming all the men who studied at the Observatory or the scientific contributions. The Observatory still stands, and will, I trust, be the rallying spot for years to come, free from routine entanglement, as its founder wished.

Letter from L.A. Wells

Fifty years! A long and honorable record of achievement in meteorological research and service to the community! The wisdom of our beloved founder, Professor A. Lawrence Rotch, has been amply demonstrated and his high aims and ideals have always been an inspiration to those who followed him.

It is fine to have here today the pioneers who have done so much for the Observatory, the first observer, Professor Gerrish, who has made a name for himself in astronomy, and Messrs. Clayton and Fergusson, who have given their lives to meteorology.

The brilliant work of the Observatory during its first 27 years, notably the cloud studies and upper-air research which had made the institution known throughout the world, was necessarily curtailed because of lack of funds in 1913, when the Observatory became part of the University.

To Alexander McAdie, Abbott Lawrence Rotch Professor of Meteorology, great credit is due for his 18 years of devoted service. He built up the endowment, a most difficult task, and his many studies and writings did much for Blue Hill. The instruction and training of 66 naval aerographers, many of whom served overseas, was an outstanding feature of the work. Meteorological work in the Navy, begun by these men, is carried on today in greatly expanded form. University students also received instruction and have shown a lasting interest in meteorology.

Great praise and support should be given the present able director, Charles F. Brooks, whose energetic administration since 1931 has greatly enlarged the work and resources of the Observatory.

Three times in the 50 years has the Observatory been expanded, and now again it has nearly outgrown its quarters.

For the writer, it was a privilege and pleasure to have been associated with the Observatory, and his only regret is that he could no longer do the work.

Thanks are due President Lowell and President Conant and the visiting committee for their loyal support through the years since 1913.

May the next fifty years show an equal record of accomplishment!

Letter from Dr. Oliver L. Fassig

At present I am engaged here in San Juan in the reduction and study of available data on the climate of Puerto Rico. My report is progressing rapidly; when completed, it will comprise about 300 octavo pages of text, tables, and charts. Dr. Carlos Charden, chancellor of the University of Puerto Rico, will include my report in his series of "Monographs of the University of Puerto Rico," some time in 1936. The report is being prepared with the aid of a grant from the Milton Fund from Harvard University. The progress of this investigation is being greatly advanced by the courtesy of Dr. George W. Bachman, director of the School of Tropical Medicine, here in San Juan, who has extended to me the facilities of the School.

A major problem to which I have been giving considerable attention during the past few years is concerned with long-term fluctuations in the annual temperature and pressure conditions. An intimate study of the correlations between these climatic factors for the same fifty-year period promises interesting results.

Letter from Dr. H. H. Kimball

The first measurements of the intensity of solar radiation made by the Observatory staff were obtained in connection with the meteorological studies undertaken during the solar eclipse of August 31, 1932. Continuous records of the intensity of the radiation received on a horizontal surface, obtained at several points in the path of totality, were summarized and correlated with the temperature changes.

Before the end of 1932, radiation apparatus had been secured and installed at Blue Hill. The routine measurements were similar to those obtained at radiation stations of the U.S. Weather Bureau. They are summarized and published in the "Monthly Weather Review," Blue Hill being the first New England Station included in the nation-wide network. During 1933, Mt. Washington, N.H., was added to the list.

At Blue Hill, the measurements include the following:

(1) Continuous records of the total solar radiation received on a horizontal surface, which are summarized to give hourly and daily totals and weekly averages throughout the year.

(2) The intensity of the direct solar radiation as received on a surface normal to the incoming rays, and summarized to five monthly averages with the sun at certain fixed zenith distances, or, reciprocally, at certain altitudes above the horizon.

(3) The intensity in selected spectrum bands, as in the visible spectrum, in the red and infra-red, and in the violet and ultra-violet. The visible spectrum is practically free from absorption bands: therefore variations in its intensity make possible a determination of the dustiness, or the turbidity of the atmosphere. With the degree of turbidity determined, the water-vapor content of the atmosphere may then be determined. These two characteristics of the atmosphere, turbidity and moisture content, are of prime importance in air mass

analysis studies. This is an international research program, participated in by most European nations. Mt. Washington and Blue Hill are the only participating stations in the United States, or indeed, in the Western Hemisphere.

Researches for improving instrumental equipment methods of observation are continuously carried on. It is hoped that in the near future the extreme ultra violet radiation, which is of special interest from the medical point of view, may be systematically measured at Blue Hill.

Professor Shapley

It is of interest that the attention of astronomers has returned to the meteorological problems in the most recent years after a considerable interval of indifference. When the Harvard Observatory was founded nearly one hundred years ago, the first director was deeply interested in meteorological observations. William C. Bond in the 1840's had a definite ambition to make scientific meteorology an important part of the new Harvard Observatory. He had the ambition of providing standards for the calibration of meteorological instruments for the country. His interest, and that of his successors at the Harvard Observatory, was almost wholly in the collection and practical interpretation of weather data. This interest was inspired by the lack of meteorological institutions elsewhere. With the growth of science, the responsibility of the Harvard Observatory decreased, especially at the home station. The founding of the Blue Hill Observatory left to the Harvard Observatory, then directed by Professor Edward C. Pickering, much less responsibility for observation and experiment. To Pickering and to Professor Rotch it appeared that the Observatory's best contribution would be in the nature of funds for the publication of systematic results. In consequence, ten of the quarto volumes of the "Annals of the Harvard Observatory" are devoted to the publication results obtained at the Blue Hill Meteorological Observatory. From 1889 to 1927 the institutions were associated in this work of publication—the observations coming from Blue Hill and the publication funds from the Harvard Observatory.

Since the time of William Bond, routine meteorological work has always been carried on in some parts of the Observatory. The largest contribution came from the Boyden station of the Observatory, especially during the 36 years when it was located at Arequipa, Peru. As in Cambridge at an earlier date, so in Peru in more recent times the Observatory's meteorology work was of a pioneer sort, and its continuance has become of less importance with the rise of meteorological stations suitably and more fully equipped for systematic investigations. Of all the meteorological observers connected with Harvard Observatory, Professor Solon I. Bailey should probably be named first. His outstanding contribution was probably the establishment in 1893 of a weather station on the summit of El Misti, at an elevation of 19,200 feet.

The recent return of interest to meteorological problems arises from the astronomical interest in stellar and planetary atmospheres. The important contribution of the meteorological physicist in recent years is leading the astrophysicist to new points of view in the studies of stellar structure. Moreover, the astronomer's interest in the shooting stars, whose average observed altitude is something like 70 kilometers, again calls his attention to the problems of the atmosphere and its constitution. The study of solar-terrestrial effects and the investigation of cosmic rays also tie up the astronomical and meteorological fields.

The newer developments at the Blue Hill Observatory are of great interest to us because of the promise of a new knowledge and new techniques in the general investigation of the most available astronomical body and its gaseous envelope.

Letter from Dr. S. S. Drury

On February 1, the 50th anniversary of the founding of the Observatory of Blue Hill, may I join in tribute to Lawrence Rotch and to his able successors. Mr. Rotch was the most unselfish and chivalrous of scientists, and was followed by Professor McAdie, true humanist among men of research. Now, with Dr. Brooks, the contacts of the Observatory with the scientific world are continuing and your future under his guidance is assuredly bright.

All honor to the past of Blue Hill—all confidence in the future.

Letter from John Woodbury

I wonder if any member of the committee besides myself visited the top of Great Blue before the Observatory was built. I did in about 1870, and as my friend and myself walked from Ashmont and back, I can still remember how footsore I was and how lame for several days. My hearty congratulations to all who arrive at the top on Friday.

G. Harold Noyes

The Meteorological Service of the Federal Government, now called the Weather Bureau, brings to you on this festival of your fiftieth year, greetings, words of hearty commendation for past and present achievements, and bespeaks courage and great urging to further advances.

Our chief of the Weather Bureau, Mr. Gregg, in Washington, has long been a friend of Dr. Brooks, and they have worked in meteorological accord successfully. Mr. Gregg bids me wish you all success, and would extend his most sincere congratulations.

The Weather Bureau takes a modest share of pride unto itself that it has had a little to do in the formation of some of the abilities which both McAdie and Brooks have brought to the Blue Hill Observatory.

McAdie and Brooks are both Harvard men, and, as another Harvard man, I bring my personal tribute to this, a Harvard institution.

I recall a visit to this place when it was about 12 years old: William Morris Davies and Robert DeCourcey Ward leading their kindergartner meteorologists from the University Museum to observe a real weather observatory. About that time kites were beginning to go up, and a little later I had the daily grind of kite flights in Topeka, Kan., forerunner of the daily airplane observation flight. The kite work done at Blue Hill afforded us both inspiration and experience.

Blue Hill Observatory is one of the great needs of present-day research. Our Federal service is so tied down to legislative limits that our excursions toward real research are usually curtailed or frowned upon. We have to show a profit to the taxpayer, and usually research takes a long time on a complicated journey to show black on the ledger.

An institution like this is always in "the red," so there are no inhibitions as to its exploration.

You would be greatly surprised to hear the frequency with which I learn that the Weather Bureau gets its reports from Blue Hill, or that we are a part of your institution. Some of our mutual neighbors, I am sorry to say, put little trust in our Bureau, and turn with complete confidence to this rocky lookout point. We are not grieved about this, for their confidence is well placed. Eventually we shall have them as clients, and we shall urge their substantial and material support to this Observatory.

The Weather Bureau extends all possible support to you, and with you. The hope and anticipation of your new progress and discoveries is a pleasure to us all.

Charles F. Brooks, Director

A fiftieth anniversary is a time to take stock with an eye to the future. We gain strength and purpose by looking back upon the intent of the founder, Professor Rotch, and upon the achievements

Figure 73. Guests and a few of the Blue Hill staff at the fiftieth anniversary meeting: I.I. Schell, I. Tucker Burr, John Clinton, C. Haramantas, Miss F. Clayton, B. Haurwitz, W.P. Gerrish, Harlow Shapley, A.E. Bent, Hurd Willett, E.M. Brooks, H.H. Clayton. American Meteorological Society—Blue Hill Collection.

of the Observatory under his direction and that of his successor, Professor McAdie. The Observatory is for research in meteorology and its incidental practical applications. The sky is the limit. But a small staff with limited financial support must, perforce, confine its attention to relatively limited problems. Today we are continuing the detailed record of the weather, now in its 51st year, not only to obtain a continuous record over a long period of years, so that fluctuations in climate—indeed actual changes in the climate of Boston in comparison with Blue Hill—may be evaluated, but also to catch the exceptional occurrences, the unusual experiments of nature, which may show more clearly than ordinary weather just how the atmosphere works.

As these occur they are investigated, for example, the eclipse weather of 1932, and we shall be ready to observe the effect of next Sunday's partial solar eclipse. We are gathering case histories of storms so that we may recognize their symptoms early enough to warn other departments of Harvard, and the neighboring towns. Dr. Haurwitz is putting tropical hurricanes into mathematical formulae to learn why they are so strong, and then he will see if any of these results can explain the enigmas of our more complicated storms of middle latitudes. We are attempting to factor our seasonal abnormalities. Perhaps we shall not have to look for the ground hog's shadow on a February 2 in order to learn whether the rest of the winter will be mild or cold.

We are keeping a continuous record of how much solar and sky radiation reach the ground, and by inference, how much is lost or absorbed en route. It is solar radiation that drives the atmosphere. We are investigating the changes in ultra-high-frequency radio transmission in relation to the weather. When we find out whether, as now seems probable, a warm wind aloft makes better hearing over the radio, perhaps we can use the radio transmission to tell us what is going on in the atmosphere. Finally, we are undertaking to summarize for all American readers the most significant work in meteorology, and to present to the meteorological world the results not only of our own investigations, but also the essence of those by other Americans.

We seem isolated way out here on a hill top, but we are not. Here we are talking to our associates on Mt. Washington and to others who may be listening on their short-wave sets anywhere from Portland to Hartford. We are in close touch daily with the Boston office of the U.S. Weather Bureau and the meteorological department of the Massachusetts Institute of Technology, with which we try to work as a unit in investigation and service. By mail we are in contact with all the principal meteorological services and establishments the world over. Our publications go out, and theirs come back in return. Our library contains the latest reports of meteorological investigations from Europe, from Asia, and from South America.

The Blue Hill Meteorological Observatory of Harvard University has a group of observers, investigators, and writers who, thanks to the courage and enthusiasm of Abbott Lawrence Rotch, his colleagues and successors, are cooperating with others in learning the ways of the weather and interpreting it for the general benefit. The speed at which we shall gain knowledge will depend largely on the funds available for bringing young enthusiasts into meteorology and providing them with the equipment necessary for their studies. It is safe to say that there will be plenty of weather in the future for all comers.

The previous speakers have told of the origin, history, and some phases of the present work of Blue Hill Observatory. I shall try to look into the next 50 years to see how this Observatory may help carry forward the exploration of the atmosphere, begun so effectively by the founder and his associates. We are no longer working in a nearly virgin field; meteorology in the United States has made great strides in the past 50 years, in the past five years, and even in the past year.

Kite-flying is practically obsolete as a means for obtaining information as to the free air. The airplane serves much better, but it is not a complete substitute—it cannot give us the wind velocity above the clouds, and, without great expense, it cannot yield a continuous record for some hours at the same altitude. The mountain station can provide these, if due account is taken of the influence of the mountain itself on the passing air. But the mountain is limited as to location and altitude.

A sounding balloon can extend to very great heights the type of record the airplane gives. But this device as now generally employed will yield its precious record, if ever, only after hours or days, when a casual finder may return the instrument. Our Observatory sent up six sounding-balloons from Pittsfield in 1910, but only three yielded records; three disappeared, two probably into the Atlantic, and the other into one of the stomachs of a voracious cow at Amherst, with fatal results. Sounding-balloons sent up every two hours can yield relatively continuous records for all levels. This has been tried recently at St. Louis by Professor Rossby of M.I.T. and his associates. Almost every day the Weather Bureau obtains reports of atmospheric conditions to heights of three or four miles. These reports come from about 20 airplane stations and 70 pilot-balloon stations scattered throughout the country. This information is supplemented by many cloud observations and by reports from two mountain stations (our own Mt. Washington and one in California). The airplane, balloon, cloud, and mountain stations are sure to multiply, to meet the demand for more complex information. But without some other means, the most important weather, namely, the most stormy will pass unobserved except at the ground and on mountains. In storms the airplanes cannot fly with safety, the pilot-balloons cannot be seen.

The next major step in the exploration of the atmosphere will, therefore, be the frequent and general use of radio sounding-balloons, a relatively recent invention. As the balloon ascends, a small radio transmitter emits signals. The intervals between signals vary with the pressure (which shows the height, the temperature, and the humidity). The position of the balloon at any time, and thus the wind direction and velocity, can be determined by radio directional receivers. If more warning of the recent snowstorm could have saved Boston any money and inconvenience, the use of radio sounding-balloons, even though they are expensive, would be justified on practical grounds alone. Blue Hill Observatory should experiment with them to test the practicability of this method as a routine procedure.

By the end of the next 50 years, I expect that radio sounding-balloons will be sent up several times daily from a hundred stations in the United States, and that our great knowledge of the atmosphere will be so greatly increased that forecasts will no longer be so uncertain as "unsettled, probably rain or snow."

(reprinted from the Harvard Alumni Bulletin)

Thus after Brooks's first four years as director, in which he restored and amplified the observational program, a specific plan for a new endeavor in observational meteorology was laid out. With meteorology's newly developed air mass analysis, a growing need for extensive and routine upper air data was evident. Data formerly obtained by kites, then balloons, and later airplanes, was to be obtained quickly and efficiently by radio from instruments attached to high rising-balloons.

The Observatory would endeavor to develop a lightweight radio-meteorograph[1] to be lifted by free balloons and transmit meteorological data to the ground as it was measured. The idea was not new but an inexpensive instrument suitable for routine use had not been developed.

Friedrick Herath and Max Robitzsch in 1917 sent signals down a kite wire (Middleton 1969). At about the same time Pierre Idrac, who introduced the term "radiosonde," used tuned radio circuity to transmit and receive the signal of each element on different frequencies. These instruments employed time cycles as the method of transferring meteorological measurements to telemetry. Varying time intervals were proportional to the varying elements that were measured. The principle, according to Middleton (1969), dates back to 1874, when E. H. Baumhauer of Haarlem, Netherlands, designed a telemeteorograph of that sort for use on land. A different apparatus on the same principle was built by H. Olland, after reading Baumhauer's paper, and from that time onward such instruments have been said to operate on the "Olland principle."

Pazel A. Moltchanoff developed an instrument that transmitted coded signals. The signals were derived from arms that made contact with metal strips, which led to the term "Kammgerat," or comb-like. This 2-kg (4-1/2lb.) instrument cost $688 in the United States. In January 1930 Moltchanoff was the first to successfully sound the stratosphere with a radio-meteorograph. It was launched from Sloutsh, USSR. Meanwhile, Paul Duckert, at Lindenberg, Germany, developed a new type of instrument. The capacitance of a condenser, and thus the radio frequency put out by the transmitter, was controlled by a temperature sensor. Pressure was indicated by interruptions in the temperature signals.

In 1932 Vilho Väisäla, of Finland, developed an instrument which was the first to telemeter all the meteorological elements in terms of radio frequencies.

---

[1] The word "radio-meteorograph" was generally used through most of the developmental period. Gradually it was replaced by "radiosonde", but not for about five years.

For the Blue Hill project, Brooks engaged Karl Otto Heinrich Lange for part time work. Lange came to MIT from Germany with experience in aerological soundings that used meteorographs on airplanes or gliders. At MIT he and an assistant, Roland Feiber, directed the airplane sounding program which Rossby started in 1931. Daily soundings were attempted to heights from 4.5 to 5.5 km (14,000 to 18,000 ft.) but often the flights had to be canceled due to bad weather. The airplanes carried a mechanical meteorograph in which measurements of the pressure, temperature and humidity were traced on a chart and read upon return to the ground.

Of the two basic types of sounding instruments, Lange elected to develop a radio-meteorograph that employed the Olland principle. Preliminary tests would be made from an instrument on the U.S. Army Air Corps airplane, which made the soundings for Rossby. A schematic draw-

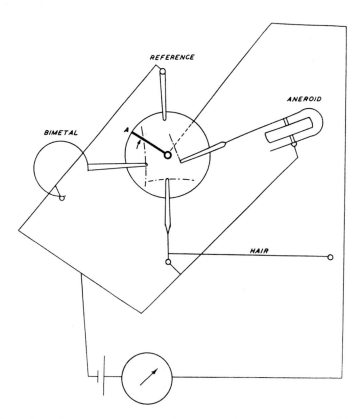

Figure 74. Schematic diagram by Lange of a radio-meteorograph planned for test from an airplane. K.O. Lange, Trans. Amer. Geoph Union, 16th Ann. Meet., p. 145, 1935; copyright by the Amer. Geoph. Union.

ing of the first instrument is shown in figure 74. The disk rotated once per minute thereby making a contact, or measurement of each element, at approximately that time interval. From the diagram it can be seen that as an element stylus moved, the time interval between its contact with the conducting straight strip across the disk and the contact of the fixed, or reference stylus on the strip, will change. Actually, the first instrument to be carried aloft by the airplane only contained a temperature element. The instrument was of necessity heavy and rugged to withstand vibration and the high wind speeds experienced on the airplane.

Meanwhile Arthur Bent designed a transmitter that would send signals, on a 5m-wavelength, from the airplane to Blue Hill. After Charles B. Pear came to the Observatory he copied, from Bent's notes a circuit diagram of the transmitter that was placed on the airplane (figure 75).

On 17 April, temperature data were successfully transmitted to the Observatory from the airplane. In the course of the flight the plane had been forced to a distance of 96 km (155 mi.) from the hill while climbing to 5.2 km (17,000 ft.) through a severe snow squall. Lange and the

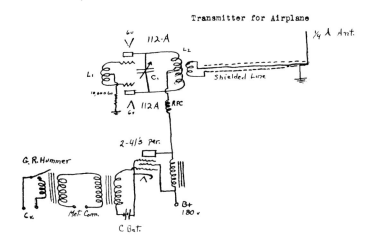

L1--8 turns #16 wire, ½in. dia.

L2--4 turns #10 wire, 1in. dia.

L3--2 turn coupling coil,one at each end.

C1--Hammarlund isolantite padding cond. 50 mmf.

Transmission Line--Shielded Auto Ignition Cable.

Figure 75. Copy of A.E. Bent's diagram of the radio-meteorograph transmitter to be used in an air plane. Courtesy C.B. Pear, Jr.

Figure 76. Fergusson and chronograph, which was used to record signals from a radio-meteorograph, 1935. Copyright President and Fellows of Harvard College. Harvard collection of scientific instruments.

transmitter were in the cabin of the plane while Bent was at the take-off field in East Boston; Towle and Ross Hull were at the Observatory to receive the signals. Signals were recorded from 1115 to 1210 EST on a special chronograph made by Fergusson. The chronograph (figure 76) made one revolution per minute, in approximate synchronization with the meteorograph. During each revolution the pen carriage spiraled down a few millimeters to allow a clear record that did not overlap the previous revolution. It is important to note that exact synchronization between the radio-meteorograph and chronograph was not required. As the meteorograph clock cooled with altitude it usually changed rate. However, since each element is measured in terms of a time interval from the reference contact the errors are minimized.

Shortly afterward, the complete system, including meteorograph and transmitter, was demonstrated at the spring meeting of the American Geophysical Union in Washington.

Later, in April, model "B" was completed. This contained all three elements of pressure, temperature, and relative humidity. It was the first complete instrument flown from an airplane.

Model "C" evolved in June. In this model the instrument was made lighter and placed in a tear-drop shaped aluminum housing for use on the airplane.

Also in June, Lange had become involved sufficiently at the Hill to require transporting his books to the small room behind the tower which he used as an office. On 1 July he was appointed research meteorologist at Blue Hill but he did not sever his relationship with MIT.

Tests of Model "C" showed frequent clock stoppages due to low temperatures, so model "D" was developed in a effort to alleviate the problem. It employed a fan to take the place of the clock. The fan was arranged to turn slowly as the instrument ascended through the atmosphere. However, due to turbulence, the turning rate could not be made constant, even for short intervals. Attempts to reduce the variations in the rotational rate were made by increasing the moment of inertia of the fan, but before significant improvement was achieved, the instrument became too heavy. Consequently the model was abandoned.

Model "E" pertained to instruments of the Moltchanoff type. It was also abandoned after a short time.

Model "F" consisted of a radically new design. The rotating disk was replaced by an insulated cylinder in which a thin wire spiraled from end to end (figure 77). Conducting arms connected to the meteorological

Figure 77. Model "F" radio-meteorograph mechanism. Clock-driven helix embedded in an insulating cylinder, center; pressure sensor, left; hair humidity and bimetal temperature sensors, right. K.O. Lange, *Bull. Amer. Meteor. Soc.*, 17, p. 10, 1936.

sensors moved up and down the cylinder surface, thus making contact with the helix at different times with respect to the contact of the reference arm.

On 18 October an "F" type instrument was used to make the first entirely successful test from an airplane. Pressure, temperature and relative humidity were recorded from takeoff to an altitude of about 5.2 km (17,000 ft.) and return over a flight time of one hour and twenty minutes.

Meanwhile Bent had been working on suitable lightweight transmitters for the instrument. Lange placed a pressure sensor in the test transmitter circuit enabling approximate altitude determinations to be made during the test flights. The first flight, on 22 October, failed, but on 23 October a transmitter weighing 890 g (2 lb.) lifted off from Blue Hill under three small balloons (see bibliography for Bent 1935). Preparations for the launch are shown in figure 78. The signal from this transmitter was heard from an estimated peak height of 13.1 km (43,000 ft.) and a distance of 96

Figure 78. Launch preparations for the first successful radio transmitter flight. Left to right: S. Pagliuca, upper unknown; lower, L.O. Lange, A.E. Bent, and C.F. Brooks, testing a small rangefinder, October, 1935. Copyright President and Fellows of Harvard College. Harvard collection of historical scientific instruments.

Figure 79. A.E. Bent with his directional receiving equipment. American Meteorological Society—Blue Hill Collection.

km (60 mi). The flight made the news through the Hearst Metrotone News Service.

Shortly afterward Bent did some directional finding on transmitters carried aloft by balloons. His receiving equipment, mounted on an old Mohn cloud theodolite, is shown in figure 79.

The need for larger balloons was great, so after considerable effort the Rubber Gel Products Corporation of Quincy and the Dewey and Almy Chemical Company of Cambridge, were persuaded to develop balloons that could provide greater lift. Soon Dewey and Almy was able to produce balloons of 1.2-1.8 m (4-6 ft.) diameter to give a free lift of 800-1200 g (1.8-2.6 lb.).

In spite of success with model "F" the effort was shifted to the "L," for Lange, model to avoid clock problems. In this model a contacting disk was rotated by the aneroid pressure sensor and the arm connected to the temperature sensor moved on the disk. In this way the temperature was measured in reference to the pressure. The instrument could not accommodate a humidity sensor but it did have the advantage of no clockwork.

The first successful radio-meteorograph ascent was made on 23 December 1935. The flight of about 1 1/2 hours reached an approximate altitude of 16 km (52,500 ft.). For some strange reason details of the sounding are meager. It appears that the "L" model with its batteries wrapped in cellophane to retain solar heat was used, since only tempera-

Figure 80. Launch preparations for the first successful radio-meteorograph flight at Blue Hill. C.F. Brooks holding instrument at right, December 1935. Copyright President and Fellows of Harvard College. Harvard collection of historical scientific instruments.

Figure 81. K.O. Lange seated before radio equipment used to receive the first successful radio-meteorograph transmission at Blue Hill. The racks include numerous receivers and transmitters and a timer, upper right, which periodically switched on radio signals, December 1935. Copyright President and Fellows of Harvard College. Harvard collection of historical scientific instruments.

ture and pressure were measured. Launch preparations are shown in figure 80.

Hydrogen-filled balloons, three white and one black, were used to lift the instrument and facilitate tracking by theodolite. Equipment used to receive the radio signals is shown in figure 81.

On 13 May President Conant of Harvard, members of the visiting committee, and others gathered at the Observatory and discussed further plans for the project. In all, twenty-two persons were present, including Charles F. Adams, chairman of the executive committee of the Board of Overseers, former observer Wells, and Clayton, who had returned from Argentina. Clayton now lived in Canton, Massachusetts, a few kilometers southwest of the hill. From there he independently studied solar-weather relationships and issued long range forecasts based on solar activity.

In June this writer came to the Observatory for the first time. I was trained briefly and proceeded on to the Mount Washington Observatory where I spent a short time as an observer and handyman. Living at the Blue Hill Observatory in those days was Spartan. The only cooking facility consisted of a small hotplate that could barely heat a can of beans or make a cup of cocoa. Drinking water, laundry and other items were brought up the hill weekly by Clinton, in his horse-drawn buggy (figure 82). The water was transferred from a huge jug to a smaller crock which was left on the kitchen counter. Milk, mail, and other necessary items were packed up by the observers and Brooks and empty milk bottles and the jug were

Figure 82. Mr. John Clinton and his delivery buggy, c. 1935. American Meteorological Society—Blue Hill Collection.

transferred by Clinton. Refrigeration at best was the cement floor in the basement. Milk often soured there in hot weather. Evening meals were taken at the Poores's home on a farm at the base of the south side of the hill.

In July the Mount Washington weather reports were sent to the Boston Office of the Weather Bureau for the first time. These reports were received by radio on the Hill and then telephoned to Boston.

In the fall, Harwood, Haurwitz and Schell left. Harwood took a teaching fellowship at Syracuse University and Haurwitz went to the University of Toronto as a lecturer. He had completed theoretical studies in dynamic meteorology which included papers on the height of tropical cyclones and the motion of air in curved isobaric fields of cyclones (see bibliography). Schell left to become a forecaster for Eastern Airlines at Newark, New Jersey. W. A. Boland arrived to serve as an observer.

Alexander McKenzie came to the Hill in 1935 and Salvatore Pagliuca, who had served at the Mount Washington Observatory from 1932 to 34, attended courses in synoptic meteorology at MIT during 1934-35. He then came to Blue Hill to become chief observer on 1 September.

In the autumn Bent, McKenzie, and Shaw organized the radio equipment on the third floor of the tower. The Observatory transmitter, W1XW, was placed in racks, with various receivers and the timing apparatus for switching on and off the tone transmissions; see figure 81.

On 10 October Wells died suddenly. He was sixty-eight years old.

Also on that date a small wind generator was erected on the kite house roof. This was to be tested to determine its usefulness in supplying electricity to a remote radio-meteorograph.

The personal library of Robert DeCourcy Ward, professor of climatology, who had died in 1931, was given as a memorial to the Observatory in 1935. It was accompanied by an endowment of $1,025 to maintain and keep it current. The books were added to the other climatological material in the Geographic Institute in Cambridge and the entire collection was then named the Robert DeCourcy Ward Climatological Collection of Harvard University. Brooks (see bibliography) wrote:

> It seems fitting that this fine collection be named after Professor Robert DeCourcy Ward, who, in the course of his 40-odd years of teaching and research in climatology at Harvard, brought to the United States from Europe the principles of climatology so well developed there, adapted them to American conditions, and, by example and precept, started an

American school of climatology which has guided governmental and private climatological work for decades.

In November Rossby visited the hill to observe progress in the radio-meteorograph development program. Also at this time it was arranged to telephone reports of high cloud measurements regularly to Willett at MIT for use in the synoptic meteorology laboratory. These reports were used to supplement sparse, high-level wind observations in their analysis and aid in the local forecasts.

Weather forecasts, mainly of snowstorms and their expected amounts, were resumed at Blue Hill after a prolonged absence. Harvard and the Boston and Albany Railroad were recipients of these forecasts. A study by Brooks and Harwood found the Blue Hill cloud motions preceding precipitation useful as predictors for snow storms.

1936

In February Pagliuca traveled to Mount Washington to conduct a series of tests to obtain rates of deposition and densities of rime on a model airplane wing. The measurements were presented at the fifth annual meeting of the Institute of Aeronautical Sciences in January 1937.

Application was made to the Federal Communications Commission for a license to transmit on a frequency of 68 Mhz from a radio-meteorgraph. The license was soon granted and call letters W1XFW were assigned.

At this time signal strength readings from radio station WTIC in Hartford, Connecticut, were underway at Blue Hill and Cruft Laboratory in Cambridge.

When bad weather threatened, Pagliuca issued local forecasts to Observatory clients. McKenzie picked up the coded synoptic weather reports from a Navy radio station in Arlington, Virginia. These he plotted while Pagliuca analyzed the country-wide maps. Blue Hill observations, reports from Mount Washington, and hourly airways reports which were received on long wave radio from Boston and Newark, New Jersey, supplemented the maps to provide an extensive collection of weather data.

In early 1936 McKenzie left the Observatory to work on cosmic ray balloon instrumentation at the Bartol Foundation in Swarthmore, Pennsylvania. His position as assistant observer and radio operator was filled by Charles B. Pear, Jr., an MIT student, radio amateur and friend of Arthur Bent. Bent officially left the staff but continued to consult on radio matters. His visits to the Observatory were less frequent however because he had moved from his home in Weston, Massachusetts, to Exeter, New Hampshire.

Early in the year Lange decided to give up the "L" model radio-meteorograph in favor of the "F" model. The "F" model was made lighter, weighing only 450 g (1 lb.) complete. Various clocks were modified to drive the helix. These ranged from dollar watches with special balance wheels made of invar for temperature compensation, to more expensive Chelsea clocks. The mechanism was constructed by the Feiber Instrument Company and balloons were supplied by the Dewey and Almy Chemical Company of Cambridge.

Bent designed a transmitter which was enclosed in a balsa wood box. The newer "F" model is shown in figure 83. The wood served as an insulator and this preserved heat, which kept the batteries from freezing and reducing the chance of clock stoppage due to low temperatures. The development was paid for by gifts from Miss Case, Mrs. Waldo E. Forbes, Henry S. Shaw, J. J. Storrow Jr., and a $4,000 grant from the Rockefeller Foundation.

Figure 83. Redesigned type "F" radio-meteorograph. Transmitter in balsa wood box, left; meteorograph, center; radiation shield, right. K.O. Lange, *Beit. zur Phy. der Freien Atmos.*, 24, 244, 1937.

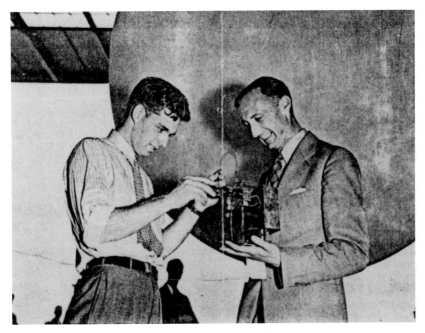

Figure 84. C.B. Pear and K.O. Lange check a radio-meteorograph before launch at Kansas City, 1936. Courtesy TWA Skyliner, August, 1936.

Soundings were demonstrated during June at scientific meetings in Kansas City and at Rochester, New York. Pear and Lange are shown in figure 84 as they check an instrument before launch at Kansas City.

Also in June Lange, an aeronautical engineering student at Carnegie Tech named Victor Saudek, and Pear went to the annual meeting of the Soaring Society of America in Elmira, New York. There, with the help of Hartmantas, they launched five radio-meteorographs, some with attached Jaumotte meteorographs, a type that traced the weather elements on a smoked glass. Two of the Jaumottes were found and the records evaluated to show excellent agreement with the radio-meteorographs.

At the time of this soaring contest Lange suggested that it would be helpful to receive temperature data from the valley to determine the stability probability of thermals. Through simple clockwork, an ordinary thermograph, and a five-meter transmitter, Lange and Pear succeeded in sending temperature data for five minute periods out of each hour. The transmitted data from 23 June to 4 July were probably the first in the country from a robot station. However, similar transmissions had been made by Moltchanoff in Russia several years earlier.

After the Elmira meeting another demonstration of the radio-meteorograph was made for the navy at Lakehurst, New Jersey.

During the summer forty-five of these radio-meteorographs were built by the Feiber Instrument Company and launched from Blue Hill. Among those who came to Blue Hill to observe soundings were Willis Gregg, chief of the U.S. Weather Bureau, Mears of the instrument division in the Weather Bureau, and Jean Piccard.

Experience showed that balloon launches were particularly difficult during precipitation or windy conditions. On 14 August a balloon launcher was erected at Blue Hill. This structure, seen in figure 85, was placed a few meters south of the kite house with the end that opened facing east–southeast. The balloon could be inflated in relative comfort, to the correct lift, inside, with the string and antenna leading out through the nearly closed sliding door. The instrument was then attached to the antenna and when the radio transmissions checked out satisfactorily the sliding door was lowered to allow the balloon to float out. As the slack string and antenna were taken up the instrument was released. This structure was built and donated by the Dewey and Almy Chemical Company.

In the fall an ingenious ground recorder, shown in figure 86, was completed. It was conveniently portable and easy to use. A sounding could be recorded on a single sheet of paper measuring 20 cm x 45.7 cm. (8 x 18 in.) (see bibliography for Lange 1937).

Figure 85. Balloon launcher at Blue Hill, c. 1935. Copyright and Fellows of Harvard College, Harvard Collection of historical scientific instruments.

Figure 86. Radio-meteorgraph recorder. K.O. Lange, *Beit. zur Phy. der Freien Atmos.*, 24, 246, 1937.

On 30 November 1936 a series of soundings was commenced from Blue Hill to compare the radio-meteorograph data with that from the airplane sounding over Boston. This required early launches at about 0400 EST to be as nearly simultaneous as possible with the airplane. After the first sounding of the series Brooks released a story to the press which was based on the following notes:

[Milton, Mass. Nov. 30, 1936]

First U.S. nighttime radio-meteorograph ascent from Blue Hill at five this morning immediately following regular weather airplane ascent showed same form of temperature curve, with 13-deg. fall to 3500 ft., then sharp 2-deg. rise, followed by 23-deg. fall to 17,000 feet, above which the radio-meteorograph, soaring upward alone, showed continued fall 7 deg. more to 20,000 where it rested at -25°F for 1000 feet, then dropped 9 more deg. to 23,000 feet, where fall stopped at -35°F for 1300 feet. The balloon radio stopped reporting at 50,000 feet at a temperature of -77°F, an hour and twelve minutes after release. The instrument was designed by Dr. K. O. Lange and A. E. Bent, research associates, Harvard, and built by R. D. Feiber. Lange and Feiber put up three balloons and instrument, and C. B. Pear, Jr., received radio signals, which were recorded by two chronographs. designed for purpose. The results, including a humidity record, will be reported to the International Commission for the Investigation of the Free Atmosphere, which designated November as a month for special effort.

These soundings were very successful; sometimes they were acquired when the plane was unable to fly and often the data, which extended many kilometers (thousands of feet) above the airplane sounding, were phoned to East Boston before the plane returned. Dr. Lange wrote, "The first American scheduled radio-sounding with immediate relay of the observed data to the central office of the United States Weather Bureau was carried out on December 10, 1936 at 3:55 a. m., by the Blue Hill Observatory." (See bibliography for Lange, 1937.)

A demonstration then followed at the American Meteorological Society's Meeting in Atlantic City, New Jersey. This sounding, on 26 December 1936, reached 21.3 km (70,000 ft). The elation experienced by this writer, who participated in the work, will not be forgotten.

Although the Blue Hill program had demonstrated the first practical system, other radio-meteorograph development continued abroad and in the United States. The instruments, according to Little (1937) were, for the most part, of the Olland type. One was developed by Dr. L. F. Curtis of the National Bureau of Standards in cooperation with the United States Weather Bureau and manufactured by the American Instrument Company. Another was developed by Army Captain O. C. Maier, and L. E. Wood and Galcit of the Guggenheim Aeronautical Laboratory of the California Institute of Technology. Piccard and others developed another at the University of Minnesota. Mears, Little, and Moon worked on another for the U.S. Weather Bureau while the Julien P. Friez Instrument Company had still another development under contract with the Weather Bureau.

An instrument following the original plan of Duckert was also under development by Diamond, Hinman, and Dunmore of the National Bureau of Standards. Temperature and humidity elements controlled resistances in a circuit which modulated the continuous high frequency transmission of the radio oscillator. The pressure element switched these resistors in and out of the circuit in a definite order as the instrument ascended to lower pressures. This instrument avoided the clock problems of most instruments employing the Olland principle.

In an effort to retain use of the Olland principle without a clock movement, Mears and Moon of the Weather Bureau Instrument Division employed an ingenious air-driven motor. Air under pressure was supplied from an inflated pilot balloon. Tests were encouraging but the system never went into production, probably because of the Diamond, Hinman, Dunmore development which was beginning to show great promise.

During the year Herb Dorsey, Boland, Saudek, and Conover served intermittently as assistants and observers. Boland left in October to take a position with American Airlines in Chicago.

Kitchen facilities were much improved by the addition of a three-burner kerosene stove and oven. The convenience of being able to cook at the Observatory far outweighed the problem of packing up all the food. Travel up and down the hill remained mostly by foot, since autos were banned on the Metropolitan District Commission ways throughout the reservation. Dr. Brooks did not like to seek overall permission for cars because he believed pedestrians and equestrians had the right not to be disturbed inside the reservation. On one occasion, Pagliuca, an energetic young man, full of fun, caught up to Clinton and his buggy on their way up the hill. He quietly grabbed the rear axle and dug his heels in to slowly stop the horse. Clinton switched the horse but it was so old and weak that it simply could not continue until Pagliuca released his hold and had a good laugh with Clinton.

On 8 May the old kite windlass was dismantled to make way for apparatus associated with the new sounding methods.

Fergusson constructed a new, mercurial barograph which was set up on the second floor of the tower. The instrument was capable of tracing pressure changes which were magnified 5.5 times on a chart.

Shaw, whose interests lay in ornithology and geodesy, as well as radio, came to Brooks for weather instruments that could be used during summers on Kent Island, New Brunswick, Canada. This island, owned by Bowdoin College, was situated in the Bay of Fundy, between Eastport, Maine, and Yarmouth, Nova Scotia. The small strip of land had been a bird sanctuary for some time and served numerous ornithologic expeditions. Shaw had become aware of relationships between climate and bird populations in the vicinity of his summer home on the Maine coast so he thought it worthwhile to expand these relationships to conditions on Kent Island. Since radio was used for communications, Shaw also desired data to augment his propagation studies over water. Brooks quickly responded to Shaw's needs by assembling thermometers, a wind recorder, a thermograph, and hygrograph, which had become surplus from the eclipse expedition. These were taken to the island by Stone and Pagliuca. Robert M. Cunningham, who became interested in the island a year later while in high school, arranged to acquire records each summer and analyze them over the years. The island site, known as the Bowdoin Scientific Station, listed Shaw and Brooks as directors and for several succeeding years as advisors, while Stone and Pagliuca were listed among the scientific staff.

On 8 November the 2000 EST observation was shifted to the enclosure shelter from the window shelter.

Also noted during the year were the regular passages of the Hindenberg dirigible on its trips between Europe and Lakehurst, New Jersey. It usually passed at about 1 km (3,300 ft) altitude.

Brooks and Baldwin published a paper on forests and floods in New Hampshire and Kimball and Baker published papers on atmospheric turbidity. A discussion of the temperature of the Schott filters which were used in the calculations was included. Stone completed a comprehensive survey of fog in the United States and adjacent regions. In addition to a map showing the frequency of fog, the country was divided into fog regions from which the patterns of seasonal and diurnal fog frequencies could be estimated.

During the year the Mount Washington Observatory was incorporated and Brooks was elected its first president. Supervision continued from Blue Hill.

On 6 December Oliver L. Fassig died as a result of injuries suffered when he was struck by an auto in Washington, D. C. Though advanced in years he had still been keen and active in research. He had completed tables and maps for his comprehensive treatment of the climate of Puerto Rico but the text was unfinished.

1937

In January Ralph W. Burhoe joined the staff as an assistant to Brooks. He also filled in as observer and usually made the base station observation each morning on his way up the hill.

Early in January a new crystal-controlled transmitter was placed in service. As before, its signal strength was recorded hourly by Ross Hull in West Hartford, Connecticut. About this time the 0130 EST Mount Washington observation was commenced, in addition to the other six hourly observations; this was placed on the national circuits via radio to Blue Hill and relay to the Boston Weather Bureau.

In late January, Brooks, Lange, Pear, Stone, Pagliuca, and Dorsey went to New York to present papers and demonstrate a radio-meteorograph at the meetings of the Institute of Aeronautical Sciences. The prestige of representation at meetings was fully restored to the glorious days of Rotch.

In February a series of thirty-one soundings was started for comparison with the Boston airplane sounding. This time the purpose was to demonstrate that the equipment could be used by those having no previous experience in radio-meteorograph soundings. The instruments were made by the Feiber Instrument Company at a cost of $29.05 each. Staff members of MIT made the launches from the roof of the Guggenheim Building; they performed the calibrations and evaluations while the Observatory supplied the instruments, balloons, hydrogen, radio receiver, and radio-meteorograph recorder. Only three soundings were lost, two due to accidents at release time, and one to an instrument malfunction. All the others rose higher than the airplane sounding, which, incidentally, cost $23 to $32 per flight, and two-thirds entered the stratosphere. A 2°C difference between the radio-meteorograph and airplane sounding, which had been found in the earlier series, was finally resolved as an error in the latter.

Brooks wrote of the following amusing experiences during the series. (see bibliography for Brooks 1937, W1XFW).

> Experience gained during the series was used to make minor improvements in instrument technique. For example, in some of the first ice storms the balloons did not rise fast enough to penetrate the freezing layer before being stopped by the weight of accumulated ice. One such turned downward at 5,000 feet, but so much of the ice had melted at 2,800 feet that it resumed its ascent, and on the second trip into the icing zone passed through and went on up to nearly 30,000 feet. Another balloon became so weighted that it failed to get through, and, after more than an hour of drifting eastward (followed by a radio-direction finder at Blue Hill) at a height of a mile or two, descended into the opposite wind below, and was seen by one of our research associates, Mr. W.G. Pickard, coming in off the sea at Seabrook Beach, N.H. Thereafter, we used more hydrogen and substituted a parachute shoved inside the balloon (to keep it dry and iceless) for the second balloon we had been sending up for parachute purposes. Two balloons gave much more surface relative to lift, and so not only ascended more slowly but also collected more ice. The rate of ascent aimed for was between 700 and 900 feet per minute. Only three of the 31 radiometeorgraphs have been returned, and two of these were from the sea. Most of the rest probably dropped back into the Atlantic. Besides the instrument recovered by Mr. Pickard, one was sent back by the captain of a fishing vessel who found it 30 miles southeast of Nantucket 5 hours after it was released in Cambridge. One of the balloons was still standing up 30 feet above the water. The other instrument struck on a steeple near Taunton, Mass., then fell to the sidewalk.

Officials from the U.S. Weather Bureau, the U.S. Signal Corps Laboratories, and the Canadian Weather Service came to Cambridge to view the work. The series, which ended 16 April, showed beyond doubt that a sounding program could be carried out by people who had not been intimately associated with development of the program.

In the spring Brooks announced to the American Geophysical Union that a radio-meteorograph network would probably be developed over the United States within two years and he listed seven studies that would benefit from the new data. He concluded by saying that more investigators must be trained for the work and especially new college men and graduate students were needed. The program was endorsed by the American Geophysical Union and brought further gifts which enabled Brooks to keep Lange on the staff.

What may have been the first work in cloud physics at the Observatory was performed by Alan Bemis, an MIT researcher, when he brought up equipment on 9 April to photograph fog droplets.

Apparently, Civil Service exams for positions in meteorology were held on 28 April; practically the entire staff took the exam while D. P. Keily and S. P. Fergusson manned the Observatory. Although good ratings were obtained, it turned out that no one left to take a government position.

On 6 May the dirigible Von Hindenberg was again seen passing southward east of the Observatory about 1045 EST. This was the flight which ended in disaster upon landing at Lakehurst, New Jersey.

The National Broadcasting Company, over its blue network, presented live from Blue Hill a thirty-minute program on 27 June. Brooks described the radio-meteorograph and Pear gave details of the transmitter and receiver. A sounding had been launched and from time to time during the program its measurements were discussed. Brooks also discussed the establishment of the Observatory by Rotch and Rotch's work. A radio contact was established with Mount Washington and weather and pleasantries exchanged. Pagliuca described the present observational programs at Blue Hill. The basic script, which had been prepared earlier by NBC writers, was complete with notes as to when "chuckles" or "astonishment" were to be heard. The broadcast was summarized in the Observatory journal by Pagliuca in two words, "great style."

Later, in August, the footrace up Mount Washington was broadcast by NBC. For this the radio link from the mountain to Blue Hill was also utilized.

In the summer Lange went to Wichita Falls, Texas, where, in conjunction with the Soaring Society of America, radio-meteorograph soundings were made. The soundings were used immediately to forecast soaring conditions and later to study convective turbulence in fair weather and in thunderstorms.

In June Stone took charge of a study of snow cover over New York and New England. The State Branch of Recreational Planning and other Federal and State agencies undertook to support the study using Civil Conservation Corps funds. Its aim was to supply basic climatological information for the developing skiing industry. Data were extracted from hundreds of "cooperative observer" records. Conover, while living and working at the Observatory, tabulated the data for New England from forms on deposit in Boston, while Sylvia McGovern obtained the New York State data at New York. The analyses were presented by Stone (see bibliography).

Also during the summer Works Progress Administration men started the construction of a stone shelter and lookout for the Metropolitan District Commission on the northeast side of the summit of Great Blue Hill.

Early in September Lange went to Europe where he presented a paper at the British Association for the Advancement of Science meeting on the comparison of sounding data from airplanes and radio-meteorographs. He then went to Salzburg for the International Commission meetings where he represented Brooks in the sections on climatology and clouds.

Alexander McKenzie returned to Blue Hill in July but on 30 September he moved on to Mount Washington where a new Observatory building had been constructed.

Edwin H. Armstrong, inventor of frequency modulation radio, was soon attracted to Mount Washington because of the height of the location, a prerequisite for long distance transmissions, the extra housing space, and the others already performing experiments in radio on the mountain (McKenzie 1982). He subsequently set up test equipment, and later an FM transmitter was housed in the Observatory. This station rebroadcast the FM programs received from the Yankee Network Station at Paxton, Massachusetts. Later the equipment was moved to a new building and FM broadcasting became permanently established on the mountain.

At the end of September Pagliucia resigned to head the newly established weather service of the Yankee Network in Boston and Charles Pear was made chief observer. At this time compensation for the job had risen to $135 a month with room, cooking facilities, etc., at the Observatory. The work period was loosely defined but specified one full day off a week. Evenings and half days "within reason" were also permitted, with a full four-week paid vacation each year.

The next test of the radio-meteorograph was to determine how well a program could be conducted by Weather Bureau personnel. The Feiber Instrument Company had contracted to supply instruments to the Weather Bureau at East Boston; Blue Hill and MIT were to provide training, calibration, and read-out equipment. At this time the radio-meteorograph's weight was down to 335 g (3/4 lb.) and the cost was only $18.00. A program of daily launches started on 1 October at East Boston to continue until 30 June 1938.

These were busy years at Blue Hill, with many projects in addition to the radio-meteorograph program. Stone, now librarian, continued to make inroads cataloging and completing serial pamphlets and volumes. In cooperation with the library of the U.S. Weather Bureau, he prepared monthly or bimonthly lists of meteorological literature which appeared in journals the world over. These were published in the *Bulletin of the American Meteorological Society*, the headquarters of which continued at Blue Hill. These were the forerunners of the *AMS Meteorological Abstracts*. A copy of an 80,000 card bibliography, compiled by the Works Progress Administration under the direction of the Weather Bureau, was received.

Dorsey continued his forecasting studies of snowstorms, using Mount Washington data. Ultraviolet measurements were resumed and the solar measurements on Mount Washington were analyzed.

Brooks was completing his comparison of shields to surround precipitation gages. Many diverse shields had been tested on Mount Washington, at the base of Blue Hill, and inside the enclosure on the summit. Two of these shields under test on Blue Hill are shown in figure 87. At times precipitation was measured every twelve hours from as many as eight gages inside various shaped shields or no shield at all.

Brooks, home at 1793 Canton Avenue, continued to serve as a base station where Observatory deliveries were made. He encouraged the parking of vehicles there but more and more trips were made to the summit by car because of the increase in loads of mail and equipment, as well as for convenience, at all hours of the day.

During the year the Metropolitan District Commission established a police radio at the Observatory. The equipment which was connected by a telephone line to the police was located on the landing just outside the tower on the second floor and the antenna was on a mast in the center of

Figure 87. Two experimental precipitation gage shields under test at Blue Hill. Fergusson weighing gage in foreground, October 1937. American Meteorological Society—Blue Hill Collection.

the tower. Its flexible pole extended well above the other masts to provide lightning protection. The Observatory and a trigonometric survey tower, at that time, are shown in figure 88.

Also during the year Fergusson developed a low-lag wind speed sensor and recorder. A Richard cinemograph was driven by an electric motor to provide rapid response to rotational speed changes from a small balsa

Figure 88. Observatory, showing masts on the tower and wooden survey tower, October 1937. American Meteorological Society—Blue Hill Collection.

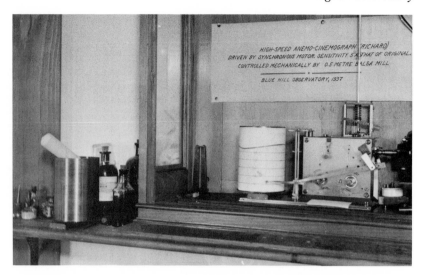

Figure 89. High-speed anemo-cimemograph, Fergusson's first model, 1937. Copyright President and Fellows of Harvard College. Harvard collection of historical scientific instruments.

bladed windmill. A plaque over the apparatus (figure 89) states that it was five times as sensitive as the original cinemograph.

At year's end Edmund Schulman, a student of Dr. Brooks, made an increasing number of trips to the Observatory. His interest was in climatological cycles and radiation measurements.

Through frugal and aggressive management, Brooks succeeded in administering these projects. He relentlessly sought gifts for immediate use or the enhancement of the endowments which slowly increased to yield larger sources of income.

1938

The East Boston series of radio-meteorograph soundings ended 30 June as planned. Lange and Pear had devoted most of their time to the program in assisting Bureau personnel, and in improving instruments and techniques to reduce the chances of failure. Pear designed a new, less expensive and lighter transmitter (figure 90). In this he adapted a single "30" tube without its phenolic base and having only a minimum of other parts. A new calibration chamber was built which could accommodate seven instruments at once. The chief problem was to keep the watch movements running at the low temperatures encountered during the nighttime ascents. Financial aid for the program came from A. Felix duPont, R. C. duPont, H. S. Shaw, J. J. Storrow, Jr., Mrs. Waldo E. Forbes,

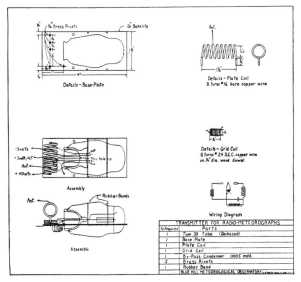

Figure 90. C.B. Pear's design of a single-tube radio-meteorograph transmitter, 1938. C.B. Pear, Jr., *Bull. Amer. Meteor. Soc.,* 19, 306, 1938.

A. C. Bemis, Ernest B. Dave, Miss Marion R. Case, R. A. Leeson, R. F. Barrett, and K. O. Lange.

A report by Knight (1938), acting chief of the Air Transport Section of the Civil Aeronautics Authority Planning and Development Division, used the Harvard radio-meteorograph exclusively to illustrate the use of radio-meteorographs before outlining procedures for transmitting sounding data over the existing communications lines. However, the radio-meteorograph which had been developed by Diamond, Hinman, and Lapham at the National Bureau of Standards was rapidly gaining in acceptance over the Harvard instrument. This instrument had the distinct advantage of no clockwork, the only moving part being a baro-switch which operated well at all temperatures. The Weather Bureau contracted for radio-meteorographs to be used at six stations beginning 1 July 1938. This marked the beginning of a network of radio-meteorograph stations in the United States. Five stations were supplied with the Diamond–Hinman type, the sixth, Boston, was supplied with the Harvard type.

Following the work at East Boston, Lange developed a heated anemometer and micro-meteorograph for use on gliders or airplanes. The anemometer consisted of a slotted vertical tube whose slots were kept pointed into the wind by a short-tailed vane for the measurement of dynamic pressure. Static pressure was approximated from a separate vertical tube that was perforated around its sides. In tests at Blue Hill the dynamic line was connected to a bellows made of latex on which a pen arm linkage rose and fell as the bellows expanded and contracted. Tests were also made on Mount Washington but the instrument was not placed in service. Lange also studied air motions on the east side of Mount Washington under lee wind conditions by observing balloons that had been inflated just enough to support their own weight. They were followed from two theodolite stations and simultaneous readings were programmed by radio from the summit Observatory.

During the year full or part time staff consisted of the following, most of whom made observations when necessary:

1. David Arenberg arrived in January to observe and later pursue ultraviolet studies
2. John Conover continued the snow cover work in New England
3. Charles Pear continued as chief observer while in charge of all the radio programs
4. Herb Dorsey studied heavy snowstorm forecasting using synoptic and Mount Washington data
5. Herb Lane assisted in the Library

Rice and Celia Beckenstein continued, in the summer, to reduce data collected during the polar year.

Schell became more and more involved in seasonal forecasting, which was based on preceding pressure patterns sometimes located over areas remote from the forecast area.

Stone and Burhoe continued to assist Brooks in American Meterological Society and library affairs. The library now honored frequent requests for loans of books or pamphlets often unattainable elsewhere.

Schulman continued with turbidity studies and after obtaining the Douglas cycloscope, a large mechanical device used to analyze cycles, he spent considerable time with the machine in analyzing tree ring growth.

Fergusson completed forty years of anemometry study, which was reported in the *Harvard Meteorological Studies* series.

To those of the above who lived on the hill, 21 June was a red-letter day when an electric refrigerator was installed.

The observational program remained essentially the same except the check or synoptic observations were changed from 0800, 1400 and 2000 EST to 0700, 1300, and 1900 EST on 1 March. This change brought the observation times into conformity with the international hours, which had been changed 1 January. On 1 July the nephoscope was turned 180 degrees to give readings comparable to other upper air data.

During the year there were times when the Mount Washington observations were sent out through Whiteface Mountain in the Adirondacks where a new observatory had been opened. From there they were radioed to Albany and then placed on the circuits.

Stone, Schulman, and Ackerman, one of Brook's graduate students, completed Fassig's text of the climatology of Puerto Rico. Stone also prepared a report with an extensive bibliography on tropical climatology and physiology in relation to the acclimatization of white settlers.

In June Schulman received his A.M. and Dorsey his S.B., both from Harvard.

On 23 March a severe brush fire swept up the southwest side of the hill on a strong, hot, and dry southwest wind. The fire burned to the Observatory walls, within a few feet of the grass thermometers and the balloon launcher (figure 91). Hydrogen tanks in the launcher were removed as the fire approached. Smoke filled the Observatory and for a time it was thought the building was on fire.

Brooks continued his non-stop pace with long work days, virtually seven days a week. In addition to routine matters, he served as Chairman of the American section of the International Commission on Climatic Variations. He frequently consulted with the Weather Bureau on the

Figure 91. Balloon launcher partially hidden by smoke from the brush fire of 23 March 1938. Photo by J.H. Conover.

preparation of instructions for cloud observations and he co-authored a paper on the great floods in the eastern United States.

The brilliant career of Ross A. Hull ended 13 September when he was accidentally electrocuted while experimenting on a television receiver (QST 1938). Hull and others had arranged the hourly radio broadcasts from Blue Hill and signal strength measurements at several locations for correlation with atmospheric weather conditions.

During the year the stone lookout and shelter with fireplace on the northeast side of the summit was completed.

On 19 September Pear and Dorsey left to study at MIT and on the 20th Conover returned to live and work at the Observatory while attending a synoptic meteorology course and laboratory at MIT.

Later Dorsey married and left immediately for Antarctica with the Byrd Expedition. Upon his return he entered the military service, but to satisfy his desire for further exploration he wheedled assignment to Greenland. There he made an unauthorized trip across the ice cap.

Beginning on 18 September 1938, the most significant weather event of the Observatory's one hundred years unfolded. A hurricane was reported north of Puerto Rico moving from the east–southeast. Brooks noted that, since the eastern United States was covered by a considerable flow of moist tropical air, the hurricane could not be expected to recurve at once, and so would probably reach the coast.

On the 19th Brooks still believed the storm would reach the coast and plans for Lange and Feiber to go to Mount Washington were made tentative. On the 20th, cloud motions at middle and high levels became available through breaks. Clouds at estimated heights between 1 and 8 km (3,000 and 25,000 ft) were moving from 190° - 203° azimuth and at speeds as high as 50 m sec^{-1} (112 mi. hr^{-1}). These motions were suggestive of a movement of the hurricane northward from its recurved position east of Florida. An arrival time was not computed simply because there was no assurance that the observed speeds existed southward to the vicinity of the hurricane. More nephoscope observations were made on the 21st and at 1350 EST the following (abbreviated) report and forecast was phoned to the Boston Weather Bureau.

Ci, Cc from 180° estimated 10 km high 30m sec.$^{-1}$

Cc " 167° " 7 km " 30 "

Ac " 155° " 4 km " 30 "

Cu " 134° " 0.6 km " 25 "

These cloud observations indicate that the whole depth of the atmosphere up to at least 10 km is moving from a little east of south at about 27 m sec.$^{-1}$ and therefore that the storm south of New England must move inland. Since the airways broadcasts show very low pressure at Lakehurst and Mitchell Field, the center will probably move northward or a little west of north through western New England or possibly eastern New York, and at high velocity. There should be intense rainfall, but not for long. The center should be well passed and the weather clearing with wind shifting to southwest by midnight.

The essentials of this forecast, including possible winds of 27 m sec^{-1} (60 mi. hr^{-1}), were given to Harvard but no forecast of near-hurricane winds was made to the public by the Weather Bureau.

After 1230 EST the wind increased rapidly from the southeast and all but residents Arenberg and Conover, went down the hill.

At 1514 EST the power failed, and at 1521 EST the windmill blew away after a speed of 37.2 m sec^{-1} (83 mi. hr^{-1}) had been recorded. As the windmill disintegrated, the noise below was described as "terrible." The vibrations shook the windvane loose on its shaft so readings below were made meaningless. A portion of the cinemograph record is shown in figure 92.

At 1530 EST it was noted that spray was being lifted an estimated 10 m (30 ft.) off Ponkapoag Pond. The odor of bruised oak and cherry leaves was very noticeable. Outside the Observatory there was little to blow away, but during a trip to the enclosure this observer was hit around the ankles by flying stones. The roar of the wind around the building and through the masts on the tower exceeded anything that the observers had experienced. At 1755 EST a window on the south side of the lower bedroom blew in. Glass fragments stuck in the pine woodwork on the opposite side of the room. After about 1830 EST the wind started to subside and slowly shift toward the southwest.

Without the windmill, wind speeds had to be read from a 3-cup anemometer that was connected to the Draper recorder. Portions of the chart records ending at 1910 EST on the 21st and 22nd are shown in figures 93a and 93b. The upper trace constituted an auxiliary record of wind direction whose scale is faintly shown at the right-hand edge of the chart dated 21 September 1938. A traverse of the lower sloping line represented wind passage of fifty miles. The slope of the trace could be used to derive speed directly. Normal use of the recorder was only for the

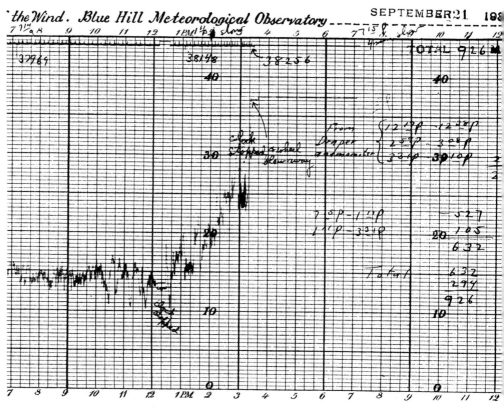

Figure 92. Copy of the cinemograph wind speed record ending at 1911
EST 21 September 1938. Speed in meters/second. The upper trace made
progressively smaller ticks at 10-, 5-, and 1-mile intervals of wind passage.
Courtesy National Archives, Boston Branch, Waltham, MA.

determination of average speed over one-hour periods. In practice the
speeds were noted at the bottom of the chart and if a cup speed correction
was required, the corrected value was noted below. Thus for the hour, 5 -
6 p. m., ninety miles of wind passed and the corrected speed was 84 mi.
hr^{-1} or 37.6 m sec^{-1}. Considerable vibration was transmitted down the
mechanical linkage from the anemometer to the recorder during the high
wind; this caused the trace to widen as shown and later a light blotch made
the trace even more difficult to read. After the chart was removed from the
instrument, some preliminary speeds over twenty-minute periods were
noted from 1440 EST to 1900 EST, but these should be ignored because
corrections were not made for the overall downward displacement of the
trace on the scale. The 3-cup anemometer and wind vane (figure 94) which
were connected to the modified Draper recorder (figure 95) were refur-

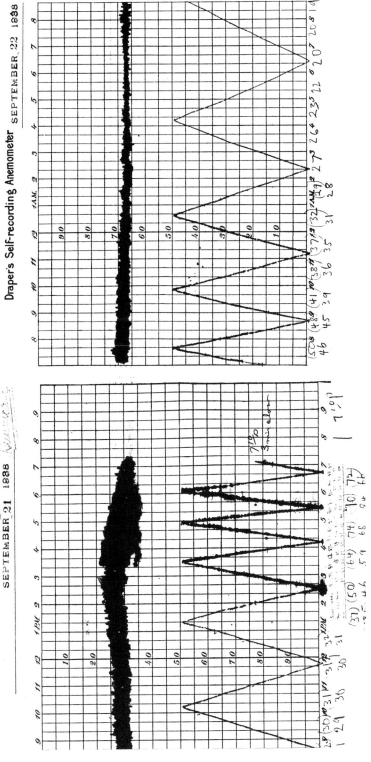

Figures 93a and 93b. Copies of the 21 September 1938 hurricane wind record obtained by the Draper recorder. Wind passage scale is in miles and the direction scale is as shown. Courtesy National Archives, Boston Branch, Waltham, MA.

Figure 94. Three-cup anemometer and wind vane constructed by Fergusson, c. 1933. Courtesy J.H. Conover.

Figure 95. Draper recorder connected to 3-cup anemometer and wind vane. Modified by Fergusson, c. 1933. Courtesy J.H. Conover.

bished and put in operation shortly after Brooks became director. The chart for the 21st was photographed and enlarged, then read by several individuals. From these readings a corrected five-minute average speed of 54.1 m sec^{-1} (121 mi. hr^{-1}) was determined. This wind, from the south, occurred from 1811 to 1816 EST.

As noted, the highest hourly average was 37.6 m sec.$^{-1}$ (84 mi. hr^{-1}), from 1700 to 1800 EST, also from the South. Numerous readings were made and a peak of 83 1 m sec^{-1} (186 mi. hr^{-1}) was established but with a uncertainty of 30 or 40 mi. hr^{-1}. The ratio of the gust maximum to the five minute maximum, or 1.54 in this case, compares well with ratios found at the Observatory during lower speeds when a gust recorder was in operation. It should be noted again that southerly winds, as in this storm, are subject to more acceleration than those of other directions as they ascend the hillside. This contributed to the extreme wind which was by far the highest ever measured at the Observatory. Rainfall during the hurricane was very light, amounting to a mere 3 mm (0.12 in.).

At the Observatory a few antennae were lost from the tower; on the south side of the building the coal bin covers were blown off and splintered. This allowed the wind to scoop out great quantities of small buckwheat coal. This writer's car had been parked in the back yard behind the wall as a safety precaution but the flying coal chipped the paint on one side.

Off the summit and in the valleys damage was tremendous. The Metropolitan District Commision estimated that 850 trees were down along its parkways within the reservation. Tree and wire damage in front of Brooks' home on Canton Avenue is shown in figure 96. Frona Brooks, shown in the picture and four years old at the time, commented "We don't have to look both ways today," as she proceeded into the street.

Power was not restored to the Observatory until 28 September.

Brooks noted that this storm, destructive as it was, did not seem to be as bad as the one of 23 September 1815. Other early notable storms in the area occurred on 8 September and 4 October 1869.

1939

An important activity of the year was the study of the September 1938 hurricane. Brooks worked diligently toward the establishment of a local forecast office of the Weather Bureau. He pointed out to the chief of the bureau and state legislatures that his local observations correctly predicted the storm's progress, while the responsible forecasters for the area who

Figure 96. Damage caused by the 1938 hurricane at 1793 Canton Avenue, Milton, MA., the base station. American Meteorological Society—Blue Hill Collection.

were based in Washington had no knowledge of these observations and failed badly in the storm's timing.

Pear returned to the hill in February after completing his degree requirements at MIT. He then worked with Lange on the development of a new automatic weather station. In this the Olland principle of the radio-meteorograph was used. Pear designed and constructed the radio equipment, which operated on a 2-1/2-m wavelength (figure 97). The system, the first of its kind in America, was demonstrated in September when it sent hourly reports from Mount Weather, Virginia, to Washington, D. C., a distance of 76 km (47 mi.). Later the meteorograph (figure 98) was used to transmit atmospheric pressure, temperatures, humidity, wind speed and direction, and a "yes or no" of bright sunshine hourly from Asnebumskit Hill, near Paxton, to Blue Hill. This path length was 72 km (45 mi.). The recorder at the receiving end of the system was known as the "barber pole" (figure 99), simply because it looked like one; in fact, the cylindrical glass cover was part of a barber pole. The mechanism was similar to the radio-meteorograph recorder, but in this case each element was plotted along a vertical line on the chart-covered drum inside the glass cover. This provided a plot of the elements that could be viewed as a conventional diagram with time running from left to right. Like all new experiments the system was subject to frequent problems which necessi-

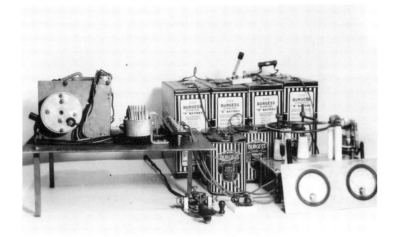

Figure 97. Remote weather station encoder and transmitter, 1938. Copyright President and Fellow of Harvard College. Harvard collection of historical scientific instruments.

tated travel to and from Asnebumskit Hill; after a short time the project was abandoned.

As part of a study of the circulation over the western North Atlantic during the iceberg season, the Observatory cooperated with the International Ice Patrol, the U.S. Weather Bureau, and the Massachusetts Institute of Technology, the U.S. Navy, J. F. Friez, the Meteorological Service of Canada, Great Britain, France, and Denmark. Radio-meteorograph stations were set up at Bermuda and Cambridge and instruments

Figure 98. Remote weather station encoder. The spiral helix slot allowed the sensor arms to drop, as the cylinder turned, to make contact on a bar. Copyright President and Fellow of Harvard College. Harvard collection of historical scientific instruments.

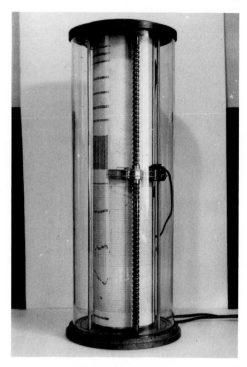

Figure 99. Remote weather station recorder. Copyright President and Fellows of Harvard College. Harvard collection of historical scientific instruments.

were released from patrol cutters and the French ship *Carimare*, stationed at 38° 30' North and 44° 00' West. Pear, on loan to MIT as a research assistant, was placed in charge of the Bermuda station. Radio-meteorographs, which employed the Diamond–Hinman System, were made by Julien P. Friez and Sons Inc., of Baltimore, Maryland. They were used at Bermuda and on the Coast Guard Cutters; Harvard instruments were used at Cambridge. Over the study period of April to June, eighty ascents were made from Bermuda and seventy-one from Cambridge; when coupled with data from the other surrounding stations the first day-to-day upper air patterns over the western North Atlantic were obtained.

This program served as a preliminary experiment of an aerological network in the interest of medium-level trans-Atlantic flights, the low-level flights by Pan American Airways having already started.

The satisfactory performance of the Harvard radio-meteorograph led the Weather Bureau to enter into a joint arrangement with the university to make certain improvements and adapt it to routine use and mass production. Some 10,000 radio-meteorographs were being used annually in the United States at that time.

During the summer A. W. Friend came to Blue Hill and Cruft Laboratory as a research fellow working on his doctorate. His work in West Virginia had been in remote sounding of the atmosphere by means of radio reflections. Pear helped in building improved equipment which was set up in October at Friend's home in East Lexington, Massachusetts. His pulsed transmitter operated on 2398 KHz; vertically directed pulses were reflected at levels of dielectric change and the echoes were measured on an oscilloscope in terms of strength and height. The levels were then interpreted as changes in lapse rate of temperature and humidity and compared with the Blue Hill radio soundings. Quantitative changes in temperature and humidity could not be obtained, but the system was of particular interest in continuously delineating changes, whereas the soundings could not show this microstructure.

Professor F. A. Brooks, brother of Charles Brooks, and research fellow in the Engineering School at the University of California, came to the hill in the fall to make fundamental measurements of atmospheric radiation. Radio-meteorograph ascents were made in conjunction with the radiation measurements.

The potentialities of deriving electrical power from the wind now came under serious study. Hourly values of wind speed at Mount Wachusett and Mount Washington were converted into power estimates for Palmer C. Putnam, a consulting engineer for the S. Morgan Smith Company, who proposed to build a large wind turbine. As this work developed, Observatory personnel were to play increasing roles in what became the first operating large-scale wind generator in the country.

During the year Brooks formed the Friends of the Blue Hill Observatory in an effort to raise additional funds. Four popular leaflets were prepared for this purpose.

Brooks also went to the University of Chicago as a visiting professor in the Department of Geography. There he lectured to twenty graduates during the first half of the summer quarter. While there he also gave three lectures on clouds before the local American Meteorological Society group and an illustrated public lecture on the New England hurricane.

In the fall six Observatory members attended the American Geophysical Union meetings in Washington where they presented five papers. Displays of the automatic weather station, a radio-meteorograph, microaerograph, heated pressure anemometer, balsa wood windmill, and data from the Bermuda and Cambridge radio soundings were shown.

Considerable work was done by Stone as editor of the American Meteorological Society's Bulletin and toward completing the catalog of serials. This work was stimulated by the need for a complete University "Union list of serials." At this time membership in the American Meteorological Society had risen to over 1,200. Stone went to the School of Tropical Medicine in Puerto Rico for about five months where he worked toward completing the climatology which had been started by Fassig.

The observational program continued with observers Max Dubin, Streeter Bass, Robert Priggen, Joseph Levine, and Dan Davis. Conover returned 1 October after leaving in December 1938 due to illness. Some of the staff are shown in figure 100. Arenberg spent most of his time on Mount Washington where he pursued studies in cloud physics.

Kimball retired from his summer work on the hill as well as from his Weather Bureau position in Washington.

It was noted that on 28 November that "the Park Commission cut brush from the fountain to the top of the hill leaving the large pine and oak."

The eminent climatologist Victor Conrad arrived in late 1939. Conrad and his wife emigrated to the United States late in 1938 when the takeover of Austria by Hitler resulted in Conrad's dismissal from his post at the

Figure 100. Blue Hill staff members: From left, Dan Davis, Ralph Burhoe, Sterling Fergusson, John Conover, Charles Brooks, Robert Stone, David Arenberg, Karl Lange, Max Dubin, Joe Levine, Irving Schell, c. 1939. Courtesy C.B. Pear.

University of Vienna. He was first sponsored by Professor Helmet E. Landsberg at Pennsylvania State College. Housing was substandard and the society of the small Pennsylvania town was not to their liking, so Brooks arranged for Conrad to join the Observatory staff. The Conrads lived in Cambridge with cherished books, furniture, and music that they had succeeded in taking from their home in Vienna. Their new home was not far from the Geographic Institute where Conrad lectured, and occasionally he and his wife came to Blue Hill, climbing to the Observatory on foot.

1940

Starting 1 January the second page of the observation forms (figure 65a) was changed slightly to include, for each day, the daily wind movement, average wind speed, and prevailing direction, maximum speed as shown by the cinemograph, and direction and time of the maximum.

In January Stone went to the Weather Bureau in Washington to prepare a correspondence course. He returned after seven months. Stone also completed the Fassig monograph on the climate and weather of Puerto Rico and the Virgin Islands.

Brooks lectured on air mass analysis at Cornell University and Lange was assigned to teach a course in aeronautical meteorology which was supported, in part, by the Engineering School and Civil Aeronautics Administration of the U.S. government.

Brooks, Lange, Pear, and Friend attended the meetings of the Institute of Aeronautical Science in New York and presented two papers.

On 14 February the famous Saint Valentine's Day snowstorm struck. Winds gusted to 36 m sec^{-1} (80 mi. hr^{-1}) at the Observatory and with the dry snow, which amounted to 38 cm (15 in.), drifting was severe. The author's car was stranded on the hill for a month in spite of efforts to drive it on heavy planks atop the drifts. Regular auto traffic up the hill did not resume until 27 March. A drift in the back yard which partially covered the shop door and windows is shown in figure 101. Snow was not always responsible for closing the road to cars. Surface water ran out on the road and froze in great sheets as shown in figure 102. Drivers occasionally attempted running starts to coast up and over the ice but if they didn't make it they sometimes ended in the woods or if lucky backed against the hillside to spin around and face downhill. In the spring thawing ground and mud also made the road impassible for brief periods.

Figure 101. Snowdrift in the backyard after the 14 February 1940 storm. American Meteorological Society—Blue Hill Collection.

In February the Englehard and Micromax recorders that were used for recording solar radiation were moved from the room behind the office to the first floor observers' office of the tower.

In March chief observer Charles Pear resigned to take a position with a firm that planned to manufacture radio-meteorographs, now more often called radiosondes. Pear had designed and constructed radio equipment

Figure 102. Ice on the summit road above the halfway turn, c. 1940. Courtesy J.H. Conover.

for the radiosondes and the automatic weather stations and kept the Blue Hill radios operational. John Conover succeeded him as the new chief observer.

In April Fergusson completed a new chart drive mechanism for his gust recorder. Speed was graphed in rectilinear coordinates on a strip chart that could move as much as 60 mm min^{-1} (2.36 in. min^{-1}). The time constant of this system was not known but it was probably in the range of one or two seconds. One difficulty with the system was that the balsa wood blades, in spite of a paint cover, chipped or even disintegrated during severe weather such as hail or sleet.

The Harvard radiosonde went through still another redesign by Lange and production was taken on by Simmonds Aerocessories, Inc. This model (figure 103) was built around a molded bakelite centerpiece. The clock and batteries were attached to one side and the pressure sensor and transmitter to the other side. Each side was covered by an insulated white cardboard box that could easily be removed to allow inspection. Fifteen of the new instruments were tested in June. One of these instruments was discovered by a hermit at Cape Pogue on Martha's Vineyard. During those days of war in Europe and increasing tensions, the rapid ticking of the clock

Figure 103. Radiosonde produced by Simmonds Aerocessories, Inc. Clock and battery removed at left; complete instrument, showing temperature and humidity sensors, at right. Copyright President and Fellows of Harvard College. Harvard collection of historical scientific instruments.

suggested "fifth column," as he noted in a letter to Blue Hill. The term, one of the period, was frequently used in reference to anything considered subversive. One ascent was made jointly with MIT at 0139 EST, on 2 September at the urgent telegraphic request of the Weather Bureau to help in the forecasting of a threatening hurricane.

Cooperation continued with P. C. Putnam on the problem of wind power. Plans were made for the erection of a 1250 kilowatt generator on Grandpa's Knob, 640 m (2,100 ft.) above sea level near Rutland, Vermont. At times the Blue Hill staff swelled by ten temporary help who reduced wind records and converted them to available power.

On 20 June Cushman of the Foxboro Company, without charge, placed an experimental resistance thermometer in the window shelter and an associated recorder in the front office. This marked the beginning of a long and cooperative liaison with the instrument company.

Occasionally the staff was treated to other visitors, like the squirrel perched on the glass covered opening to the pole star recorder in figure 104.

In the fall Fergusson's office was moved from the old library to the room behind the tower. This move was made preparatory to the coming of the U.S. Weather Bureau Solar Radiation Supervising Station in November. In making the change it was considered that the observing conditions

Figure 104. Squirrel on pole star recorder window. Courtesy K. McCasland.

on Blue Hill, which is relatively free from smoke, and the intensive solar radiation work being conducted by both Harvard and the Massachusetts Institute of Technology, under grants from Godfrey L. Cabot of Boston, would be advantageous. A concrete pier was erected in the back yard, to accommodate normal incidence pyrheliometers during calibration (figure 105). Irving F. Hand was placed in charge of the station with Helen Cullinane as assistant. The Bureau paid $700 per year for space and use of the facilities.

On election day a contingent of army men and Boston Post newsmen moved in to flash election returns by means of a rotating antiaircraft searchlight. The Observatory staff and these men were treated to a turkey dinner which was served in the Metropolitan District Commission lookout. After dinner the huge light (figure 106) was turned according to a prearranged code to signal the results, which were received by telephone. The beam was reported as having been seen from Montpelier, Vermont, 290 km (180 mi.) distant, and other closer points, such as Holyoke, Massachusetts, Concord, New Hampshire, and North Harpswell, Maine.

In the fall a district Weather Bureau forecast office was opened at the East Boston Airport. The great Saint Valentine's Day snowstorm, which had not been properly forecast from Washington, played an important role in Brooks' efforts toward obtaining the office. Charles Pierce, former

Figure 105. Backyard of the Observatory, showing concrete pier and sliding instrument cover used for Weather Bureau solar equipment. Courtesy J. H. Conover.

Figure 106. Searchlight used to signal election returns. Courtesy J. H. Conover.

assistant at Blue Hill and forecaster at Washington, was made chief forecaster at the new station.

Activities at Mount Washington remained under the direction of Brooks. At this time icing measurements were made at least every three hours. These were coded and transmitted with the synoptic messages and sometimes given directly to aircraft by radio.

The Observatory assisted the American Meteorological Society in preparing the 232-page, fifth edition, of the booklet, Air-Mass and Isentropic Analysis, by Jerome Namias and others. This edition, edited, like the others, by Stone, included a 50-page bibliography by him for synoptic meteorologists. Over the past years the society had distributed some 10,000 copies of the booklet.

F. Burhoe assisted part time in the library work.

John Clinton, caretaker of the Observatory and porter of equipment, died in November.

The radio signal strength measurements that had been collected for propagation studies by the late Ross Hull had been turned over to Brooks and Pear for analysis, but this work was not completed at the time. After the war and after leaving Blue Hill, Friend (1945) published an analysis of the data. In this he showed the importance of the role played by Observatory staff members in this work, and showed that Hull was well on the way to explaining the variations in signal strength by means of the atmospheric sounding data that he had collected.

Kenneth G. McCasland arrived 29 August as observer and radio man. He had served in the same capacity at the Whiteface Mountain Observatory in New York. A month later McCasland made all the observations, including early sunrise measurements and radio schedules for fifteen consecutive days, while this writer married Ethel E. Chase and took a honeymoon. The year ended with a Christmas party on 24 December. Toy gifts were exchanged. Dr. Brooks received a corn-cob pipe (he never smoked) and Mr. Fergusson a toy windmill. Those present with their toys are shown in figure 107.

Figure 107. Christmas Party, 1940. From left: Irving Hand, Robert Stone, Ethel Conover, Francis Burhoe, Edmund Schulman, Edward Brooks; back row: Victor Conrad, Ida Conrad, Helen Cullinane and son, David Arenberg, Sterling Fergusson (kneeling), Ralph Burhoe, Eleanor Brooks, Charles Brooks, and Karl Lange. Courtesy J. H. Conover.

1941

Starting 1 January a new observation was entered in the books each day at 1300 EST. The day was classified as to whether it was one on which the observer preferred to be outdoors or indoors and for each of these classifications three degrees of level were assigned. The scheme probably originated with Conrad, who planned to use the data in his climatological studies. Meanwhile, Conrad, who had been appointed Robert DeCourcy Ward research associate, embarked on numerous climatological studies in which Blue Hill and Mount Washington data were utilized.

Staff meetings were started and held about every three months in order to bring all members up to date on current work and future plans.

Fergusson rebuilt the ombroscope, which remains in use at this writing. He also made a new wind vane which was attached to the anemoscope (figure 113). Aluminum was used for the split tail surface; otherwise it was very nearly the same as the old vane which had been in use from the earliest days of the Observatory.

A baffle in the form of a cross was added to the receiving bucket of the weighing precipitation gage. This was designed to help retain snow in the receiver under windy conditions. However, during wet snow with low wind speeds the snow built up on the baffle, decreasing the size of the gage orifice.

At the spring meetings of the American Meteorological Society, R. W. Burhoe noted the rapid increase of meteorological literature. He proposed that the meeting "make a resolution to the effect that it recommends to the various meteorological institutions and scholars of America that they cooperate toward the establishment of an adequate bibliographic service for this science in this country." Resolutions to this effect were adopted by the Society and the meteorological section of the American Geophysical Union. This was another step toward the eventual publication of the *Society's Meteorological Abstracts and Bibliography*.

A. W. Friend resigned in mid-year to take a position in what became the Radiation Laboratory at MIT.

Walter M. Elsasser, of the California Institute of Technology, joined the staff in July. With Arenberg's assistance he completed a sensitive radiometer which was used at the time of radiosonde ascents. Elsasser soon completed a treatise titled "Heat Transfer by Infrared Radiation in the Atmosphere," which was published as *Harvard Meteorological Studies*, No. 6. This well-known monograph was used into the 1960s and the radiation chart that was included in it served to illustrate the effects of upward and downward flux in the lower atmosphere.

On 29 June the Observatory was, once again, struck by lightning. The new grounding cables and pipes which had been installed after Brooks's arrival apparently carried off the charge. The only damage consisted of burned-out fuses in the delicate pyrheliometer circuits.

In July, after some redesign of the Harvard radiosonde, another series of test soundings was made from the roof of the Geographic Institute in Cambridge. The changes and tests were required as problems developed during quantity production of the instruments for the U.S. Weather Bureau and Chinese government.

Putnam's wind power project demanded additional wind data from the Grandpa's Knob site outside Rutland, Vermont. A so-called "Christmas tree" was erected which was festooned with cup anemometers. The 56-m (180-ft.) high "tree" consisted of a vertical mast and long cross arms extending from midlevel in four quadrants. Data from this array of instruments was to give information on speed gradients over the proposed 55-m (178-ft.) diameter of the turbine blades. Lange was in charge of this instrumentation. He also erected a gas-heated anemometer to obtain speeds during icing conditions on nearby Pico Peak.

The wind turbine was first turned on in August of 1941 and underwent modifications while under test for about 1000 hours of operation over 4 1/2 years (Putnam 1948). Finally it went into regular operation in February 1945. Only a month later one of the two blades, weighing 8 tons, let go and flew 229 m (750 ft.). The turbine was not rebuilt and the project ended. The meteorograph on Mount Wachusett was also reactivated 9 December to supply further wind measurements.

Course theses in Brooks's climatology class centered about the climate of New England. Edward Sable of the Weather Bureau compared monthly Blue Hill temperatures with other surrounding stations. Sable also tabulated many wind records that had accumulated without reduction during McAdie's time to form a table of monthly speeds for the entire period.

At the beginning of Brooks's second decade at Blue Hill, he proposed that research be directed toward solar, terrestrial, and atmospheric radiation, vertical turbulence and horizontal circulations. He no doubt had Elsasser, Lange, and Hand in mind for the work. Most of these studies were never undertaken.

Mrs. Abbott Lawrence Rotch, widow of the founder of the Observatory, died in May. It was only by virtue of her interest and generosity that the

Observatory was kept going during the difficult years following her husband's death. She contributed from $1,500 to $5,225 each year for thirteen years until the income from endowments made this no longer necessary. Subsequently she contributed about $40,000 and $15,000–$20,000 in real estate.

In September Conover started teaching the laboratory course that was associated with Willett's course in meteorology at Harvard. Five sessions were required for each laboratory to accommodate the large number of students, many of whom were in a navy program.

The prominent hill-top location now brought a state police radio relay station to the Observatory. The antenna was placed on the northwest side of the tower roof (see figure 108).

The events at Pearl Harbor on 7 December quickly led to traumatic events at Blue Hill. On the following day Lange, who had filed final papers for U.S. citizenship nine months earlier, was one of a group of 2,000 enemy aliens taken into protective custody by the Federal Bureau of Investigation. After 6 years of distinguished development of aerological apparatus, especially the first routine operational American radiosonde, and four years of teaching aeronautical meteorology to several hundred aviators—mostly students—he was forced to terminate his work at Blue Hill. He was released later to do special work for the government and still later moved to Kentucky where he taught at the University. Informal cooperation with the Weather Bureau and Signal Corps, largely in the testing and improve-

Figure 108. State Police antenna and horizontal dipole antenna used to communicate with Mount Washington, 1941. Note glaze. Courtesy J. H. Conover.

ment of the Harvard radiosonde, was continued by W. K. Coburn, Jr., and Jean Booth, with occasional supervision by D. P. Keily of MIT.

On 13 December, arrangements were made to supply the local anti-aircraft battery with upper air winds for ballistic purposes. On 14 December amid frequent false air raid alarms, eight army men from the 208th anti aircraft division of Quincy arrived to spot and report airplanes. They strung their own communication lines beside the road. Two pibals, using Observatory equipment, were made daily at 0930 and 1500 EST. Observers and staff literally had to climb over these men who manned the post twenty-four hours a day. The road soon deteriorated due to heavy use, but on the 20th, on short order, it was repaired and placed in good condition.

On 25 December the exchange of weather information by radio to Mount Washington was discontinued by government order.

During the year observer and radio operator, McCasland divided his time with the Mount Washington Observatory. Ethel Conover spent many days at the Observatory working on the Putnam wind project, cooking, and staying overnight when her husband could not leave. Ed Maher, William Wadell, and John McNaye served for brief periods as observers. Robert Elsner also spent a short time at the Observatory as a trainee for Mount Washington.

References

Friend, A. W. 1945. A summary and interpretation of ultra-high-frequency propagation data collected by the late Ross A. Hull. *Proc. Inst. Radio Engs*. 33: 358-373.

Knight, R. W. 1938. The radiotelemeter and its importance to aviation. Rept. No. 1. *Civil Aero. Authority Planning and Development Div*. 35 pp.

Little, O. M. 1937. Contribution to the development of the radio-meteorograph by the United States Weather Bureau. *Trans. Amer. Geophys. Union*, 138-141.

McKenzie, A.A. 1982: The Observatory and Dr. Armstrong. *Mount Washington Obs. News Bull*. 23:3-7.

Middleton, W. E. K. 1969. *Catalog of meteorological instruments in the Museum of History and Technology*. Washington, D.C.: *Smithsonian Institution Press*, 128 pp.

Putnam, P. C. 1948. *Power from the wind*. New York: *Van Nostrand Reinhold, Co*. NY, 224 pp.

QST. 1938. It seems to us —. Obituary of Ross A. Hull. *QST*. Nov. 1938: 7-10.

Chapter XI

The War Years and Rising Costs: 1942-47

1942

In January the army departed for a more permanent airplane spotting location and the Observatory was able to return to normal operation. In March blackouts were held in various areas. It was an eerie sight to see the valley lights go out and total darkness envelop large areas of the countryside. Blackout curtains were installed throughout the Observatory in July; no outside lights were permitted, except flashlights, and the top halves of automobile headlights were painted. These precautions were necessary, otherwise vessels close to the coast were silhouetted against sky-light from large metropolitan areas and thereby could become targets for submarines. Living at the Observatory was made more comfortable by the installation of a used electric cook stove.

Brooks, as usual, attended the spring meetings of the American Geophysical Union and American Meteorological Society in Washington. In July he lectured for four days at Clark University and on 7 August he was appointed consultant to section D-1 of Division D of the National Defense Research Committee. Later in the month his geography course started with twenty students. Beginning in June, Conover resumed teaching the laboratory course for Willett's elementary course in meteorology. More than 100 students attended. In September Brooks had five additional students in physical climatology. He made two international radio broadcasts over WRUL on "Clouds" and "March Weather." For the illustrated Red Cross book, *Science from Shipboard*, that was prepared by the Boston and Cambridge branch of the American Association of Scientific Workers, Brooks contributed the opening chapter titled "Waves, Wind and Weather." The book was designed for use by men traveling to war by ship transport. As a member of the Committee on Geography of the U.S. Office of Education, Brooks outlined wartime courses of study in meteorology and climatology for U.S. colleges.

Conrad published his 121-page book titled *Fundamentals in Physical Climatology* and in the fall he lectured on physical and war climatology at the California Institute of Technology.

Hand, J. and E. Conover, and Boland made a joint expedition to Mount Washington in the fall. Simultaneous observations of direct insolation were made at the top and base of the mountain when the density and water-vapor content of the intervening atmospheric layer was measured.

The radiosonde development which began in 1935 was concluded. The manufacturer had taken over completely; in its final form, cost of the instrument was $15.

The observational program was increased somewhat by supplying pibals or cloud observations, on demand, to the Weather Bureau of Boston.

In February Stone went on leave of absence to New York University. This leave proved permanent but Stone had served the Observatory well, playing an important role in reestablishing the Observatory's international prestige. He had been involved in the tremendous progress toward restoring the library. In the course of this work he had developed a reputation as a walking encyclopedia for his ability to recall material and references on a wide range of meteorological and climatological subjects. His reputation also included an interesting habit of arriving on the hill about 1300 EST and working straight through until about 0100 EST the next day. He stood most of the twelve hours while working, interrupted only by breaks which consisted mainly of the consumption of large quantities of milk.

Ralph W. Burhoe was made acting librarian in Stone's absence.

The times made it very difficult to keep assistants and observers. Hand had a series of assistants; then W.A. Boland, who had been a research assistant at the Observatory in 1935-36, arrived to assist Hand.

Rose Harrington worked full-time as a secretary to Brooks. She was the first full-time female Harvard employee at the Observatory. Several female weather Bureau assistants of Hand's had served earlier. On 4 November Sarah H. ("Sally") Wollaston arrived as an observer. She soon learned the schedule and methods. Usually she arrived early on foot starting with the 0700 EST observations and worked until midafternoon. Elsasser left in February for work in the Signal Corps Laboratories and at year's end McCasland left to go to the Radiation Laboratory at MIT Alexander McKenzie, who left the Observatory in 1936, and Arthur Bent also joined the Radiation Laboratory.

At about this time a number of the rarer early works in the library which had been collected by Abbott Lawrence Rotch were transferred to the Houghton Library in Cambridge. In July Rotch's son, Arthur, and daughter, Margaret, presented to Houghton, in memory of their parents, most of their father's private library of meteorology and aeronautics books.

The gift amounted to over four hundred early books and manuscripts, the latter including letters of Benjamin Franklin to Sir Joseph Banks regarding balloons, and letters from the Wright brothers and Graf von Zeppelin. The remainder of his library was sold through a New York book dealer.

1943

Research during this wartime year was halted while the staff was kept busy teaching and responding to a wide variety of requests. These ranged from simple library loans to small bibliographies on specific subjects. Climatological data from all over the world was sought by many agencies and often this required special interpretation, such as the development of probability tables of certain weather conditions over specified areas.

Brooks and Conover continued teaching the Harvard courses until summer.

At the request of the Weather Bureau, Conover and Brooks participated in an experiment to instruct weather observers throughout the United States on cloud observing and techniques for making these reports more valuable to forecasters. Brooks made the first country-wide trip in the fall, spending about a week at each regional office located in Flushing, New York, Chicago, Kansas City; Seattle; Pacific Palisades, California; Ft. Worth; and Atlanta. Lectures and outdoor observing periods were given to Weather Bureau and Civil Aeronautics Administration trainees, Weather Bureau observers, liaison officers, and army and navy personnel. Conover made the second trip, starting in December, and Brooks made the last trip in 1944.

In March a new high-voltage power line was placed underground on the south side of the hill, thus providing more consistent voltage on the summit. A transformer house was built in the backyard of the Observatory from which power was fed to the Observatory and to the new metropolitan police and state police transmitters on the north and southwest sides of the summit. The old transmitters and the Metropolitan District Commission antenna on the center mast of the tower were removed from Observatory property in April.

Harvard and the Weather Bureau entered into joint investigations of long-range forecasting for the army and navy. A new office for this work was opened in Cambridge in March. The studies were under the direction of Mr. R. A. Allen of the Weather Bureau's Central Office. Clayton was appointed research associate 15 March–30 June and Mr. Schell assisted along with nine Weather Bureau statisticians. The navy sent Commander Anderson to observe. Clayton's solar weather relationships and Schell's foreshadowing techniques were under study for possible wartime application.

Conrad's manuscript, *Methods in Climatology*, was submitted for publication in November. It would be the first American book of its type and brought to U.S. climatologists the wealth of techniques developed abroad, and gathered together effective methods developed in the United States.

In early winter the Observatory participated in a program of snow cover observations that were designed to provide relationships between traffickability for the military over snow and synoptic weather reports. A number of measurements, including penetrability, density, specific heat, and temperature, were made at frequent intervals. These were difficult on the summit, especially at night, in spite of a portable canvas wind shield which was employed. Heavy snow in January amounting to 79 cm (31 in.) all but buried the Nipher shield and standard gage, as shown in figure 109. The recording Fergusson gage, also shown in the picture, had been mounted on a concrete pillar which elevated it well above the snow surface. Snowfall for the season 1942-43 totaled 155 cm (61 in.).

In February the MIT radiation laboratory mounted two small radomes on the north parapet of the tower. These housed secret antennas presumably to measure signal strength at radar frequencies over the path to

Figure 109. The author standing beside the weighting precipitation gage; nipher shield, and standard gage in the background, January 1943. Courtesy J. H. Conover.

Cambridge during variable weather conditions. Tests showed that the domes were not high enough to disturb the solar or wind measurements.

About this time David Arenberg, former Observatory employee, suffered a humiliating, but, in retrospect, amusing experience. He had joined the MIT radiation laboratory and was experimenting with voice transmission by modulating a light beam. The time came for field tests, so, being familiar with the Observatory and surrounding area, it was decided that a colleague go to the Observatory tower while he set up his equipment in a field south of the hill. Around midnight Arenberg was engrossed in sending signals from his base location when two policemen unceremoniously descended upon him and hustled him and his equipment off in a cruiser. A member of the spy-conscious populace had reported his flickering light. In vain he explained the experiment but was held in the Canton jail as calls to MIT went unanswered. Finally, toward morning, the authorities at MIT were contacted and he was released. Meanwhile, his friend at the Observatory knew only that his signals had ended abruptly, and upon investigation found that Arenberg and his equipment had simply vanished.

In the summer Conover modified the Mount Washington heated number 2 anemometer. A new calrod heating element with thermostat was installed in the top, shown in figure 110. A small cooling fin was also made which attached to the upper end of the anemometer shaft. This was intended to carry off heat from the upper bearing which had previously overheated and failed. Following these changes the calibration was checked at a Harvard wind tunnel. Use of the thermostat was abandoned because vibration caused intermittent operation and excessive radio interference.

Conover also made a heated anemometer for use under icing conditions at Blue Hill. The anemometer (figure 111) consisted of an open slit

Figure 110. Mt. Washington heated anemometer number 2, with rotating cap removed to show new heating unit and thermostat. Courtesy J. H. Conover.

which was kept facing the wind by a small attached vane. From this slit the approximate dynamic pressure was obtained. Below the slit a series of holes around a cylinder provided a negative pressure as the wind passed. Two pressure lines led to a Friez pressure-type wind speed recorder on the third floor of the tower. This recorder was loaned to the Observatory by the Yankee Network weather service which had been forced to close at the start of the war. A 750-watt calrod unit was mounted vertically inside the wind sensor and controlled by a variac indoors. Since the transmitted pressures were not truly dynamic and static, the recorder required calibration against the other anemometers. Use of this instrument assured records during icing conditions, which mainly resulted from freezing rain, and saved climbing the masts to de-ice the instruments with hot water.

In January the base station thermometers were returned from a location near the Trailside Museum to a shelter in Brooks's yard at 1793 Canton Avenue, the location that they had first occupied.

Also in January tabulations of the chief weather elements by the hour were resumed.

Brooks acquired a rangefinder which rested on a tripod located on the bridge just outside the library door. He made cloud height measurements from this short, 1-1/2-m (59 in.) long instrument.

Figure 111. Conover's heated anemometer on the tower at Blue Hill, 1 December 1943. Courtesy J. H. Conover.

In the fall dim-out restrictions were lifted, although the Observatory continued to be blacked out at night. The Observatory was permitted to send its weather by radio to Mount Washington but Mount Washington's weather remained encoded.

Frances Wollaston, sister of Sally Wollaston, arrived in July. She assisted with the preparation of a current bibliography and abstracts of new work which were published monthly in the *Bulletin of the American Meteorological Society*. Conrad Chapman helped with this library work. Burhoe organized the publication in Spanish of a Latin-American section of the *Bulletin of the American Meteorological Society*.

During this period it was very difficult to keep the observational program covered by personnel, especially overnight. G. Oleshefsky became the regular observer in residence. Sally Wollaston and Conover filled in the remaining hours. All of the Brooks children were trained in observing and, as they grew up, occasionally filled in for a time that was particularly difficult to cover. At this time Edith Brooks took many observations. Much time was spent training and checking out observers, only to have them leave, often for military service. A group picture (figure 112) shows most of the staff and others.

However, not all was hard work. A highlight of the winter for this writer was a trip down the hill, after dark, belly down on a sled. At the time the road was covered with rough ice, a result of many footsteps in slush and

Figure 112. Group picture, c. 1943. Left to right, back row: Commander Anderson (USN), John Conover, Ethel Conover, Sally Wollaston, Jim Taylor, Rose Harrington, Frances Wollaston, Irving Hand (USWB), William Boland (UWSB), Sterling Fergusson. Left to right, front row, Lydia Timmons (UWSB), Irving Schell, Bernard Haurwitz, Henry Clayton, George Olehefsky, and Ralph Burhoe. Courtesy S.H. Wollaston.

snow. The trip was made at what seemed to be fantastic speed. My wristwatch stopped from the vibration and I developed a headache which lingered for hours.

During the year much effort went into teaching military and Weather Bureau meteorologists methods to maximize the use of cloud observations in the preparation of forecasts. Following the second trip around the country during which observing and forecasting techniques were taught, Brooks and Conover spent five days in Washington discussing the trips and planning further teaching aids. Conover was offered a position with the Weather Bureau but he declined and arrangements were made to do much of the work from Blue Hill. A 35-mm Leica camera with a wide angle supplemental lens was purchased to start a collection of 35-mm colored slides depicting clouds and states of the sky in general.

About this time, the first whole-sky time lapse color photography in this country became available to the Observatory. H. R. Condit of the Eastman Kodak Laboratories in Rochester, New York, had mounted a pulsed 16-mm movie camera at the side of a spherical mirror; the camera looked at a small plane mirror mounted above the spherical mirror thus seeing the entire sky, horizon to horizon, as reflected from the spherical mirror. Condit used these films to prepare exposure guides for different sky conditions. These films also proved very interesting in depicting changes in the sky and cloud motions. Some of these dawn-to-dusk cloud sequences were shown at the many lectures on clouds which were given by Brooks or Conover.

Brooks published a paper on international cloud nomenclature and coding in which he proposed a fourth cloud group, to be called CV, or "clouds vertical," in addition to the CH, CM, and CL, high, middle, and lower groups, already in use. This group was to include many layers at "low" levels and clouds of vertical thickness that extended from low or middle levels to high levels. The proposal would have solved a dilemma for observers and added more meaning to coded synoptic reports but it was never adopted by the international commission.

In midsummer, daytime hourly airways observations were made by Sally Wollaston and Conover for one week and phoned to the airport Weather Bureau station. During this experiment every effort was made to obtain cloud heights by rangefinder, which was used by Brooks, or by triangulation between Blue Hill and the South Weymouth Naval Air Station, 15.56 km (9.7 mi.) to the east–southeast. The experiment was continued through the summer but only at the hours of 0700, 1000, 1300, and 1600 EST.

In September Conover took a leave of absence to prepare cloud observing instructions for the Weather Bureau. Since the Blue Hill observing techniques embodied so much more detail than the bureau's, he was assigned to the Weather Bureau airport station at East Boston in December to become more familiar with their methods and needs. While there he took routine observing and then flight advisory weather service (FAWS) shifts.

Other work during the year involved Brooks's participation in an aircraft icing conference in Washington. As director of the Mount Washington Observatory, this helped to keep him informed of icing research, and its future needs, while determining the role that the mountain station might play in the work.

Brooks consulted frequently with the military during the year. He was especially helpful to the Quartermaster Corps in supplying "cooling power," now referred to as "wind chill," information in respect to the human body and in planning new measurements.

Conrad also prepared for the Quartermaster Corps a 202-page gazetteer of the climates, soils, orography, and biogeography of the climatic regions of the world.

The Harvard chronometric radiosonde was still in competition under at least two contracts with manufacturers. W. K. Coburn, Jr., of the Observatory staff, was chief technician during performance tests against the Diamond–Hinman–Dunmore frequency modulation type in Washington.

As a result of the previous year's concentrated study of Clayton's forecasting techniques, which was based on sunspot numbers, he went to Washington for further discussions. Owing to the wartime classifications of forecasts it is uncertain whether his work was used.

Irving Schell provided the army, navy and Weather Bureau with experimental long-range fall and winter forecasts for northwestern Europe. These were based on the extent of ice in the Greenland Sea from April to August. The reasoning was that with above-average ice, the sea remained cooler than average and storm tracks were shifted southward with their associated precipitation for ensuing months. Such a condition increased rainfall over Great Britain and decreased it over Scandinavia. Sixty years of records showed this sequence to be 70 percent correct. In the spring and summer of 1944 the ice presence was reported as being considerably less than normal, favoring less-than-normal precipitation over northwest Europe in November and December. It turned out that a break in the wet weather did not materialize until December.

Chapman prepared a descriptive list of the Observatory's collection of weather maps. A total of over 70,000 maps, some series dating back to 1871, from all parts of the world were listed. Many of these were borrowed for the study of future military operations.

S. Wollaston, acting chief observer in the absence of Conover, helped Hand in the calibration of new solar instruments.

As a result of studies by Gorczynski, Conrad and Haurwitz, relationships between bright sunshine or cloudiness and its type to total insolation were under preparation.

In midsummer another serious brush fire swept up the southwest side of the hill on 14 m sec^{-1} (31 mi. hr^{-1}) winds. Approximately forty acres were burned.

Use of the twenty-four-hour clock was started in the observing program and Fergusson made a completely new windmill and mechanical transmission system to the cinemograph. The new aluminum windmill was rugged but still as responsive as the more delicate windmills of the past. The new windmill and vane of 1941 are shown in figure 113.

The new windmill was soon tested by the hurricane of 14-15 September 1944. This storm tracked northward along the East Coast of the United States, moved inland over Rhode Island, and passed about 16 km (10 mi.) east of Blue Hill about 0010 EST on the 15th. Once again Brooks and the Observatory staff made frequent observations of cloud motions preceding the storm and sent them to the district forecast office at Boston. With additional data, such as winds aloft measured by radar reflection from drifting balloons (RAWINS), the Bureau made good forecasts which were disseminated quickly, thereby saving many lives and much property. At the Observatory the wind reached 34.5 m sec^{-1} (77 mi. hr^{-1}), as recorded on the cinemograph, a two-minute average, from the east–southeast. As the center passed the winds diminished markedly, only to accelerate to about 22 m sec^{-1} (49 mi. hr^{-1}), from the northwest in the storm's southwestern quadrant. Heavy rain, amounting to 93 mm (3.66 in.) fell over a four-hour period preceding the lowest pressure. The deluge washed out the summit road leaving gullies up to 0.6 m (2 ft.) deep. This writer and his wife spent the wild night at the Observatory making hourly observations. A State Police officer was also stationed on the summit in his cruiser to relay messages in case of a power failure, which indeed occurred. He ex-

Figure 113. Wind vane of 1941 and windmill of 1944 atop the southern-most posts. Courtesy J. H. Conover.

perienced a wild night rocking in his cruiser and finally drove a short distance from the summit fearing his car would be overturned. As we walked down the next morning drag marks on the road clearly showed his difficulty in negotiating the deep gullies in the road. Dr. and Mrs. Conrad were vacationing at a beach house on Cape Cod at the time. They were persuaded to leave for higher ground in the early evening of the 14th. The next morning they returned to the site of their house to find nothing but the refrigerator. He was so impressed that he published a short paper in which he described his observations. (See bibliography for 1945.)

The Observatory was made more cheerful by the removal of blackout paint on 13 November.

The observing staff was more stable in 1944. Jim Taylor left but F. Wollaston and Norman Brooks took his place while S. Wollaston was acting chief observer. Oleshefsky continued as resident observer and Fred Lund III filled in occasionally at night. The staff was saddened to learn on 1 May that former staff member Salvatore Pagliuca had been killed in a jeep accident on Mount Mitchell, North Carolina, while serving in the military.

Margaret R. Storrow (Mrs. J. J. Storrow, Jr.) made a gift of $5,000 to establish the Abbott Lawrence Rotch Fund. The fund was to provide for Abbott Lawrence Rotch Fellowships or Associateships. This and other gifts totaled about $7,500 for the year. Payments from the Weather Bureau and Quartermaster Corps significantly helped to ease expenses over the year.

Conover continued to work at the airport office of the Boston Weather Bureau through the first half of the year. While there he expanded his interest in depicting clouds and weather in three dimensions. Utilizing the continuous flow of available weather data at the airport, efforts were made to construct cross-sectional diagrams, including fronts, on a real-time basis and use these to make airways and terminal forecasts over New England. The work was instructive but not practical. Good forecasters could visualize the changes quite well without the use of charts; furthermore, the need for detailed cloud layers and weather aloft for general aviation had already diminished due to navigational aids and instrument flying techniques which were in common usage.

Meanwhile Brooks and Conover prepared a guide to cloud nomenclature for the Weather Bureau. This guide, which was constructed similar to an organizational chart, started with the simple choice of whether the clouds were "lumpy" or "not lumpy," and led to a final name which gave the genera and species of the cloud. The authors felt that proper nomenclature was required for the best coding of the clouds by observers and interpretation on the part of forecasters. However, this guide was not made a part of the Bureau's official instruction book because it was thought excessive in detail and more applicable to the research community.

Conover was once again offered full time work, this time at the airport, but he declined in favor of a return to the Observatory and work on a new and urgent cloud study. This study, which was initiated by Harry Wexler, was to prepare three-dimensional cloud diagrams around cyclones and, in particular, hurricanes. The project was sponsored by the Weather Service of the Army Air Force. Work was done in a temporary penthouse on the roof of the meteorology building at MIT. Lt. Edward G. King and Conover made the analyses while Isabella Hynes and Hazel Miller plotted the data.

The detailed three-dimensional cloud structure and its evolution associated with the hurricane of June 1945 was completed and presented at an MIT seminar on 27 August, a few days after the end of the war. With the end of hostilities immediate need for the study ended and the contract terminated a month later.

On 1 February Dr. and Mrs. Brooks, in the tradition of Rotch, gave a tea at their home commemorating the sixtieth anniversary of the Observatory. Fifteen guests, including Arthur Rotch, attended.

Margaret R. Storrow, sister of Abbott Lawrence Rotch and wife of J. J. Storrow, Jr., died 21 March. She and her husband were active benefactors to the Observatory for many years. A new effort to raise the endowment was begun in the spring by a subcommittee headed by Henry DeC. Ward, vice chairman of the Visiting Committee.

In October Brooks gave a fifteen-minute talk over the Yankee Network home radio station, WNAC. His subjects were aeronautics and the work at the Mount Washington Observatory. During the year he spent considerable time on the design and construction of new facilities to study icing on Mount Washington. This involved the coordination of four government agencies, three universities, one airline, and two manufacturers. Much of this conferring was carried out between the two Observatories by radio. In preparation for a more active winter, the mountain staff was increased from six to nine.

The Mount Washington Observatory heated anemometer number two and a pilot-static tube were tested in a Harvard wind tunnel to determine the effect on the of winds having various upward components.

In mid-May Sterling Fergusson and an old colleague from the University of Nevada, J. E. Church, decided to visit the White Mountains. Church, an expert on the Sierra snowpack, wanted to observe snow conditions in Tuckerman's Ravine so they climbed easily to that point. Since the weather was good they ascended a trail beside the ravine and continued up the cone to the Observatory. When they knocked at the Observatory door the residents were astonished to see these two men, dressed in overcoats and wearing rubbers over their street shoes. At that time Fergusson was 67 years old. Their unheralded arrival was even more amazing when it was learned that they had simply "walked up" from Pinkham Notch via the steep Lyons Head trail.

The observational program continued at Blue Hill as before except for termination of the classification of the character of each day, an observation which was started in 1941. Gast gave a micromax recorder to the Observatory for recording total radiation. This replaced the old Englehard recorder.

The state police antenna on the northwest mast was removed and Ken McCasland visited to construct and install a new dipole antenna for communication with Mount Washington. This antenna was mounted horizontally outside and below the north side of the tower parapet. At the same time the wooden mast on the north side of the tower was sawed off.

In October the recording of diffuse, or sky, radiation was commenced by the Weather Bureau. A conventional Eppley total radiation pyreheliometer was mounted so the sensor would always be shadowed from the direct rays of the sun by an adjustable ring. Figure 114 shows the apparatus, which was mounted on the top of the north side of the parapet, where the MIT radomes had previously been located.

A new series of seasonal weather summaries was started during the year. These mimeographed summaries which included a discussion of the weather, and tabulated data were sent to interested institutions and individuals.

Louvan Wood of the Friez Instrument Division, Bendix Aviation Corporation, designed a new anemometer–wind vane modeled after the pattern of the Blue Hill windmills. This instrument, called the Aerovane, had three molded plastic blades which turned a magneto, thus generating a voltage which could be recorded remotely. The streamlined vane that kept the windmill facing the wind was connected to a selsyn generator. This transmitted wind direction to remote locations. Wood brought one of the first models to Blue Hill, where it was left for evaluation. It was mounted on the central tower mast and first connected to indicators on the third floor, and in 1955 was connected to a Friez recorder on the first floor of the tower. Conover devised a lightweight attachment that could supply

Figure 114. Diffuse radiation sensor. The ring adjusts along the two parallel bars to allow for solar elevation changes throughout the season. Courtesy J. H. Conover.

contacts for each revolution of the mill. In this way response times were recorded and other tests made (see bibliography for a report on the tests). With slight modification this new wind sensor proved to be an excellent instrument for station use on land or sea; thousands, undoubtedly, have since been manufactured. The wind indicators for the Aerovane and radio equipment are shown in figure 115. Instrumentation on the tower can be seen on figure 116.

In December Fergusson installed an electric motor in the cinemograph. This provided power to the two rotating disks of the mechanism. The old recorder in which the disks were spring driven was made interchangeable for use during power failures.

In February resident observer George Oleshefsky resigned and Lund took his place for a brief period. Francisco deCosta took the resident observer position in August. deCosta also gave violin lessons and in the course of his practicing he occasionally played while we worked. S. Wollaston continued to maintain accuracy and homogeneity in the observational program.

Figure 115. Wind indicators connected to the aerovane and radio equipment. Receiver on the left and transmitter panel for W1XW on the right. The loop of patch cord above the receiver was used to hang the telephone mouthpiece over the speaker, thus easing the burden of relaying both sides of conversations. C. 1950. Courtesy J. H. Conover.

Figure 116. Tower instrumentation looking toward the north–northeast. Left to right: Cinemograph windmill, aerovane, seen through the ladder, anemoscope wind vane, 3-cup anemometer and vane connected to the modified Draper recorder, lightweight windmill, and heated pressure tube anemometer. Below: Normal incidence pyreheliometer and other pyreheliometers on the platform. C. 1946. Compare with figure 21. Courtesy J. H. Conover.

1946

Systematic icing observations had been made on Mount Washington for about two years by this time. The wealth of data revealed a large potentiality for icing and cloud physics research which eventually was realized in the form of a $37,000 contract with the Air Material Command. The Blue Hill Observatory undertook the contract, arranging for research space in Cambridge and communication via Blue Hill, while exposure facilities, living quarters, and working space were provided on Mount Washington. Wallace E. Howell was appointed Research Fellow in Icing Research and made scientific director of the project while Brooks continued as director of the mountain Observatory. Victor Clark, who had been in charge of the icing studies on the mountain from 1942-46, became a voluntary assistant.

For years Clayton had claimed statistical relationships between solar activity and surface weather. In an effort to show some physical meaning behind his relationships, Haurwitz prepared a brief paper in which he suggested a connection between solar activity and the lower atmosphere through the absorption of radiation and heating of the ozonosphere. He postulated that this heating could cause divergence at midlatitudes and convergence over the equator which conceivably could lead to changes in the weather of the lower atmosphere. Later, in the words of Haurwitz, "my theory was shot down by Professor London who pointed out that certain values that had been used were in error."

Conrad continued his work on such diverse subjects as trigger causes of earthquakes and the physical development of polygon nets by freeze and thaw cycles in soil.

Brooks, in addition to his directorship duties of Blue Hill and Mount Washington, also prepared papers ranging from a note on the fluffiest snow to a comparison of sunshine recorders. The latter he prepared jointly with Mrs. Brooks. He also wrote on the spacing of stations in climatic networks, and aided in the publication of three books. He gave tutorial instruction in meteorology and climatology to two undergraduate students.

During the war years the American Meteorological Society membership expanded greatly and consequently the office work associated with it. Three to four women employed full time plus the services of treasurer Ralph Burhoe, Frances Burhoe and Brooks, were required to handle the work. Due to the lack of space at the Observatory most of the work took place at the Burhoe residence in Dorchester. Also during the period a move, headed largely by C. G. Rossby, developed to publish a new journal which would contain articles more scientific than those generally found in the Bulletin. Kenneth Spengler was appointed executive secretary of the society in April. Office work was consolidated in new headquarters at 5 Joy Street, Boston, and the job of organizing what became the *Journal of Meteorology* was undertaken.

S. Wollaston, in addition to her observing duties, worked on the comparison of the diffuse radiation data obtained with the new Weather Bureau instrumentation versus the subtraction of direct radiation from total radiation to give the diffuse quantity.

Conover made two trips to the Signal Corps laboratories at Belmar, New Jersey, in an effort to obtain a small weather radar. For a while prospects for obtaining the radar were good and a platform on which it was to be mounted was erected on the west side of the parapet. However, plans fell through and it was not obtained.

An extension of about two meters (6 ft.) was added to the south side of the tower roof. This extension, made of open steelwork, was put up primarily for the Weather Bureau's solar radiation work. It provided new space for their pyrheliometers, two of which were mounted vertically to measure radiation on surfaces facing east and south. The extension and instruments are shown in figure 117.

Figure 117. Steel extension on the south side of the Observatory tower. Equatorial mount and normal incidence pyreheliometer (2) and (4) were replaced by a Weather Bureau instrument which was mounted on the extension. Steps were then added for access to the extension, Campbell–Stokes sunshine recorder (3), total radiation (1) and east vertical surface radiation sensor at (5). Courtesy S. H. Wollaston.

Cloud reports continued to be sent to the Weather Bureau and gradually the observers were given government "certificates of authority" to make these observations.

The various Observatory anemometers, which are shown in figure 118, were wind tunnel tested during the year.

In June Willis Hicks of the Foxboro Company brought a vapor pressure sensor, called the "dewcel," and recorder to the Observatory. He had developed the instrument for use in air conditioning systems but thought that with proper shielding it might be adopted for meteorological use. The sensor which operated at a temperature above ambient could not be wet or subject to sudden drafts so it was necessary to devise a shield that would prevent these problems yet sample the air without undue lag. When the sensor was properly housed, kept clean, and reconditioned with chemicals every month or so excellent continuous records of the dewpoint temperature could be obtained remotely. Eventually many of these instruments were sold to airport observing stations but improper care gradually led to replacement by newer instruments (see bibliography).

At the same time the Serdex Company constructed a relative humidity recorder in which goldbeaters skin replaced human hair as the sensor. Instruments were tested at the Observatory but they were not useful in the rugged outdoor weather.

During the year Fergusson constructed a new pole star recorder that used 35 mm film, thus allowing a month's record on one film strip.

Friend had returned to Harvard where he carried on new work in sounding the atmosphere by radio. He required a calibration target for his antenna array so a corner reflector of heavy screening was constructed and

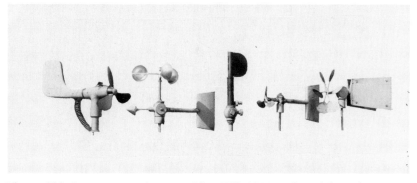

Figure 118. Anemometers in use at Blue Hill, 1946. Left to right: Friez aerovane and attached contacting mechanism, three cup and vane, heated pressure, Fergusson balsa-bladed windmill, cinemograph aluminum windmill and vane. Courtesy J. H. Conover.

fastened to the northwest side of the tower facing his antennas in Lexington. This reflector is shown in various pictures of the tower.

In connection with balloon ascensions which carried cosmic ray counters, the Observatory facilities were used by the Laboratory for Nuclear Science and Engineering at MIT. A small prefabricated metal building was erected just south of the old balloon launcher site. Balloons were inflated inside the building and then launched in clusters numbering as high as fourteen to lift the 11 kg (5 lb.) apparatus. An attached radiosonde was used to compute the height of the equipment. This launching site was used for about a year; then it was given up because of the infrequency of winds light enough for successful launches.

The observing program was simplified somewhat by reducing the times of detailed cloud observations to 0700, 1300, and 1600 EST, and in addition continuous records at the base station were terminated 31 December.

Requests for consultation or other services continued to come in but all of them simply could not be honored. Brooks generously allowed some of these to be taken on by personnel working at home. He viewed this work as instructional as well as of monetary benefit to the individual. During the year this writer consulted for a company in Woburn that received complaints from nearby neighbors of objectional odors. The meteorological conditions of a temperature inversion and light winds, that caused entrapment of the odors was shown to be the cause of the problem. The writer also prepared short "weather fact" stories for use in radio commercials. Altogether 225 stories were prepared and broadcast over Boston stations.

Henry Helm Clayton died on 26 October 1946. He had been associated with the Observatory from 1886 to 1909 and from 1943 to 1946. Some of the most important work carried out at the Observatory was performed by Clayton in the early period. After leaving in 1909 he served as dean of the School of the Aeronautics Association Institute in Boston for one year. He then went to the Oficina Meteorologia Argentina to study methods of forecasting and to inaugurate a kite station in Cordoba. From 1911 to 1912 he was engaged in private practice in America. Then he was invited to return to Argentina as chief of the Forecast Service, a position he held for nine years. While there he inaugurated forecasting based on solar variations. Upon his return he published *World Weather*, a general work on meteorology. In cooperation with the Smithsonian Institution, he engaged in further researches on world weather, especially in relation to solar changes. One of the by-products was three volumes of *World Weather Records* which consisted of monthly temperature, pressure, and precipitation records from stations scattered over the globe. He then headed the Clayton Weather Service from his home in Canton, Massachusetts. Only a few years before his death he presented his solar–weather relationships at an MIT seminar. He was rudely criticized although he defended his work with dignity. Ten years after his death, at the 250th memorial celebration of Benjamin Franklin's birthday at the Academy of Arts and Sciences, these same individuals had reversed their opinions and publicly paid tribute to Clayton for his initiative in this work.

Chapman and Harrington, secretary to Brooks, resigned during the year. Schell also resigned to start his own long-range forecasting service.

The subcommittee which was set up in 1945 to raise funds reported that $7,650 had been subscribed, a significant amount, but inadequate for expansion.

During the year J. Murray Mitchell, Jr., an M.I.T, student, first came to the Observatory as a relief observer. He continued to help through his postgraduate years at MIT. Thereafter, throughout service in the Air Force and most recently as senior research climatologist in the National Oceanic and Atmospheric Administration in Washington, he has remained a staunch supporter of the Observatory.

The December meeting of the American Meteorological Society was held in Cambridge. In addition to papers which were presented by Observatory personnel, various activities at Blue Hill and Mount Washington were displayed.

1947

Work under the icing research contract continued to occupy the Blue Hill staff. At Blue Hill a new bench lathe, drill press, and grinder were purchased. The old foot-powered lathe was sold. Conover made a fog drop ranger and Fergusson assembled a set of windmill anemometers (figure 119). The anemometers were designed to measure wind speeds from a few centimeters above the ground to a height of about two meters. Howell designed an icingnometer, or icing rate meter. Conrad prepared many statistics on the frequency of different icing rates, ice densities, and accompanying synoptic situations. The staff on the mountain was augmented by Weather Bureau personnel through this period.

In August the International Meteorological Organization met in Toronto. This was the first assembly in twelve years. Brooks was elected

Figure 119. Six sensitive windmill anemometers made by S. P. Fergusson, and recorder for use on Mount Washington. From left to right: Ann Reiter, Sally Wollaston, John Conover, Frances Wollaston. Courtesy S. H. Wollaston.

president of the subcommission on station instruments and exposures. He also accepted membership in the permanent subcommittee on actinometry.

After adjournment of the meetings, Harvard was host to some 30 foreign meteorologists at a joint meeting of the American Geophysical Union and the American Meteorological Society at the Institute of Geographical Exploration.

The three-week Conference of Directors which followed in Washington was notable for its transformation of the sixty-nine-year-old International Meteorological Society into the World Meteorological Organization. The authorized representatives of thirty-one sovereign nations signed immediately and most of the 400 resolutions presented were approved. While in Washington, as advisor to the President of the Commission on Instrumentation and Methods of Observation (CIMO), Brooks met with others to work out and recommend improvements in the international cloud code and cloud atlas.

Thirty-six foreign meteorologists visited the Observatory from August to October and the Argentine delegation honored the late Henry Clayton by placing a wreath on his grave in Canton.

Brooks continued to serve as Secretary of the American Meteorological Society and during the year was elected president of the Association of American Geographers.

On 14 January Sally Wollaston fell on glazed rocks just outside the Observatory. She apparently had suffered a fracture so she was bundled up and protected from a near freezing easterly wind by a large cardboard box while Burhoe and Conover went down the hill for his van and a doctor. Both made it up the hill with chains and Sally was taken to the Milton Hospital one hour and forty-five minutes after the accident. It was a difficult ordeal but within three months her leg had healed and she was back at work.

Wollaston's enforced vacation required quick recruitment of a substitute observer. In this instance a burly ex-Marine took the job. A few days after he had started to sleep there alone, he told of strange noises at night and his call to the police. He suspected someone was trying to jimmy a window. That night the same noises were repeated and the next morning he vowed, once again, that the place was about to be burglarized and he quit. This writer filled in that night and he too heard noises after retiring; however, investigation showed that a skunk was rummaging through a metal trash barrel—his sounds were enough to scare an ex-Marine off the hill.

In the spring Tor Bergeron and his wife made several visits to the Observatory during a series of lectures which he delivered at MIT. Their presence was always a source of pleasure at the Observatory and we welcomed Bergeron's interpretations of our detailed observations in terms of the larger scale synoptic picture.

Observer Mitchell rendered the pencil sketch of the Observatory shown in figure 120. This drawing, facing toward the west, accurately portrays the Observatory and apparatus attached to the tower at that time. The steel parapet extending to the south is shown, the old window shelter in the north window is shown and beyond and above it a portion of the corner reflector is visible. The anemometers and wind vanes are all correctly positioned. Below the aerovane, on the central mast, a dewcel housing is shown, and outside the Weather Bureau's diffuse radiation

Figure 120. Pencil sketch of the Observatory drawn by J. Murray Mitchell, Jr., 1947. Courtesy J. M. Mitchell.

occulting ring another stovepipe is shown which housed a dewcel. Shrubbery in the right foreground marks the original site of the ground thermometer shelter. Later it was paved with asphalt.

On 31 August the Magnetron Development Laboratory of the Raytheon Company used Blue Hill for experiments in two-way voice communication. A new world record, using 5-cm wavelength equipment, was set for the 72-km (44-mi.) distance to Mount Wachusett.

Also during the summer Conover started experiments on methods to measure precipitation remotely using a weighing device. He believed this method superior to other techniques that required the melting of snow and volumetric measurement. An attempt to transmit using a selsyn generator was made. It was found that to provide the necessary power to move the selsyn, a large and therefore inefficient collector under windy conditions, was needed. The method was therefore abandoned, but the work continued.

Conover and Howell were on loan for part of the year to the Ultrasonics Corporation. Experiments in the dissipation of fog with sound were underway and a portable wind tunnel and instrumentation was needed to characterize the fog particles before and after tests. A tunnel was designed which had a throat of about 20 x 20 cm (8 x 8 in.) and air speed of roughly 30 m sec^{-1} (67 mi. hr^{-1}). This was constructed by Conover and later transported to Mount Desert Island on the Maine coast for use during ultrasonic tests.

In the spring Haurwitz resigned to chair the Department of Meteorology at New York University. Haurwitz had served part time for 15 years. Over that period, as shown by the bibliography, he had published papers on a wide variety of subjects.

The Burhoes also resigned in the fall. Ralph Burhoe became executive officer of the American Academy of Arts and Sciences and his wife, Fran, assisted him. The Burhoes had given the Observatory exceptional service in a wide range of affairs and effected many improvements in the library. Ralph Burhoe's institution of a monthly abstract and bibliography section of the *Bulletin of the American Meteorological Society* was an important contribution.

By late October a severe drought had developed in the northeast, fires raged around Bar Harbor, Maine, and at Blue Hill smoke was strong enough to smart the eyes on the 24th and 25th. On 29 October the drought was broken when 11 mm (0.43 in.) of rain fell, the first measurable amount since 30 September.

In December, at the American Meteorological Society's expense, an annotated list of monthly library accessions was started and sent to the American Meteorological Society bibliographer in Washington. From there it was distributed to libraries of the U.S. Weather Bureau, the Air Weather Service, and the meteorological departments of six universities. On 23 December 44 cm (17 in.) of snow fell and from the 26th to 28th another 38 cm (15 in.) fell. At year's end 51 cm (20 in.) lay on the ground, marking the beginning of the snowiest season in Observatory records.

During the year, in the face of rising costs, the university reduced its use of unrestricted funds to some departments. One of these departments was the Observatory. Over the past sixteen years the university had made up an average deficit of $6,000 per year. The amount was cut to $5,000 and by year's end it was eliminated entirely. The work over recent years had not been significantly productive for dollars expended, at least in the eyes of the University. President Conant, a chemist, was more familiar with exact science. He was not interested in climatology and desired more "hard core" research. In September a $7,000 contract with the Air Material Command for research in cloud physics at Mount Washington, Blue Hill, and in Cambridge was completed. Howell and Brooks had negotiated for the continuation of the work, beginning in December, with a new contract for $55,000, but the corporation deferred acceptance on the ground that it was too late to accomplish much before 30 June of the next year. The more likely reason was that the university expected the Observatory to close and approval would imply that it could continue a year or more. The action left the Observatory, whose budget was about $30,000, to fend for itself using endowment income, gifts, and other sources of income. The year ended with this financial blow. Closure of the Observatory was a definite possibility and, at best, a reduction in personnel seemed imminent.

Chapter XII

The Observatory and Government Contracts: 1948–1957

1948

No time was lost in the search for new funds to carry on Observatory programs. Brooks departed for Washington on 2 January, where discussions of Blue Hill and Mount Washington affairs were held with Francis Reichelderfer, Chief of the Weather Bureau, and assistant chief, Little. Brooks pointed out the many ways in which the Observatory had helped the bureau by supplying cloud reports and discussing forecasts with the Boston office, in addition to the testing of instruments that proved satisfactory for station use. This was the beginning of negotiations which developed in the form of a $7,000 contract with the Bureau for one year. The contract was chiefly for researching snowstorm forecasting for Boston and testing apparatus. With this contract and $11,000 in gifts for the year, $7,000 of which was for capital funds and the remainder for immediate use, the Observatory was as financially secure as it had been over recent years.

Under the contract a statistical study was begun in which the Mount Washington and Blue Hill weather elements preceding winter storms were used as predictors of snowfall amounts.

A new 4-m (13-ft.) rangefinder was secured from the Weather Bureau and placed on a platform inside the enclosure. This and the smaller 1-1/2-m (4.9-ft.) rangefinder were compared as instruments to measure cloud heights.

Various bimetal thermometers, some of the maximum and minimum type, were also exposed and compared with the standard thermometers. The bureau had hoped that these might be used to replace the thousands of glass thermometers, which often broke, located at cooperative stations throughout the country. Tests, however, showed that the bimetal thermometers were not durable and their calibrations shifted with time.

Conover made two new nephoscopes: one for the Belgian government, the other for Mount Washington.

And finally, under the contract certain cloud observations were routinely telephoned to the forecast office at the East Boston Weather

Bureau Forecast Office. Most of these pertained to the motions of high clouds and the advent of sudden low ceilings.

Arthur Rotch, son of the Observatory's founder and frequent contributor to the Rotch endowment, wrote to Harvard's President Conant on 26 January. In his letter he expressed

> . . . a disappointment that Harvard did not take the opportunity, that was hers when she received the Observatory, of assuming leadership in a new and growing science. However, she did not, and I can understand that under the present conditions, all expenses that do not yield valuable results must be reduced and that therefore the question of holding the Observatory arises.

His concern in regard to the endowment was noted by, "Endowment was given primarily to Blue Hill, not to Harvard." He ended his letter by saying that he ". . . believes Harvard has a moral obligation to continue the climate record."

Meanwhile, the visiting committee, chaired by Robert E. Gross, made a comprehensive study of the situation. It noted that the university had already refused to execute a proposed $55,000 contract with the Watson Laboratories and that $500,000 in new capital money could not be raised to continue advanced operation. Alternative arrangements were explored and their feasibility determined as follows:

1. Merge with the Harvard Observatory Council as a step toward decreasing administrative and financial Observatory matters in Cambridge. To this possibility Shapley replied that the Council was not ready to do so.
2. Merge with MIT. They did not wish to take on the responsibility.
3. Reorganize as a separate corporation like the Mount Washington Observatory. This was considered not practical because of the problems entailed in shifting the Observatory endowments to such a corporation and in obtaining a new lease for the site.
4. Arrange to have the Weather Bureau take over the observing program. The bureau declined this possibility.
5. Seek outside sources of funds, namely, in the form of contracts to carry on current and new work. This turned out to be the immediate solution to the problem.

The committee also considered the effects of closing the Observatory. Representatives of the American Meteorological Society, MIT, the Weather

Bureau, the military, and meteorologists at the University of Chicago and New York University all stated that the results would be harmful.

The report to Harvard University, dated 10 June 1948, concluded by saying: "The Committee feels therefore that it would be premature to conclude that Blue Hill cannot continue on some acceptable basis, and that today is not the time to suspend operation."

The Harvard–Mount Washington icing research contract was completed and results were detailed in an 802-page report. At this time reorganization of the Mount Washington Observatory Corporation allowed it to deal directly in contractual matters and Wallace E. Howell became acting director. The research then continued with new government sponsors.

Early in the year Howell took a five-week leave to complete work for his doctorate in meteorology at MIT.

Brooks also took leave for three months to prepare a revised edition of the Weather Bureau's circular S, or cloud manual. He was given the title of meteorological consultant; he did most of his work at the Observatory. Brooks, with Mrs. Brook's assistance, also prepared a thirty-nine-page summary of techniques for measuring sea surface temperatures and recommendations for U.S. Weather Bureau and international experiments.

The second and greatly enlarged edition of *Methods in Climatology* by Victor A. Conrad and L. W. Pollack, the latter of the University of Dublin, was completed (see bibliography, 1950).

The Observatory contributed substantial portions of Chapters 2-5, pages 15-97, of Palmer C. Putnam's book *Power from the Wind* (1948). These pages contained data that had been collected and analyzed in connection with the wind power project.

Conover and Wollaston finished a study of the cloud systems of a winter cyclone from the time when the storm first formed in the southern Great Plains until it passed over the North Atlantic seaboard as a heavy snowstorm. The analysis vividly displayed large departures from the textbook examples of clouds around such storms. In this case the first precipitation fell from middle-level clouds that apparently had been seeded by higher cirrus clouds.

The heavy snows that fell in late December 1947 continued throughout the winter and by 25 January 106 cm (48 in.) was recorded as the depth on the ground. The depth figure was the result of a survey by Murray Mitchell over the summit of the hill. A snowcover of 102 cm (40 in.) was measured at the base station. All travel up and down the hill was done on skis or snowshoes. Skiing was good on the open southwest slope of the hill where the scrub oak was completely covered. Earlier, on 18 January, the standard precipitation gage was snowed over, necessitating its excavation and the raising of its orifice and shield to one meter (3.3 ft.) above the ground. The road was closed on 23 December and did not reopen to automobile traffic until 22 March. Snowfall for the season amounted to 345 cm (136 in.), the largest for any season on record.

The new lathe, which had been obtained with Harvard–Mount Washington contract funds, was transferred to Mount Washington and another surplus machine costing $400 was obtained.

In July the Weather Bureau took over the recording of total radiation on a horizontal surface and direct radiation.

After much experimentation, a suitable shield for the Foxboro dewcel was found. It was self-ventilating due to convectional currents set up inside the shield by the warm cell. The shield was modified further and tried on Mount Washington, but during conditions of blowing snow it was impossible to prevent all ice crystals from reaching the cell and spoiling the readings and still have it sample the air adequately.

The three-bladed, balsa-wood windmill, which had been built by Fergusson for use with his low lag wind speed recorder, was modified with a new helical bladed mill made of sheet magnesium. This lightweight mill was durable and not subject to damage as was the balsa mill.

Sterling Price Fergusson, research fellow, retired 30 June at the age of eighty. He had worked in meteorology nearly sixty-one years. He was a member of the Observatory staff from 1887 to 1910 and returned in 1931 after service at the University of Nevada and at the U.S. Weather Bureau. His total of forty full time years at Blue Hill exceeded in a sense that of Wells, who had worked nine years part time and thirty-one years full time. Fergusson was noted chiefly for his outstanding work in the design and testing of apparatus.

The Fergusson weighing precipitation gage was and still is used by thousands of observers all over the world. His kite and balloon meteorographs were widely used before the radiosonde replaced them. For years he was the authority on anemometry. Accounts of his experiments were published in *Harvard Meteorological Studies*, No. 4 (see bibliography).

Fergusson enjoyed participation in the Milton Choral Society and, wherever he was, when appropriate, withdrew a small tuning fork from his vest to strike the note of C.

Fergusson was rugged, as noted at other times in this history. After returning to Blue Hill in 1931, he rode the bus from his Milton home to the base of the hill and often climbed in the worst of weather. He always dressed with a stiff collar, tie, suit, a felt hat, and overcoat; in cold weather he never wore gloves. The observers and assistants who were some forty to fifty years younger were amused when he occasionally came in the office to call his wife. The conversation went something like, "Mrs. Fergusson? . . This is Mr. Fergusson . . . I am calling to ask you to dinner . . . Yes, well that will be excellent, I will be home in about one hour." We never really knew whether this dialogue was put on for our benefit or whether it was just another mark of the true southern gentleman he was throughout his life.

Alan Faller, an MIT student in meteorology, served part time as observer. In the summer he took leave to go on a supply ship to Weather Bureau stations in Greenland. Harry Vaughan replaced Faller. Francisco da Costa resigned to be married after three years as resident observer. He had made a number of improvements to the Observatory including a burglar alarm, a rain time recorder, and a walkway leading from the gate to the front door. While at the Observatory, da Costa studied electronics by correspondence course. With this new knowledge he obtained work with the Philco Corporation. Paul Oyler replaced da Costa for a brief time then moved on to Mount Washington. Charles Harrington, who had experience in the Air Weather Service, replaced Oyler.

A picture of the Observatory and its surroundings as seen from the air at about this time is reproduced in figure 121.

1949

By this time a Mount Washington Observatory office had been set up, thanks to the generosity of A. Hamilton Rice, in the penthouse of his Institute for Geographical Exploration in Cambridge. This was a cooperative effort with Blue Hill. Howell, Boucher and Braun worked on the icing studies and Catherine M. Whalen served as secretary and library assistant. At the same time new instrumentation for the mountain was under development in the Blue Hill shop. Conover constructed a new anemometer, the second model for the mountain which used a heated pilot tube to measure wind speed. The position of the vane which kept the pitot tube pointed into the wind was transmitted to an anemoscope to record direction (see bibliography). Timothy Hanley, an MIT student, worked on an

Figure 121. Aerial view of the Observatory and surroundings, facing west–southwest, 1948: 1) enclosure thermometer shelter, 2) old kite house, 3) balloon launcher, 4) site of the temporary MIT cosmic ray building, 5) State Police radio shack and antenna. Courtesy J. M. Mitchell, Jr.

icing rate meter. A small rod that could be heated was made to oscillate axially; as ice accumulated the frequency lowered to a value where heat was applied and the ice melted; the sequence then was repeated. The frequency of this sequence could then be used to indicate icing rate. The instrument was never perfected but years later the principle was used in a commercial instrument.

Once again, emulating early experiments with kites, an antenna was lifted above the hill in an effort to obtain stronger radio signals. This time, in June, the Dewey and Almy Company used one of their kytoons, a "flying balloon," to lift an antenna that was designed to pick up distant television signals. The experiment partially failed. Sound was heard without video signals.

On 31 May shields that had been designed and constructed by Conover were installed, for the Weather Bureau, around their vertical pyrheliometers facing north and south. The shields prevented reflections from finding their way to the thermopile when the sun was behind the facing surface.

Beginning in July, the Weather Bureau began publishing the Observatory's daily weather record in their monthly *Local Climatological Data* format. An annual summary, titled *Local Climatological Summary*, for current and past years was also published by the Weather Bureau. In addition, in October the Bureau started to include Blue Hill monthly averages in their *CLIMAT* reports, a summary report of data from stations located throughout the world. These summaries, along with the seasonal summaries, restored international distribution of meteorological observations at Blue Hill to that of the years prior to 1930.

The summer was unusually hot with a new all-time maximum of 38.5°C (101°F) recorded on 10 August. This was measured in the enclosure thermometer shelter.

In November the old base station thermometer shelter, which was located near the present Trailside Museum, was sold to Harry Larkin, Jr., of Buffalo. The shelter, made in 1886 according to the specifications of H. A. Hazen, was still in good condition.

Also in November a new ski trail was laid out by Hanns Schneider, designer of the North Conway, New Hampshire, slopes, with their mobile. A series of two rope tows pulled skiers to the top of two new trails on the northwest side of Great Blue Hill. These trails and tows were maintained and operated by the Metropolitan District Commission.

Progress on the development of statistical methods for forecasting snow, using Mount Washington observations as predictors, had leveled off. Brooks' successes over the pure statistical methods were due to the inclusion of numerous other factors, so it was decided to classify cases using six surface weather map types preceding the start of precipitation. Boucher helped in developing the types which, when employed, improved the forecasts slightly.

A partly annotated bibliography of some 500 papers in the last dozen years dealing with meteorological instruments was prepared, mainly by Howell, at the Weather Bureau's request.

Howell was appointed director of the Mount Washington Observatory and Brooks continued as president. Howell was also appointed to membership on the subcommittee on icing problems of the National Advisory Committee on Aeronautics.

Brooks participated in the preparation of a new *International Atlas of Clouds* and submitted a total of eighty-one cloud photographs for possible inclusion in the atlas. He also published a paper on the revised cloud codes.

In the course of the year Brooks participated in seven examinations relating to the doctor's degree and provided divisional examinations for two undergraduates. P. K. Chang, whose graduate work Brooks had supervised since 1947, completed his study of cold waves and prepared to return to China.

The staff continued with four professionals, Brooks, Howell, Conrad, and Conover. Conover was made technical manager and Abbott Lawrence Rotch research fellow; Sally Wollaston was appointed chief observer. Paul C. Dalrymple and Tim Hanley jointly filled the resident observer position. Arthur Adams, a Harvard student, William L. Gates, Murray Mitchell and Alan Faller, all of MIT, along with students Charles Harrington and Harry Vaughan, served as part time observers.

1950

In addition to the routine observations, new instruments were tested and constructed. The catches of five commercial rain gages were compared. The smallest one, having a 2.54-cm (1-in) diameter orifice, clearly was the most efficient in collecting drizzle under windy conditions. Gages with larger orifices diverted the droplets, giving lower catches.

Harry Larkin presented the Observatory with a new nephoscope of his own design. The movable eyepiece was attached below the mirror, making the instrument entirely self-contained for use in any of the three tower windowsills. Previously the eyepiece was attached to a stand that was rolled along the floor, thus tending to introduce errors in the height of the eyepiece above the mirror.

An electrical·indicator of the occurrence of rain was loaned to the Observatory by the French Meteorological Office for testing. The instrument, named "Fifi" by the observers, relied on a heater to dry the sensing element during the recycling sequence. This design was found to be inadequate at the windy hill location.

A new cloud droplet collector was made for use on Mount Washington. It consisted of six porous cylinders of different diameters. The system was water-filled; then as cloud water struck the cylinders, the equivalent amount exited through the system into graduated test tubes to allow calculation of liquid water content. The different diameter cylinders made median-volume drop size computations possible.

In an effort to maintain a consistent circulation of air around the enclosure thermometers, major brush-cutting took place in the fall.

During the year computation of daily and monthly 24-hour mean temperature was completed. This standard of computing mean temperature is important for the study of small changes. When the method of one-half the maximum plus minimum temperature is used, different averages result when the beginning time of the 24-hour period changes, as it did several times since the Observatory records began. This revised temperature data allowed an improved analysis of annual temperature change, which Conover presented at an American Meteorological Society meeting in Washington. The Observatory record was extended back thirty-six years using the Milton Center record, as with the analysis by Sweetland in 1901. This time, however, corrections were applied to the old thermometer that was used to obtain the early data, making the record more significant. From these data a rise of about 1.8°C (3°F) in annual temperature was evident over the past 100 years.

In February Howell was asked by the city of New York to develop a procedure aimed at increasing the precipitation over the Catskill watersheds to ease a desperate shortage of water in the reservoirs. A program of cloud-seeding by dropping dry ice from airplanes and discharging silver iodide smoke into the atmosphere from the ground was developed. During the spring and early summer the reservoirs nearly filled and Howell was widely acclaimed. In spite of many statistical analyses, the portion of rainfall that fell as a result of artificial stimulation was never resolved to the satisfaction of some scientists.

In June Brooks attended a meeting of the Committee for the Study of Clouds and Hydrometeors in Paris. He then translated the French text of the new cloud atlas into English and returned a detailed critique of the final manuscript for the new *International Cloud Atlas* for consideration by the editing committee. While in Europe Brooks visited the British Meteorological Offices, the Kew Observatory, the Department of Meteorology of the University of London, and the Royal Meteorological Society.

Meteorological Abstracts and Bibliography formally began publication by the American Meteorological Society in 1950; for this the Observatory library contributed material on all its new accessions. Classifications were done weekly by Howell and Gates with assistance by F. Burhoe and Miss

Wollaston. The second edition, revised and enlarged, of the book, *Methods in Climatology* by Drs. Conrad and Pollak, was published.

On 26 and 27 September a great pall of smoke originating from forest fires in Alberta, Canada, passed over the area. Surface visibility from the hill dropped as low as 7 km (4 mi.) and during a cloudless sky at noon on the 26th the direct radiation from the sun was only 67% of normal.

Work on the snowstorm forecasting project was mostly directed toward the preparation of new predictors for different weather map types. In November Vaughan and Conover installed a 30-day Foxboro thermograph on Mount Greylock in northwestern Massachusetts. This was done to determine the usefulness of continuous temperature data from the 1,064-m (3,491-ft.) high mountain for the snowstorm forecasts. The winter's data, however, proved of little use because of the low elevation and distant location of the site.

On 25 November a severe coastal storm caused wind gusts to reach 42 m sec^{-1} (94 mi. hr^{-1}), the highest since the September 1944 hurricane. Power failed on the hill and Mitchell made the evening indoor observations by kerosene lamps. Salt from the ocean, 13 km (8 mi.) distant, was found in small deposits on the Observatory.

In December David Davidson of Harvard's Cruft Laboratories directed the placement of a van, housing radio equipment into the backyard. This project, sponsored by the Office of Naval Research, was concerned with the measurement of the movement of clouds of electrons in the upper atmosphere. The station on the hill was one of several that received pulsed radio waves from reflections aloft. By triangulation the height and motion of the reflecting areas could be determined.

Operation of the Observatory now cost about $33,000 per year and only 40% of this was covered by endowment income. Contract money and much effort in obtaining gifts was required to raise the remaining $20,000. The visiting committee suggested that undergraduate instruction in meteorology and climatology be restored and that Brooks prepare such a plan for the University. In this way college funds would partially support a competent researcher and teacher.

Photios Karapiperis, chief assistant of the Meteorological Institute of the National Observatory of Athens, arrived and was appointed to the staff as a Research Fellow. Under a scholarship from the University of Athens he planned extended studies in climatology. The meteorological and climatological researches of four Harvard graduate students were supervised and several MIT students were assisted. One, Roger Bond, prepared a thesis in architecture on the Observatory and supplied blueprints for a larger plant better adapted to needs of the day.

Dalrymple served as observer in the early winter; in the summer he went to Mount Washington, but returned in the fall to attend classes at Harvard and MIT.

In a recent letter from Dalrymple the following episode and tribute to J. Murray Mitchell came to light:

> There is one Murray Mitchell story which shows what kind of a guy he was and is. For some reason, I wanted out of Blue Hill on a Friday night (probably some basketball game or hockey game). So I called Murray during the middle of the week and asked if he could come out early that weekend so that I could take off Friday evening. Murray never said a word beyond the fact that he could come. When he showed up he was in a tuxedo. It seemed that he was going to a formal dance that night. I felt so damn bad that I had asked him that I could have crawled into a crack in the floor. But it is just an example of the kind of fellow that Murray Mitchell was and is. It was probably the first time in the history of Blue Hill that a sling psychometer was read by an observer in a tux.

In January the university "dropped the other shoe," so to speak. Financial support had already been cut off, leaving the Observatory to survive on endowment income and whatever other means of support it could find. Now the overseers planned to close the Observatory in five years in spite of its proven ability to survive without Harvard support. Although documentation is meager, the announcement may have been precipitated by an application for an associate professorship and a position of assistant director of the Observatory for Howell. Such an appointment would have made him a likely heir to the directorship after Brooks retired. Although considered by the corporation, it was declined, since closure of the Observatory was anticipated.

Presumably, as in 1947, it was felt that the research was not sufficiently basic to warrant continuation of the facility and its attendant administrative work in Cambridge, which had been complicated further by the use of space there for the Mount Washington Observatory contract work. Again, there was substance for the reasoning, with the exception of some work by Schell and Haurwitz there was practically no real research, only applied research at best, over the preceding ten years. Teaching continued and textbooks were written, services of all sorts on a scale from local to international were performed. New instruments for research on Mount Washington and for the observational programs at both the mountain and

Blue Hill were constructed. The observational data were accumulated but not used extensively, so there was justification for termination if the lack of hard-core research was indeed a factor for the decision. On the other hand, removing all financial support had virtually eliminated any possibility of the Observatory's entrance into more basic research.

In response to this new development, the visiting committee met several times. Proposals were drawn up for the Weather Bureau to assume the expense of maintaining the climatological record and for liaison with the Division of Applied Science with the thought that it might make greater use of the Observatory and that the Observatory could benefit from expertise in electronics.

The staff remained loyal. Other positions were not ignored, but all recognized the fact that the work place was pleasant and most of all fulfilling under the honorable personality of Brooks.

Two significant weather events occurred early in the year. Freezing rain accumulated to a thickness of 1.3 cm ($\frac{1}{2}$ in.) on the ground on 30 January. This was one of the heaviest ice storms on record at the Observatory. Eight days later, on 7 February, wind gusts from the south–southeast reached 38.9-m sec^{-1} (87 mi. hr^{-1}).

Conover analyzed the new "24 hour" temperature data to prepare a paper titled "Are New England Winters Getting Milder?" for which the answer was "yes" at the time. The paper, which was presented at the January meeting of the American Meteorological Society in New York, was preceded within the hour by a telegram from observer and jokester Paul Dalrymple. The message had been prompted by the morning observation at Blue Hill. It read: "Flash Winters Now Getting More Severe One Below And Snowing."

In April preparations were started on the hill for a new, noncommercial FM station owned and operated by the Lowell Cooperative Broadcasting Council. The Observatory lease with the State was modified to permit this endeavor. The storage room under the library was cleared to house a transmitter, a small shop, and a technician; a 29-m (95 ft.) high transmitting pole was erected on the rocks northwest of the tower. A new power line was laid underground going up the south side of the hill and transformers were placed outside the Observatory wall to the south. The station WGBH-FM, one of the first educational FM stations in the country, went on the air 6 October 1951 with a power of 20,000 watts. To assure high fidelity of the station's Boston Symphony Orchestra broadcasts, a microwave link was set up from Symphony Hall to the Observatory in November. A 1.7-m (5-1/2-

ft.) diameter dish was mounted above the window shelter on the north side of the tower to receive the signals. The Observatory received $1,000 per year for the use of its space.

On 1 November Brooks made the first of many broadcasts titled "Why the Weather" over the station. He spoke directly from the transmitter room on current and popular weather subjects.

An amusing story, as told by Dalrymple, comes from one of his letters:

> One morning Brooks called up to tell me they were sending up some large balloons from Lexington to study cosmic rays or something, and he asked me to put the range finder out and he would be right up. So I put it out and he arrived. It was spring time and he wanted to zero the instrument on a landmark but the trees had grown so much that he could not see the pole in the valley. So he sent me out there and I would shake a tree and he would say yes or no. If it were a yes I had to chop it down. Then he would tell me to go 20 feet to the left and he would go through the same procedure again. I guess I was out there over a half hour chopping down trees. Then I came in and he told me to go up in the tower and get some cloud movement values on the nephoscope, so up I went and I recall there were multiple levels. By the time I got through, it was late in the morning, so I grabbed this fried egg which I had on the stove when he called and gulped it down. Returning something to the refrigerator, I thought that it was Sally sitting at Brooks's desk and that he was still outside with the range finder. So I blurted out "that was the most miserable damn breakfast that I have ever had." Brooks was all taken back by the fact that I had missed my breakfast. But it was comical.

In the fall an oil burner and new furnace were installed. A 2,000 gallon tank was placed in the eastern most portion of the yard. This supply of fuel was considered adequate to heat the Observatory through the most protracted period when the road would be closed to trucks.

At this time the Air Force Weather Radar Branch of the Geophysical Research Directorate was located at Hanscom Field in Bedford, Massachusetts. The group planned to acquire a CPS-9 radar, one of the first of a production run, and they convinced the Air Weather Service to establish a radar unit attached to the Weather Radar Branch. With this arrangement five or six enlisted personnel would report their observations to the weather station at Hanscom Field while the research team could use the radar between observations.

Since this was to be a semioperational endeavor a good site with an excellent radar horizon was desired. David Atlas, head of the research group, approached Brooks with a proposal that the operation be set up on Blue Hill. The plan was obtained, permission was granted by the Metropolitan District Commission and in July, land clearing for a tower near the center of the hill commenced, and vans that were to house electronic equipment arrived.

Dalrymple completed the large task of tabulating daily maximum and minimum temperatures for the entire sixty-five-year record. The data were used immediately by him in a course at Syracuse University.

The American Meteorological Society, since its founding by Brooks in 1920, had been collecting meteorological and climatological publications from many sources, both in this country and abroad. As secretary to the society, the material came to Brooks at Clark University prior to his move to Blue Hill. After his move it continued to accumulate at Clark and was deposited in its library. In 1951 the entire collection was transferred to the Blue Hill library, with the climatological contents added to the Ward Collection at 2 Divinity Avenue in Cambridge.

Following the recent rainmaking experiments much concern developed regarding legal aspects of the work. In March Howell, at the invitation of Senator C. P. Anderson, testified before several subcommittees of the U.S. Senate with regard to bills on weather control and augmented water supply.

Consultation demands and rainfall stimulation requests became so great that Howell and others formed the Wallace E. Howell Associates, Inc. in May. Howell was then granted extended periods of leave for studies and for operations in Cuba, Peru, and Canada.

The Mount Washington Observatory office space in Harvard's Geographic Institute had to be vacated because Hamilton Rice, director of the Institute, retired and the building was temporarily closed. Work was continued in nearby offices.

Storms that caused heavy snow in Boston were now under study from the synoptic aspect. For these studies snowfall and precipitation were mapped throughout New England in detail.

Methods for remotely recording precipitation were resumed by Conover. Again, the gage would weigh the precipitation, thus avoiding the problems of melting it before measurement. This time the weight was determined by a strain gage and its indications were recorded on a Foxboro dynalog; both were supplied by the Foxboro Company.

Fergusson published an historical paper on nephoscopes and their use at Blue Hill (see bibliography).

Brooks received a $1,500 grant from the American Academy of Arts and Science toward the expense of publishing Ukichiro Nakaya's book on the physics of snow crystals.

On 8 October an unusually long period of fog ended; 82 consecutive hours were recorded. Another long period of fog occurred 22-27 May 1940 with 115 hours and also 144 hours out of 155 hours in the same month.

Frances Wollaston married William Lawrence Gates in the fall, and Charles Cunniff took her position in the Weather Bureau's Solar Radiation station. Karapiperis departed in November after finishing a comprehensive investigation of the climate of Blue Hill according to air masses and winds. His work was published in *Harvard Meteorological Studies*, No. 9. He is shown with others in figure 122. Raymond Wexler was added to the Mount Washington scientific staff. Dalrymple left in the summer to take a position at the Woods Hole Oceanographic Institute. Tom Nastor, Richard Gordon, and Robert Vogel served various periods as resident observers. In those days observers were paid about $115 per month with living quarters included.

1952

At the start of the year the Mount Washington Observatory assumed the responsibility for transmitting the 3-hourly synoptic observations directly to the Weather Bureau in Portland, whence they were placed on the teletype circuit. The Blue Hill radio station was changed from KC2XBS

Figure 122. A favorite place for lunch in the sun atop the coal bin. Left to right: Sally Wollaston, Frances Wollaston, Dr. Brooks, Frances Burhoe, John Conover, and Dr. P. Karapiperis. Courtesy F.W. Gates.

to KAE41 having a government frequency of 30.34 MHz. The Observatory continued transmission of a forecast from Boston to the mountain for the White Mountains and an exchange of local data each morning at 0720 EST. With Brooks's continuation of the presidency of the mountain observatory, scientific direction and other affairs were scheduled over the air almost daily. Blue Hill maintained some of the mountain station's equipment and handled the secretarial work, which included its membership list and the mailing of notices and publications.

For the first time, as a service to WGBH-FM as well as Observatory personnel, the summit road was plowed regularly by the Metropolitan District Commission during the winter of 1951-52.

In June the old balloon launcher was torn down and removed.

In September the Weather Bureau Solar Radiation station resumed recording radiation on vertical surfaces facing east and west. (These measurements were also made from April to September 1946.) From this time onward, measurements included radiation on vertical surfaces facing the cardinal directions, total and diffuse radiation on horizontal surfaces, and normal incidence radiation. The radiation on vertical surfaces was sought by architects and later used to derive relationships between total and vertical surface radiation at different latitudes and time of day.

In November, evening observations in the enclosure were made easier by the addition of a light inside the shelter and a floodlight mounted about 1-m (3 ft.) above the shelter.

During the year Conover had been granted leave to consult regarding a microclimatological study which was part of a project being carried out by Boston University under contract with the Air Force. Instrumentation was prepared for several sites on the North Slope of Alaska and set up there by Conover during a three-week period in June. A spare instrument from this project, which recorded temperature at the ground surface and 10 cm (4 in.) below the surface, was operated inside the Blue Hill enclosure for a short time beginning in December.

In July Brooks attended two commission meetings of the World Meteorological Organization in Europe. The commission for marine meteorology was held in London and that for the Study of Clouds and Hydrometeors in Paris. Brooks actively participated in these meetings, devoting long hours toward final approval of the cload atlas, which contained several cloud photographs by Observatory personnel.

Conover was invited to discuss the interpretation of meteorological data in a 2-day "Symposium on Climatic Change" held by the American

Academy of Arts and Sciences in May. He not only showed changes observed at Blue Hill but discussed wind, temperature, and precipitation changes over much of the United States.

As part of the Weather Bureau contract, the job of obtaining and preparing data for the publication of *World Weather Records 1941-1950* was undertaken. This was a continuation of the series started by Clayton in the 1920s. It contained monthly average pressure and temperature and total precipitation for selected stations throughout the world. For the new volume new stations having good records were sought and included in full. One of these was Blue Hill Observatory. In some cases station locations had changed, such as from city to airport. When overlapping data were available, conversion factors to the original station were developed.

For the snowstorm study a handy manual was prepared for use by Boston forecasters to estimate probabilities of class intervals of snowfall amounts from various predictors.

Snowfall from several storms was surveyed over the Blue Hills, eastward and south to a distance of several kilometers (miles) to gain further insight into orographic effects of the hills and the ocean–land convergence zone. The Blue Hill height of some 190 m (635 ft.) relatively close to the coast appeared to be as effective in increasing snowfall as two or three times that height in the interior.

Demands upon the library made it one of the most active units of the Observatory. It was used daily to answer queries that arrived by mail or telephone. Researchers frequently came for periods of days, and books were loaned to American Meteorological Society members without charge. Occasionally large amounts of data were loaned, as when the Quartermaster Corps borrowed 200 - 300 pounds of monthly climatic summaries.

At this time the library was in five places. Blue Hill housed the books and pamphlets most needed for reference: mostly those published during the preceding twenty-five years and the most used journals and monograph series, the original Blue Hill and Mount Washington records, an extensive card file, and an author file of the contents of the entire library. Older books, pamphlets, lesser used and duplicate files of journals and climatic data from all over the world were in the erstwhile Institute of Geographical Exploration. A few very old books, antedating 1800, were located in Houghton Library and the American Meteorological Society's offices. Other publications, not yet turned over to the Observatory, were in use at the Meteorological Abstracts and Bibliography offices in Washington.

Total library expense ran around $2,200 per year, excluding the costs of Observatory publications and their mailing for exchange issues.

During the year the visiting committee met again and pledged their support to the Observatory. Forty-two friends, including several members of the committee, contributed $6,500 during the year.

Wexler temporarily joined the professional staff part time as a cloud physicist while the semiprofessional staff remained the same.

1953

In the spring the overseers' committee favorably reviewed operations over the past five years at the Observatory. These were years when no university unrestricted funds were granted. Income was about $30,000 annually; fifty percent came from endowments, eighteen percent from gifts, and thirty-two percent from reimbursements for staff time or facilities supplied for cooperative projects. These consisted of $6,500 from the Weather Bureau, $700 from Mount Washington Observatory, $1,000 from the American Meteorological Society and $1,200 from WGBH-FM. Expenses were: eighty percent for salaries, fifteen percent for general expense and five percent for the library and publications. Aside from Observatory activities the committee again strongly urged reestablishment of instruction in meteorology, preferably a cultural course on the subject. Instruction for undergraduates at the time was incidental; occasionally students assisted at the Observatory, and Brooks continued his ten-minute talks over WGBH-FM.

Brooks attended meetings of the Commission for Instruments and Methods of Observation (CIMO) of the World Meteorological Organization in Toronto. There he presented recommendations on the exposure of anemometers, precipitation gages, and thermometers and on improved apparatus for the measurement of precipitation, dew, cloud motion, sea surface temperature and air temperature at sea.

Following the CIMO meeting a session was held at Blue Hill on the construction and standardization of pyrheliometers.

Brooks was invited by the Department of Commerce Advisory Committee on Weather Bureau Operations to present the Observatory's views on how the bureau's service could be improved.

These activities clearly illustrate Brooks's influence on the international and national scene, which was much like that of Rotch, but dissimilar to that of McAdie, who was openly hostile to the Boston Weather Bureau office.

The statistical forecasting studies showed that in predicting snow vs rain, simple use of half the Mount Washington temperature plus the Blue Hill temperature in Fahrenheit gave a good index. When the sum did not exceed thirty-seven, snow occurred in most cases and when the sum exceeded forty, rain was highly probable.

During the year Wexler prepared a chapter on "The Physics of Tropical Rain" for Herbert Riehl's book, *Tropical Meteorology*, and embarked on numerous studies in which weather radar was used.

With the hope that less time could be spent on the observational program, Conover went to Oak Ridge, Tennessee, to evaluate a system for automatically tabulating the weather as it occurred. It appeared that a method could be developed in two to three years for about $15,000, an expense beyond the means of the Observatory. Such systems were unique at that time. In retrospect, if the proposed and old methods of observing could have been compared, in view of present-day automatic systems, no doubt the use of manpower would have proved less expensive, more accurate, and reliable.

In June an anemometer was installed six to seven feet above the ground in the enclosure. This was done to determine if the coming WGBH-TV building would influence circulation inside the enclosure. After building construction and analysis of the data, no change as referenced to tower winds could be detected.

Also in June a concerted effort was started to keep the natural vegetation at the 15-30 cm (6-12 in.) height inside the enclosure and to 0.6 m (2 ft.) outside to the southwest of the enclosure. Trees in other directions were discretely cut. A comparison was made of temperatures at the enclosure and window shelter at 0200 and 1400 EST for two years before and after the cutting. No change before and after cutting was found at 0200 EST, but at 1400 EST the temperature averaged 0.2°C (0.4°F) higher before the cutting. A study by Mitchell (see bibliography for 1953) concluded that about a third of the 0.8°C (1.4°F) increase in summer temperature from the 1900-1919 decade to the 1920-1939 period was caused by progressive growth of the vegetation.

Storm precipitation amounts in the spring were measured at two nearby locations: one on the Rotch estate, 1.4 km (0.9 mi.) to the north, the other 1.7 km (1.1 mi.) to the east at the Metropolitan District Commission police station. Large differences for individual storms, presumably induced by orography, were found, but the data were inadequate to derive significant conclusions.

The weather event of the year was the Worcester tornado which occurred on 9 June. When the first debris fell at Blue Hill the Weather Bureau was notified and they immediately put out a severe thunderstorm and tornado warning for the area west of Boston. Debris fell at Blue Hill for thirty minutes; first to fall were oak twigs with frayed leaves, dark green from freezing at high altitude. These were followed by papers, rags, splinters, shingles, roofing paper, insulation, clapboards [up to 2-m (6-1 2-ft.) long] and pieces of walls or roof a meter square (3 x 3 ft.). This material had traveled 40 - 60 km (25 to 38 mi.). Some of it is shown in figure 123. Spectacular cumuliform clouds in tall, narrow columns were seen at Blue Hill and in the evening the receding clouds formed an apparently continu-

Figure 123. Sally Wollaston holds an embroidered dresser scarf found on Blue Hill as a result of the 9 June 1953 Worcester tornado. Courtesy J.H. Conover.

ous wall that was unceasingly illuminated by lightning in the southeastern sky.

In midsummer Bergeron spent several weeks at the Observatory. He gave four lectures at Cambridge on the analysis of local precipitation patterns. Professor Bergeron and Mrs. Bergeron (figure 124) joined some of the staff for an evening picnic and swim at nearby Ponkapoag Pond. The observers took turns manning the Observatory during such events, which were frequent at the end of hot summer days.

The first of a series of contracts with the Air Force's Geophysical Research Directorate was obtained in October. This pertained to the study of weather radar data that was supplied by the Air Force.

Ground was broken for the WGBH-FM radio and TV building on 5 November. This building, located east–southeast of the Observatory, was

Figure 124. Professor Tor Bergeron, standing, and Mrs. Bergeron, at Ponkapoag Pond. Mrs. Jack Webber is seated near the tree and young Gerry Conover's back is toward the camera. Courtesy J. H. Conover.

located on land leased to Harvard by the Metropolitan District Commission.

In November Conover set out a shielded minimum thermometer in the bog area at the western end of Ponkapoag Pond. It was read intermittently throughout the winter. The greatest temperature inversion between the bog and the summit of the hill occurred on 14 January 1954. The bog minimum was -27.4°C (-17°F) while on the summit it was -17.8°C (0°F).

The remote precipitation gage had reached a stage of development where a recorder was placed in the observer's room and the gage inside the enclosure.

In April Jack Webber arrived to work on the Weather Bureau contract. The work had now shifted to experimental hand computations of vertical velocities under the advice of W. L. Gates. The patterns of vertical motion were then compared with precipitation patterns.

At this time Joachim Kuettner of the Geophysic Research Directorate frequently worked in the third floor of the tower and experimented with a few time-lapse cloud pictures.

Irving Hand, supervisor of the solar station, retired in the summer and Charles V. Cunniff was placed in charge.

Roy Hines, Dave Kelley, and Thomas Riley served as plotters, computers, and relief observers. Resident observer Larry Howard resigned in February. He was followed for a brief period by H. A. Oluwasanme, then by John Skarbek.

The new WGBH building was finished in January and the FM transmitter was moved from the sub library room to the new building on 21 April. This transmitter was built by the late Edwin Armstrong, considered to be the father of FM broadcasting. It was used later by a station in Washington and ultimately it was accepted by the Smithsonian Institution. Ground was broken on 15 September, just east of the new building, for a TV tower. The tower was completed 11 December. Total height of the tower and antenna was 71.6 m (235 ft.). The structure was designed for 80 lbs. ft.$^{-2}$ wind load which figures out to be a wind of 61.7 m sec^{-1} (140 mi. hr^{-1}) at a factor of .0004 and 72.9 m sec^{-1} (163 mi. hr^{-1}) at a factor of .0003, the coefficients for flat and cylindrical surfaces.

Brooks resigned as secretary of the American Meteorological Society on 30 June. Since the society's founding in 1919 he had held the position and donated his time unstintingly to it. While the American Meteorological Society was headquartered on the hill, many operating costs were absorbed by the Observatory.

The research publications and climatological material that the society had deposited in the library were officially presented to the Observatory at this time.

Brooks continued his work with the International Commission for Synoptic Meteorology by preparing legends for their new selection of sixty-three photographs for the International Cloud Atlas. He checked and edited the final 230-page text of the atlas.

Conover presented details of his remote recording precipitation gage at the American Meteorological Society meeting in New York in January. It was now capable of recording five inches of precipitation on an indoor recorder. Its 17.9 cm (7.0 in.) diameter orifice could be electrically heated and controlled by a thermostat during icing conditions. Later in the year a modified Shasta wind shield was added. The gage is shown in figure 125 and a sample recording in figure 126. A unique feature of the gage was its accuracy of 0.25 mm (.01 in.) collection over the entire range. This was achieved by a balance mechanism having a flexible bar and strain gage attached to one side and the precipitation collector on the other side. When one inch of precipitation accumulated, a weight corresponding to that weight of precipitation was automatically dropped on the opposite side of the balance, so the sensor worked over the original one-inch range again.

Figure 125. Conover precipitation gage with cover removed. Heat from a light bulb was first used to melt snow in the collector. Courtesy J.H. Conover.

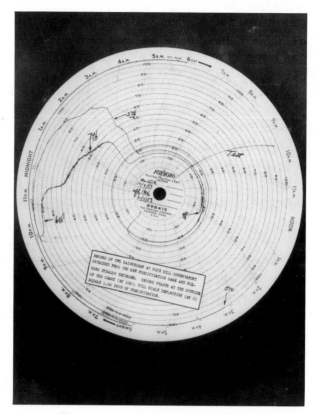

Figure 126. Sample recording from the Conover gage. Courtesy J.H. Conover.

This could be repeated four times. Changing resistance of the strain gage was recorded by a Foxboro Dynalog to give a continuous trace. Snowfall was reduced to water and kept from evaporating by a charge of oil and calcium chloride in the collection bucket. The gage worked well except under very windy conditions when "pumping" of the receiving bucket caused a "painted" trace. This was never eliminated in spite of the application of mechanical and electronic damping systems.

On 1 June Roland J. Boucher and Paul H. Putnins joined the staff. Boucher's primary work was the analysis of weather radar data that was collected under different synoptic conditions; he also worked on the forecasting study for the Weather Bureau. Boucher had wide experience in forecasting, as well as research in cloud physics. Putnins was appointed Abbott Lawrence Rotch Research Fellow. He began a detailed study of the Blue Hill precipitation record and later, after Webber's resignation in September, helped with the work for the Weather Bureau. Webber's phase of the work for the Weather Bureau was carried out using computer-generated vertical motions that were obtained from the Geophysical Research Directorate. Humidity values were combined with vertical motions for correlation with precipitation. Good qualitative but poor quantitative relationships were obtained.

Much activity took place on the hilltop in the fall. With the increased Observatory staff, Air Force, and WGBH personnel present, it was noted that on 8 November 24 cars and one crane were on the summit at one time. The Air Force CPS-9 weather radar was positioned on a steel tower about 90 m (295 ft.) north of the Observatory. Connecting, cables ran to a console inside a Jamesway Hut in the Observatory back yard. The radar and tower with newly prepared space for the vans that housed equipment are shown in figure 127. The insulated temporary hut having a 4.8 x 4.8 m (16 x 16 ft.) floor area neatly housed the console and a heater. A plane position image (PPI) repeater scope and intercom was located in the observer's first floor room of the tower. Kenneth Ultsch, of the Observatory staff, was placed in charge of the radar. It was first started up on 16 November.

Also at this time the shop was renovated and enlarged by removing the partition between the old shop and a darkroom.

The Observatory experienced three hurricanes in the fall. Carol, which passed on 31 August, was much like the 1938 hurricane but less intense. At Blue Hill a maximum gust of 56.0 m sec^{-1} (125 mi. hr^{-1}) from

Figure 127. Weather radar and work area in the center of the hilltop.
American Meteorological Society—Blue Hill Collection.

the southeast, and 62.5 mm (2.4 in.) of rain were measured. The highest hourly speed, 36.3 m sec^{-1} (81 mi. hr^{-1}) occurred between 1100 and 1200 EST. An interesting note told of bruised leaves that blew in around the kitchen window. It did not seem possible that these leaves could have come through such small cracks between the window sash and frame but this writer witnessed the phenomenon. No damage occurred to the Observatory but the paint and windows of several cars were chipped due to blowing gravel from the roof of the new WGBH building. Power was restored on 4 September but at the base it was off for nine days.

Hurricane Edna passed to the east on 11 September. Strongest winds were measured at the rear of the storm. A gust of 45.3 m sec^{-1} (101 mi. hr^{-1}) from the northwest was clocked at 1547 EST. 132.8 mm (5.23 in.) of rain fell.

Hurricane Hazel passed to the west on 15-16 October. A maximum gust of 30.5 m sec^{-1} (68 mi hr^{-1}) and 24.4 mm (0.96 in.) of rain were measured during that storm.

In spite of the greatest level of activity ever achieved at the Observatory, significant scientific research, and financial independence, the threat of closure of the Observatory after his retirement hung over Brooks. He presented the problem to Conrad Chapman, a scholar who came to the Observatory almost a decade earlier to "help out" after his interests were stimulated by a course in meteorology at MIT. "Chappie," as he became known, occasionally filled in as an observer and performed various duties in the library; more recently he had become an annual contributor to the Observatory. Chappie wrote to Harvard president Pusey only to hear the University position repeated, but he continued to take an active interest in the problem and later was made a member of the Visiting Committee.

In anticipation of Brooks' retirement the Visiting Committee met again. As recommended earlier it was thought best to expand teaching and possibly arrange for two permanent professional positions and one technical position. This would provide for teaching in Cambridge while maintaining the Observatory. The plan did not bode well for the large and increasing staff but the group remained loyal to the Observatory and interested in their work. A start to raise more endowment for these positions was made when Marian DeCourcy Ward, sister of Robert, presented Harvard with $25,419.71 to establish the "Robert DeCourcy Ward Memorial Fund." The fund was subject to an annuity and therefore was to be used in such a manner as the President and Fellows of Harvard University shall think most suitable . . .

Additions to the staff were Charles C. Shackford, an assistant on Air Force work in Cambridge, and Anastasije Cakste, assistant librarian in Cambridge and research assistant at the Hill. Skarbek continued as resident observer while Arthur Adams, Roy Hines, Richard Gordon, and Jim Cameron assisted Sally Wollaston with the observing duties.

Naturally, some of the observations were made under difficult conditions. Chief observer Sally Wollaston is shown in figure 128 on one of these occasions as she returned from the enclosure during a snowstorm.

1955

Chapman wrote to the new chairman of the Visiting Committee, Thomas Cabot, urging that a plan for continuing the Observatory after Brooks's retirement in 1957 be quickly settled. He also reiterated to Brooks that the administration was united in its plan to close the Observatory. However, after study, Cabot "believed that the Observatory was so much better off than it was 2 years earlier that thoughts of abandonment should be completely reconsidered." By May the corporation had softened its earlier plans. It had agreed to extend the use of the Observatory provided:

Figure 128. Sally Wollaston returning from the enclosure during a snowstorm. Courtesy J.H. Conover.

(1) new endowment became available; (2) the Astronomical Observatory took over responsibility of Blue Hill; and (3) a new director be appointed under joint sponsorship of the observatories and that he serve as a responsible member of the research staff.

Meanwhile, to save the Observatory from financial embarrassment at this critical time, when it was doing so much better, Chapman quietly contributed sufficient funds to put it in the black.

During the year Putnins and Boucher worked on the Weather Bureau precipitation project. Boucher was loaned for brief periods: first to the President's Advisory Committee on Weather Control, and then to the Mount Washington Observatory for the committee's "Operation Overseed."

In May four meteorologists, David Atlas, Albert Chmela, Ralph Donaldson, and Edwin Kessler, of the Weather Radar Group, Air Force Geophysical Research Directorate, moved into the sub library room. During the next several months they transferred much of their radar and meteorological equipment to Blue Hill from Lexington. A long-lasting impression of that room came to the attention of the author in a letter from Atlas, who recalled, ". . . temperature remained about 60° or colder at the floor level all year. In those days I was a smoker and my pipe smoke settled at an inversion in this office just about the height of my head at five feet."

The Air Weather Service sent five men to the hilltop to operate the CPS-9 radar twenty-four hours a day. Beginning on 14 July two-hourly radar weather reports (RAREPS) were telephoned to Hanscom Field at Bedford, Massachusetts. These reports were used immediately by forecasters and then placed on a national teletype circuit. Radar coverage from the unobstructed hilltop was unique; in one case a high thunderstorm top echo was monitored at a distance of 520 km (323 mi.). Lines of thunderstorms were often detected west of Syracuse, New York, or more than 400 km (249 mi.) distant.

With this latest complement of staff, a total of 26 persons were involved rather directly at the Observatory; the group was comprised of: nine meteorologists, nine part-time research and secretarial assistants, three technicians, and five radar operators.

On 1 June the Air Force compound which surrounded the radar tower was enlarged by extending it about 15 m (50 ft.) toward the east–southeast. The new area housed additional trailers and equipment.

A second contract with the Geophysical Research Directorate was entered into on 1 March. This was to provide the directorate's weather radar meteorologists space, access to the library, library consultation and bibliographic services.

At the end of June a contract was negotiated with the Quartermaster Research and Development Center at Natick for a thorough study of ground temperature, mostly of the literature and records already obtained over the world, and to tabulate and map the world with respect to hot and wet climates.

During the summer, volunteer observers started to collect data on severe local storms. These data were for correlation with radar data that were collected on film.

During the year five members of the staff presented a total of fifteen papers. Brooks continued his course in meteorology and climatology at Harvard. Other Harvard students participated in Observatory activities and at the observation station atop the Geographic Institute building, 2 Divinity Avenue, Cambridge. Brooks also lectured on climatic change at the 50th Anniversary of the Babson Institute and over TV for the National Broadcasting Company.

By year's end plans were underway for a method of recording all the regular surface weather observations in a single format suitable for easy comparison with vertical radar observations. Placement of the collated weather data on punch cards at the same time was considered but this idea never materialized.

Other Observatory affairs of lesser importance started with a weekly cleaning service which was welcomed by the resident observer who had been saddled with the job for many years.

On 10 March the enclosure was enlarged toward the southeast and a new fence with gate erected around it.

WGBH-TV went on the air 2 May.

The Bendix Corporation gave the Observatory a wind direction and speed strip chart recorder which was connected to the Aerovane. The Foxboro Company was loaned the original Aerovane, dial indicators, and a Negretti and Zambra recording rain gage. These instruments were to be used in the collection of severe storm data in Foxboro.

The Foxboro Company set up a radioactive point collector and recorder which indicated atmospheric potential gradient and its sign. These data were to be correlated with vertical radar data.

In midsummer, new above-ground power lines were strung up the west side of the hill near the ski slope. The new lines then ran to transformers behind the Observatory wall and to the Air Force compound.

It was noted in midsummer that the Weather Bureau Solar Radiation Supervisory Station at Blue Hill was recording solar radiation as follows: at normal incidence, total on a horizontal surface; on vertical surfaces facing, north, east, south and west; and diffuse on a horizontal surface.

Hurricane Diane passed 18-19 August. During the storm lightning struck the State Police tower on the evening of the 18th and the TV tower early in the morning on the 19th. On the second occasion two weights were tripped in the Conover precipitation gage. The following extraordinary rainfall rates were recorded:

25.4 mm (1.00 in.) in 25 min
44.2 mm (1.74 in.) in 1 hr
103.4 mm (4.07 in.) in 3 hr
252.2 mm (9.93 in.) in 24 hr
324.4 mm (12.77 in.) in 57 hr

An additional 25.1 mm (0.99 in.) of rain fell on the 17th in three hours to make a total of 349.5 mm (13.76 in.) for the three-day period. The 3-, 24- and 57-hour amounts were the highest for those time intervals in the Observatory's 100-year record.

The Blue Hill temperature record was greatly lengthened with the discovery of a record by Bazin in Ponkapoag, 2.0 km (1.2 mi.) south of the Observatory. It was checked for homogeneity with records from Milton Center, Cambridge, New Bedford, and New Haven and adjusted to Blue Hill to extend the record back to 1831. Monthly and annual averages are therefore available for a total of 154 years.

In September the Jamesway Hut was moved from the backyard to the Air Force Compound; the remote PPI scope remained in the observers' office.

12 October was a particularly busy day. Brooks held a consultation with members of the Harvard College Observatory; Bent and Keutner conferred on Mount Washington affairs; and a boy scout troop, cub scouts, girls, and foreign students from Tufts University visited, making a total of about 120 visitors.

The Observatory as it looked from the weather radar tower during these busy days is shown in figure 129.

In the year 1955 Brooks was given the American Meteorological Society's Award for Outstanding Service to the Society. He was cited as follows: "Charles Franklin Brooks, for a lifetime of devoted and untiring service to the Society, notably as its organizer in 1919, as its Secretary for

Figure 129. The Observatory looking south from the weather radar tower. WGBH building on the left; thermometer shelter and flood light, Conover precipitation gage, shield, and heating controls in the box, standard precipitation gage inside the modified Nipher shield, Fergusson recording precipitation gage and shield on concrete pillar, and the ombroscope shown at base, 1957. Courtesy R.J. Newcomb.

the next thirty-four years, and as Editor and mainstay of the Bulletin for most of its first twenty years."

Brief notes in the observatory journal read, "9 June—Kangueser, Weather Bureau Section Climatologist, recommends that Blue Hill be made a Bench Mark station for New England. . . . August—Blue Hill chosen tentatively as one of twelve bench mark stations in the U.S." These notes refer to a program originally formulated in 1954 by Helmut H. Landsberg, director of the newly created Office of Climatology of the Weather Bureau. At that time it was known as the Climatological Benchmark Program. Cooperative climatological stations were to be sought out to form a benchmark network and used for purposes of monitoring climatic fluctuations. Criteria for selection were: (1) long-existing records of an apparent homogeneous nature; (2) good prospects for continuation in the same location; (3) prospect of unchanging environmental conditions at the station; and (4) reasonable geographical distribution of about five stations in the conterminous United States. Stations outside that area were to be added later.

In the late 1950s the climatological benchmark station concept was introduced at the international level through the Commission for Climatology of the World Meteorological Organization (WMO), of which Landsberg was president. This led to a WMO initiative in recommending to all WMO members that such stations be identified and maintained as newly named Reference Climatological Stations (see World Meteorological Organization 1981 and 1983).

According to Mitchell (1985) the analysis for homogeneity of temperature records was completed in 1960. Blue Hill was found to rank two or three out of twenty-two candidate stations.

In 1974 the network was stabilized around twenty-one stations nationwide. Two other stations, Presque Isle, Maine, and Norfolk, Connecticut, accompanied Blue Hill in the New England area.

Not until 1 January 1976 was Blue Hill was officially made a station. This followed after the resultant wind measuring system had been made operative.

Data from the stations are carefully monitored for errors and over the years 1981-83 Blue Hill proudly topped the list with no errors in 1981 and 1983 and one in 1982. This gave the station top ranking in the network.

No world-wide selection of reference stations has been achieved at the time of this writing. However, a project is being initiated by the WMO to develop standard methods for selecting stations with the aim of establishing at least one station for each 5° latitude–longitude box.

Jen-Hu Chang joined the staff as a research assistant in climatology on 3 January. He was in charge of the Quartermaster contract and helped immensely in checking data for World Weather Records. Richard Ashley also joined the staff as a research assistant in climatology on 25 July. Raymond Wexler became a full time researcher 1 July and George Kenneth Thompson joined 27 December as an electrical specialist. Elizabeth Carlson was hired as a secretary, and Lorna Ridley was hired as an observer. Albert Kaufman served as resident observer for the summer, succeeding Roy Hines; later Stanley Yarkin became resident observer. Wollaston continued as chief observer with Richard Gordon and Richard Parsons as part time observers.

Activities at the Observatory continued at an increasing tempo.

Wexler studied the melting layer using new radar data. Boucher was involved in radar meteorology, forecasting, cloud seeding, and the use of snow crystal forms as indicators of the cloud and temperature structure aloft. The seeding and snow crystal studies were Mount Washington Observatory projects.

Conover, with the assistance of Ashley, completed the collection of data for World Weather Records, 1941-50. The new volume was nearly twice the size of the former 1931-40 volume and was comprised of 678 stations, many new and with long records Conover also directed a new Air Force contract which was called Cloud Photo. Time lapse photography was employed to show and measure motions of cirrus, especially near the jet stream, cumulus "streets," and the roll cloud of thunderstorms. All the observers assisted in the detailed photography. Two to four daily rawin-sondes, which were made at Fort Banks, in Boston Harbor, were used in the study. Through the cooperation of the U.S. Navy special cloud reports were often obtained from high-flying jets based at nearby South Wey-mouth.

Thompson was engaged in the preparation of new instrumentation that would feed into an eight-channel oscilloscope. The oscilloscope was then photographed on 16 mm film to give a record of the weather. A precipitation-rate computer, fed by the Conover precipitation gage, was designed and constructed by WGBH, MIT and Raytheon Manufacturing Company personnel. A Hudson–Jardi precipitation-rate recorder was set up north of the enclosure. This gage had a collection area of 1.8 x 1.8 m (6 x 6 ft.). It could be heated when snow fell. A dewcel, for deriving dew point temperature, was set up in the enclosure and a temperature sensor placed in the enclosure shelter. A combination dew point and temperature recorder, connected to the sensors, was installed in the first floor office. The enclosure instrumentation is shown in figure 130. Another picture, figure 131, taken from the TV tower, shows the aerial positions of these instruments and a part of the weather radar enclosure.

New closets were installed against the northwest curved wall of the first floor tower. The new instruments as they were developed, were mounted above the closets with ancillary equipment contained within the closets.

The Observatory radio transmitter and receiver that were used for communication with Mount Washington were moved by Thompson to the

Figure 130. Enclosure looking northwest. Dewcel housing in white shield at left; thermometer shelter and flood light, Conover precipitation gage, shield and heating controls in the box, standard precipitation gage inside the modified Nipher shield, Fergusson recording precipitation gage and shield on concrete pillar, and the ombroscope shown at the base, 1957. Courtesy R.J. Newcomb.

downstairs closets and shelf. The radio racks on the third floor were removed, thus providing additional desk space. Provision was made for talking to the Mount Washington Staff from either the first floor observer's desk or third floor desk, an excellent improvement and time-saver. Other improvements consisted of a second telephone trunk line, two additional extensions, and an interoffice communication system. Living conditions at the Observatory were also improved by connecting the water system to the WGBH pressure tank, and the installation of a shower and new kitchen stove.

In the spring a series of heavy snowstorms struck the area. On 16-17 March 33 cm (13 in.) of snow fell with wind gusts of 35 m sec^{-1} (78 mi. hr^{-1})

Figure 131. Aerial view of the enclosure and Hudson–Jardi precipitation gage. May 1956. Courtesy J.H. Conover.

from the northeast. This was the worst storm, as far as snowfall amount and wind were concerned, since that of 14 February 1940. On 19-20 March a 49 cm (19.3 in.) snow fall with a peak wind of 28 m sec^{-1} (63 mi. hr^{-1}) from the northeast occurred which resulted in a total of 61 cm (24 in.) on

Figure 132. Sanding truck stuck at the halfway turn on the summit road, 9 April 1956. Courtesy J. H. Conover.

the ground. On 24 March another 31 cm (12.2 in.) fell with lighter northerly winds. One hundred thirty two cm (52 in.) of snow fell during the month. Atlas wrote of the difficulty in hiking up the hill after such storms: "Indeed, I recall having hiked up with B.J. (now Sir John) Mason in about 1957. It was always a source of embarrassment to 'us men' when Sally Wollaston and Lorna Ridley beat us up the hill regularly on snowshoes."

Most accidents on the summit road occurred in winter. However, on 9 April, after a late snowstorm, a heavy sanding truck broke through frost in the ground at the side of the road on the halfway turn. It is shown in figure 132 preparatory to towing by a heavier vehicle.

Radar technician Ultsch and Thompson worked with Air Force technicians Graham Armstrong, Ruben Novak, and William Lamkin and Airman/2d Robert Whittaker in improving the AN/CPS-9 radar (3.2 cm wavelength), the vertical beam radar AN/APS-(1.25 cm), the zenith scan radar (1.25 cm) and the cloud base and top radar, TPQ-6 (0.86 cm). Most of these technicians and other personnel are shown in figure 133.

Professor Frederick A. Brooks, brother of Charles Brooks, spent part of his sabbatical from the University of California as a Blue Hill research fellow. He worked on a method of separating solar radiation attenuation by haze particles from that by water vapor.

Figure 133. Weather radar group. Top, left to right: Graham Armstrong, Edwin Kessler, Mary Weaver (BHO), Stanley Yarkin (BHO), Kenneth Ultsch (BHO); second row: Dave Atlas, Bill Lamkin, Al Chemla; lower row: Ruben Novak, Ralph Donaldson, and Robert Whittaker. Air Force personnel except as noted. 1957. Courtesy J.H. Conover.

At this time the Observatory started to receive groups of meteorologists from all over the United States for participation in their refresher courses. Two groups of Air Force Officers one group of navy reserve officers and two U.S. Weather Bureau Advanced Study Groups each spent the greater part of a day at Blue Hill during the year.

Roger Brown, David Williams, Cakste, and Ridley all worked to bring the climatological data for the station up to date. Much checking of back data was required.

On 8 November fire destroyed the Metropolitan District Commission building at the base parking lot located on the west side of the hill. The building had served refreshments through the summer and provided shelter since its construction in 1905.

Work started 27 November on the erection of a temporary insulated and heated building in the back yard of the Observatory. It provided a darkroom in one corner and space for desks and the projection of time-lapse cloud films. Hot and cold water lines above ground were insulated and kept from freezing with electrical heating tape. The building was located just outside the shop and adjacent to the large fuel tank.

The tower was once again waterproofed with a plastic-like material and the roof covered with felt paper and tar in December.

A bronze plaque showing a bust of Abbott Lawrence Rotch, as he appeared shortly before his death, was presented by his son, Arthur Rotch, at a dedication ceremony on 26 June. The work was done by A. Q. Ladd. The plaque (figure 134) which was mounted over the fireplace on the southeast wall of the first floor of the tower, was unveiled by Lawrence Rotch, grandson of the founder, while his father made the presentation. Harvard president Pusey accepted it on behalf of the President and fellows of the university. The plaque now hangs in the Center for Earth and Planetary Physics, Pierce Hall, Harvard University. After the dedication ceremony, President Pusey, Mrs. Pusey, and Donald Menzel of the Harvard College Observatory were shown the Observatory and results of some of its work. Menzel was favorably impressed and promised to increase the amount of money sought for additional endowment according to the university plans of last year.

Robert Stone, who had worked at the Observatory as an observer, librarian, and editor of the Bulletin, received an award in 1956 for his outstanding service to the American Meteorological Society. The award read: "Robert Granville Stone, for his many years of faithful editorship of

Figure 134. Plaque of Abbott Lawrence Rotch by A.Q. Ladd, 1956. Copyright President and Fellows of Harvard College. Division of Applied Sciences.

the *Bulletin of the American Meteorological Society* and his active participation in committee and council work."

Greenleaf W. Pickard, research associate in the 1930s, died 8 January. He was a past president and Fellow of the Institute of Radio Engineers (IRE), and received the IRE Medal of Honor in 1926. He was a fellow of the Radio Club of America and received their Armstrong Medal in 1941. His other society memberships included the American Association for the Advancement of Science, the American Academy of Arts and Sciences, and the American Meteorological Society. He was active in amateur radio during the last 40 years, holding the call letters W1FUR.

On 1 June Shirley J. Richardson joined the staff as a full time assistant librarian, and Ralph Newcomb joined as a map plotter and assistant in the cloud photo project.

Over this busy time gifts amounting to $29,533 were received for the fiscal year 1955-56. The endowment rose to $400,000 and the Observatory did not require university general funds.

1957

By January an active search was underway for a successor to Brooks. This period was especially difficult for Brooks. He attended numerous conferences with Dean McGeorge Bundy and others in an effort to determine just what would happen to the Observatory. Details, however, were to be decided by the new director and a candidate had not yet been found, let alone appointed. At one point Bundy was negotiating with the Air Force Geophysics Laboratory to take over the observing program, but Brooks quickly vetoed this, saying it would be the first program to be cut when a period of austerity set in.

The new temporary building in the back yard of the Observatory was ready for use by mid-January. A desk and tables were brought in and Newcomb processed film and supplied numerous prints in the darkroom for project Cloud Photo.

Arnold H. Glaser arrived 1 February to direct work on Project Satellite Cloud Photo. This project, sponsored by the Geophysical Research Directorate of the U.S. Air Force, was to evaluate the potential meteorological utility of televised images from a satellite vehicle and devise techniques by which such information might best be put to use.

The outstanding scientific event of the year was the sponsorship, jointly with the Air Force Cambridge Research Center, MIT, and the American Meteorological Society, of the Sixth Weather Radar Conference. Meetings in Cambridge on 26-28 March, drew an attendance of nearly 800, including meteorologists from the United States, Canada, Japan and the United Kingdom. A 372-page preprint volume was published jointly by MIT and Blue Hill and distributed by the Air Force. The Observatory and the Air Force brought Professor Frank H. Ludlow, of the Imperial College of Science and Technology, and Professor Chojii Magono, of Kokkaido

University, to the meetings. A group of attendees and hilltop staff members are shown in figure 135, following tea at the Observatory.

Boucher's radar weather research consisted of the grouping of precipitation processes in the atmosphere into four main categories, each of which exhibits characteristic patterns on the vertical radar record. Boucher, also, with the help of Goldman and Penn of the Boston Weather Bureau, constructed detailed time cross sections of the radar-echo-intensity for a number of winter storms.

Thompson put "Weather Log" into operation in April. Eight weather elements, atmospheric pressure, air temperature, dew point, wind direction, wind speed, atmospheric electrical potential gradient, precipitation rate, and accumulated precipitation amount were logged. The weather log control panel and other Observatory recorders, some of which fed into weather log, are shown in figure 136. Values of the various elements were transferred to an oscilloscope as displacements from a zero level and photographed on 16 mm film at a rate of 30.5 cm (12 in.) per hour or exactly the same speed as the recording of the vertical radar, thus providing easy correlation.

Figure 135. Visitors and hilltop staff members at the observatory. Left to right: Mrs. Brooks, John Conover, Shirley Richardson, Raymond Wexler; in rear: Elizabeth (Dib) Carlson, Ken Thompson, Frank Ludlum, Jen-hu Chang, Edwin Kessler, Arnold Glaser, Robert de Chancenotte (back), Chojü Magono, Albert Chmela, Dr. Brooks, Charles Cunnif and Roland Boucher. Courtesy J.H. Conover.

Figure 136. Weather log monitor meter panel and other Observatory instruments and recorders along the northwest wall of the observer's room, 1957. From left: Conover accumulated precipitation, aerovane wind speed and direction, weather log monitor panel, Foxboro temperature and dew point recorders. Below: Radio receivers and auxiliary monitors. Microphone was for communication with the Mount Washington Observatory. Courtesy R.J. Newcomb.

In the first phase of the study of severe storms Boucher, assisted by Ashley, analyzed mesoscale conditions preceding and during the storms.

The development of objective methods to forecast snow vs rain using Mount Washington and Blue Hill data was brought to an end.

Tabulation of hot–wet climatic data for the Quartermaster was also completed.

The Cloud Photo project was expanded to include stereo photography of cirrus clouds. Large aerial cameras having a 23 x 23 cm (9 x 9 in.) film format, aimed vertically and triggered simultaneously by radio signals, were used (figure 137). Baselines of several kilometers extending from the summit were used. Three whole-sky cameras were also constructed and set

Figure 137. Ralph Newcomb with a vertically aimed aerial camera and walkie-talkie radio. The adjustable panel provides a sun shield for the camera lens. Courtesy J.H. Conover.

up at the Hartford and Worcester Weather Bureau stations and atop Blue Hill. During jet stream cirrus cloud conditions these cameras were triggered at intervals of twenty minutes by cooperating Weather Bureau personnel and observatory staff members. From these pictures successive maps of the cirrus clouds covering an area about 250 x 350 km (155 x 218 mi.) were constructed. These were probably the first detailed cloudscape maps of this size, prior to the reception of cloud pictures from satellites. The temporary building in the back yard, where much of this work was carried out, was occupied in late February.

Details of all of these studies, as with all other Observatory work, may be found in reports and papers listed in the bibliography.

For a period of several months the PPI scope of the radar was transmitted by microwave radio directly to the Boston Weather Bureau District Forecast Office, the forecast office at the navy's South Weymouth Air Station, and the Air Force forecast office at Bedford.

Housing for all these activities and the WGBH facilities is displayed in an aerial photograph taken about this time (figure 138).

Figure 138. Aerial view of the summit buildings. From left: weather radar tower, kite house, Observatory, WGBH building in background with TV tower, WGBH-FM tower in front of Observatory. c. 1957. Courtesy J.H. Conover.

Elizabeth Carlson, secretary, resigned in the spring. Carlson, or "Dib" as she was known to the staff, was an excellent secretary; she had done an outstanding job of typewriting all the preprint papers for the Sixth Weather Radar Conference.

Plans for Brooks's retirement commenced, and on 10 May the Division of Geological Sciences honored Brooks with a dinner at the faculty club. About forty people, including most of the Observatory staff, attended.

Brooks was given a farewell party at the estate of Arthur Rotch on 26 May. Past and present Observatory personnel, as well as other friends and family members, attended. Brooks was presented a scrapbook of letters and photographs, particularly from those who could not attend, and a $1,070 purse for travel purposes if he so wished. Presentation of the gift is shown in figure 139.

Among the many tributes to Brooks from associates and those whom he employed, the following are reproduced.

> Atlas wrote, in relating some of his memories of his stay at Blue Hill: ". . . most of all it was the association with Charlie Brooks that made it so interesting and exciting. He stimulated our interest in clouds and storms and most of us owe our continuing intense interest in these phenomena to him."

Figure 139. Presentation of check to Dr. Brooks at his retirement party. Left to right: Dr. Charles Brooks, Mrs. Eleanor Brooks, Mr. John Conover, and Mr. Ralph Burhoe. Courtesy A.J. Faller.

Dalrymple wrote: "I think I probably speak for every observer who ever walked through the front door when Brooks was there when I say that we all left much better for our association with Blue Hill and Brooks."

Ashley wrote: ". . . most of all I remember Brooks, the perfect gentleman, more of a father-figure on the staff than 'the boss.' Adjectives such as kind, understanding, and caring applied to him as they do to very few others."

The following is excerpted from a letter by Ralph Burhoe.

I suppose many other things could be mentioned, but I close with a word about the impact of Charles Brooks as Director during my years. Basically trained in the early part of the century as a geographer and climatologist, he was nevertheless very open to the increasing importance of theoretical physics for the field, and ready to cooperate with and foster the work of those who knew how to handle it in ways he only partly understood technically even though he was keen on its application in meteorology. I perceived his gracious character, his open and searching mind, and his vast energy and enthusiasm for progress in the field, making him, and hence Blue Hill, a primary influence upon the advancement of Meteorology (his helpful correspondence was international as well as American) from the First through the Second World War.

Furthermore, Brooks' open, fair, and kindly personality as well as his enthusiasm made him a productive and honored leader for most of those who worked with and under him. I had the privilege of observing many aspects of his home life—where he was obviously a wonderful father and husband—as well as his notable professional life. Nearly all of the diverse persons who worked with the Observatory during those years seemed to me to exist in an unusually pleasant inter-personal environment, and the one or two who had some serious built-in problems probably never had it so good. I count myself privileged to have had those twelve years' experience of a blossoming applied science with such good companions.

On 30 August another small party to mark Brooks's retirement was held at the Observatory. His retirement became effective 1 September, after twenty-six years as director of the Observatory and professor of meteorology at Harvard. He was appointed Professor Emeritus at that time.

The base house, in which the Brooks family lived, at 1793 Canton Avenue, was rented by Brooks subject to a three-month termination notice.

In summary, the directorship of Brooks can best be summarized as productive, very active, and turbulent. He first refurbished the Observatory, initiated new detailed weather observations, and restored and expanded the library. He was instrumental in establishing the new Mount

Washington Observatory and acted as its director for many years. The Observatory at Blue Hill participated in the development and use of UHF radio. The Harvard radiosonde was conceived by Brooks and developed by Lange, Bent, and Pear. They demonstrated the first routine use of such instruments in the country. This was probably the most important contribution to the science under the directorship of Brooks. Experiments were made in tracking the heights of levels of dielectric change by the vertical reflection of radio waves. The Weather Bureau supervisory solar radiation office was brought to the Observatory, and Brooks was an important factor in bringing a Weather Bureau District Forecasting Office to Boston. During World War II Brooks and Conover taught the rudiments of meteorology and observing to hundreds of students. Throughout his tenure Brooks taught graduate students, usually one or two each year. A lull in important activities followed the war years, although consultation flourished, including detailed forecasts each year for the Harvard commencement, and new instruments were developed and tested. In 1947 the use of unrestricted university funds was discontinued, leaving the existence of the Observatory in a precarious condition. However, Brooks quickly negotiated the first of many government contracts to keep the Observatory in operation. The opening of one of the first educational FM radio stations in the country was supported by Brooks. He arranged for transmitter space within the Observatory and later consulted on the construction of a station building and TV antenna tower at the hilltop. He made weekly broadcasts, without pay, over the FM station.

Brooks was an avid observer of all natural phenomena. Figure 140 shows him at the rangefinder which he used to determine cloud heights. He often filled columns in the journals with detailed notes. These included cloud and snowfall observations, snow types, ice columns in frozen soil, optical phenomena, thunderstorms, phenological and ornithological notes, to mention a few. He discussed these observations with his observers and encouraged them to follow with similar notes. The routine detailed observations and frequent training of part time observers required much time, and he received some criticism for expending so much time and money on their continuance. However, he viewed some of the effort as educational and, indeed, it was to the young observers, although some stayed only a few months. Many went on to careers in meteorology and have since vouched for the training which reminds them of what really lies behind the data which they so often analyze.

Figure 140. Dr. Charles F. Brooks at the rangefinder, c. 1957. Courtesy J.H. Conover.

It is surprising that, as a climatologist, Brooks did not seek to reduce and compile data from the numerous charts that were taken from the instruments each day. For years, after he became director, the only summary data consisted of a single page submitted to the Weather Bureau once a month. He implied that if additional data were desired it could always be obtained from the autographic charts. However, when records are not routinely reduced, small inconsistencies in the recordings develop; these may continue unnoticed for a long time, then if the record is needed, the problem is discovered and reduction is more difficult, if not impossible. For this reason, and a growing demand for hourly values of temperature, humidity, precipitation, wind direction and speed, he was persuaded to have these data tabulated as the charts came off the instruments. At first they were entered on separate sheets of paper but beginning in 1943 they were entered on new forms in the observation books.

In 1951 Harvard University made it known that it no longer wanted the Observatory after five years. However, in 1953 a report by Robert E. Gross, chairman of the Visiting Committee, to the overseers was viewed with some favor, but the university's plan for eventual closure did not change. Research was beginning to significantly increase with the addition of one new professional meteorologist. The Observatory demonstrated its ability to operate financially independent of the university. The report recommended that teaching be expanded, but this could not be done without university assistance.

Two more professionals were added in 1954, in 1955 other professionals arrived, as well as the Air Force's weather radar group and its radars. At this point Thomas Cabot, the new chairman of the Visiting Committee, noted that improvement over the past two years had been so great he would recommend that the Observatory not be abandoned. In this light the university reconsidered its position and agreed to extend the Observatory's use under three provisions: new endowment, that the Astronomical Observatory take responsibility, and that the new director be under joint responsibility of the Astronomical and Blue Hill Observatories.

By 1956 the work of the Observatory had advanced to a point where groups of practicing meteorologists placed the Observatory on their travel itineraries several times a year. These advanced study groups comprised Air Weather Service, navy and Weather Bureau meteorologists.

Menzel of the Astronomical Observatory visited and was impressed with the new work, compared with an earlier visit, so he agreed to assist in the raising of endowment. A plan was developed whereby Blue Hill would have two professors: one to work with the astronomers on solar–weather influences and aerodynamics of the upper atmosphere, and the other to continue the cooperative researches with the Weather Bureau and Air Force on problems of the lower atmosphere.

The Observatory remained financially secure; in fact, $76,000 was added to the endowment for the fiscal year 1956-57. This was from accumulated profits from endowment and the reduction of expenses by government contracts. Nevertheless, the endowment could not fully support new plans without a closer tie with the university.

At this time the American Meteorological Society, which perceived the jeopardy of the Observatory, contributed $1,000 for current expenses "in token of the esteem in which the Society holds Charles Franklin Brooks, and, particularly, for his selfless and untiring devotion to the building-up of this outstanding meteorological library, with the hope that Harvard will find a means for preserving the library on a permanent basis."

Thomas Cabot deserves much credit; without his dedicated work the Observatory would have surely closed. He had succeeded in developing a compromise between those who fought for the Observatory status quo and the university which had adamantly sought its closure. A search for a new director became a reality by 1957.

References

Mitchell, J. M. 1985. Notes of the U.S. reference climatological station program, its history and present status, with special reference to Blue Hill Observatory. Personal communication.

Putnam, P. C. 1948: *Power from the wind*. D. Van Nostrand Co., 224 pp.

World Meteorological Organization. 1981 ed. Manual on the global observing system. I (Global aspects), *World Meteorological Organization* 38ub. No. 544. p 6, Part III, Sect. 2.3.3.3 and Part III, Sect. 2.3.6.3.2.

——1983. Guide to climatological practices, 2nd ed. *World Meteorological Organization* Pub. No. 100, lB.7, Sect. 3.3.2.1.

Chapter XIII

Selection of a New Director and the Interim Directorship: 1957–1958

1957 (continued)

In the spring McGeorge Bundy, then dean of the Harvard Faculty of Arts and Sciences, invited Professor Richard M. Goody to become professor of dynamic meteorology at Harvard and director of the Blue Hill Observatory. Goody was reader in meteorology at the Imperial College of Science and Technology, University of London. He was educated at the University of Cambridge where he received his Ph.D. in 1949. During World War II he served in the British Government Scientific Service. At the end of the war he was put in charge of the Max Planck and Prandtl Institutes in Götingen and was instrumental in preventing their dissolution.

At the time Goody was 36 years old, knew very little about the United States or its educational system, and knew nothing about the Blue Hill Observatory. The attraction to the offer, in Goody's words, was " ... the growing ascendancy of U.S. science and Harvard's international reputation."

Characteristic of the appointment of full professorships at Harvard, the university made no attempt to inform the appointee of anything. In this case Goody was given no instruction regarding the Observatory; the only requirement was that he join one of four departments: physics, geology, astronomy or engineering and applied physics.

Cabot called upon Goody in London to explain something about the workings of a privately endowed University. He understood the difference between American and British systems of university support and governance and was concerned that Goody's ignorance might lead him into difficulties. Cabot scarcely mentioned Blue Hill, but instead related the recent history of Harvard's Arnold Arboretum, which was in a situation similar to that of Blue Hill.

Goody was best known for his theoretical work and the book *The Physics of the Stratosphere*, published in 1950. He had always been interested in the direct measurement of the atmosphere and had spent much of his scientific career working in observatories and even setting up a new observatory for Imperial College, so the use of a new observatory offered additional appeal.

If he made use of the observatory the only condition was that he stay within the limits of its budget.

In April Goody accepted the offer and an effective date of 1 July 1958 was set for his appointment.

Goody found the Observatory bursting at its seams with personnel and all available space filled to capacity in the library. Although the projects brought in sufficient income to carry the facility, there was no space for Goody's use, nor did the projects serve the Harvard faculty or students in the manner that he believed they should. Therefore, it was decided to terminate all the projects and move the library to Cambridge after some consolidation. Income from new projects, which were relatively easy to obtain during that period of rapid expansion in science, would continue to support the Observatory.

John H. Conover served as acting director from 1 September 1957 to 30 June 1958. It was his job to carry out these changes as far as possible before Goody's arrival.

Roger M. Lhermitte, though he held an appointment until the end of the year, was continued on leave of absence because he was unable to return from France.

Jen-hu Chang, after completing the preliminary draft of his book on ground temperature, left in September. Tabular data were completed, with the assistance of de Chancenotte and Ashley, and the final papers and reports were finished while Chang pursued postdoctoral studies at the University of Wisconsin.

Anastasija Caskte left in October after more than three years of service as library and general assistant.

Kenneth Ultsch also left in October to join the MIT Radiation Laboratory.

Arnold H. Glaser left in October to direct a meteorological research team at Allied Research Associates, Incorporated, in Boston. His in-depth evaluation of the potential use of photography of the earth from a satellite was the first of its kind. Many of the uses were realized with launch of the first satellite and within a short time additional applications of the televised pictures followed.

The library had been greatly increased in size. Some 450 periodicals and serials were received per year with daily and monthly weather maps and data from many countries. Loan requests numbered over 100 in a nine-week period and many visitors came to use the extensive bibliographies and references. A concerted effort was made to eliminate duplicate material. Services on an appreciable scale had been rendered to the American Meteorological Society, the Army Quartermaster Corps, the Air Force Cambridge Research Center, the Sandia Corporation, other agencies, and various departments of the university.

On 22 November a portrait of Brooks, executed in oils by artist A. Gourine, was unveiled and hung on the northeast wall of the first floor tower room (figure 141). Twenty-seven people attended the ceremony. Conrad Chapman, long-time friend of the Observatory and part time assistant in the library and observing program, commissioned the work. The painting now hangs in the Center for Earth and Planetary Physics, Pierce Hall, Harvard University.

Throughout Brooks's retirement he continued his work, much of it for the International Meteorological Organization. He frequently came to the Observatory where he worked as usual at his roll-top desk in the library.

1958

On 8 January Brooks died suddenly of a massive heart attack. It had snowed the previous night and he was outside his home knocking snow from the lilac shrubs when the attack came.

A memorial service was held at the Friends Center, in Cambridge, on 12 January. In addition to his lifetime of scientific work, much of which has been set forth in this history, he will be remembered by all who knew him, because during all this activity, he showed a warm sympathy for all men and he gave his time and energy unstintingly, day or night, to all who sought his advice.

Brooks held membership in the following scientific organizations and honorary societies: American Academy of Arts and Sciences, American Geophysical Union, American Geophysical Society, American Meteorological Society, Royal Meteorological Society of Great Britain (fellow), Phi Beta Kappa and Sigma i.

Following Brooks's death, the American Meteorological Society named its award for outstanding service to the Society the "Charles Franklin Brooks Award for Outstanding Service to the Society."

Coincidently, Edmund Schulman, a student of Brooks who worked with the Blue Hill studies in radiation and tree ring data from the west in the late 1930s, died, also of heart failure.

Beginning 16 January, 56.4 cm (22.2 in.) of snow fell in 22 hours. This was the largest amount measured from a single storm to date. The previous

Figure 141. Portrait of Dr. Charles Franklin Brooks by A. Gourine, 1957. Note the old Observatory and cloud pictures in the background. Copyright President and Fellows of Harvard College. Division of Applied Sciences.

record of 50.8 cm (20.0 in.) occurred on 31 January–1 February 1898. The dry snow of the 1958 storm was accompanied by winds of 13-18 m sec^{-1}

(29-40 mi.hr^{-1}), which caused heavy drifting. A maximum gust of 31 m sec^{-1} (69 mi.hr^{-1}) from the northeast was measured during the storm. For several days following the storm, personnel travelled up and down the hill by foot, on snowshoes, or on skis.

The base house was sold in the spring and Mrs. Brooks moved to Greenwich, Connecticut, on 29 May. This necessitated the quick disposal of vast files of papers and stores of reprints and weather maps in the barn. These were hastily screened and divided into lots, which were sent to the archives, various institutions, agencies, individuals, or were destroyed. The large thermometer shelter on the property, known as the "base station," was sold and all observations there terminated.

The Cambridge climatological station at 2 Divinity Avenue was closed and the instruments which belonged to the Observatory were sold.

In June another coat of waterproofing was applied to the tower.

Shirley J. Richardson, who had become librarian in September 1957 in succession to Wexler, resigned 1 February. Her report on the Observatory library, which was published in 1958 under joint authorship with Brooks (see bibliography) summarizes activities, in detail, since 1885.

Roland Boucher resigned 30 April. He took a position in the group headed by Arnold Glaser at the Allied Research Associates Corporation.

Robert de Chancenotte, resident observer since January, resigned in June; Richard Royal took his place and James Walker assisted with the observations.

Mary Weaver resigned in June and Richard Ashley left for summer school and a fall teaching position in Bridport, Vermont.

All contractual research for the Air Force and Weather Bureau was completed and terminated.

For the fiscal year 1957-58 a balance of $16,000 was shown and gifts amounting to $10,500 were received in spite of no fund-raising program.

Chapter XIV

Directorship under Richard M. Goody: 1958–1971

1958 (continued)

Richard M. Goody, whose portrait is shown in figure 142, had decided to join the department of Engineering and Applied Physics and an agreement was made by the corporation that the Observatory become a part of that division. After consultation with the primary benefactors of the Observatory's endowment, the university allowed funds to be transferred for use in Cambridge under continued responsibility of the director. As Goody saw it, "My task had two parts: first, to consolidate atmospheric science as a proper activity for the University; and second, to incorporate the Observatory into that activity in a constructive way, if at all possible." He concluded early "that there never would be support for a traditional Department of Meteorology or even a broader Department of Atmospheric Sciences with the competition of nearby MIT. The only worthwhile effort would be to build something complementary to MIT that would be able to bring into the atmospheric sciences some of the excellent graduate students at Harvard who, otherwise, would not know anything about the field."

Courses in geophysical fluid dynamics, embracing physical oceanography and dynamical meteorology, were initiated. By emphasizing that these courses concerned some of the many applications of physics, some students entering graduate school to study physics became interested enough to pursue further studies in atmospheric sciences.

Work at the Observatory became centered around two new programs.

One project dealt with the measurement of suspended particles in the atmosphere up as high as 70 km (43 1/2 mi.). Goody had been interested in the problem for some years because of the potential chemical and thermal effects of high-level dust and its possible use as a tracer of atmospheric motions. Important work on the subject was taking place at the Air Force Geophysical Research Laboratory under Dr. Christian Junge, and because of his interest, support for a program was obtained. Through Cabot's generosity, it was possible to appoint Frederick Volz of Mainz, Germany, as research fellow to work with Goody on the project.

Figure 142. Professor Richard M. Goody, Director, 1958-1971. Copyright President and Fellows of Harvard College. Division of Applied Sciences.

The second project was the development of a new technique for measuring the concentrations of certain minor atmospheric constituents in particular nitrous oxide, a gas that Goody had studied and which he believed to be a major link in the atmospheric nitrogen cycle. Since then, study of minor atmospheric constituents has become a central concern in studies of atmospheric chemistry. The oxides of nitrogen and, particularly, nitrous oxide, have emerged as subjects of heated controversy.

In October removal of material from the Observatory library and its transfer to the Gordon McKay Library in Pierce Hall at Harvard University commenced. Transferred books were distinguished by a bookplate with the words "Blue Hill Observatory Collection." Forty-eight sets of serials, some of very great length, and 576 books were retained. Subscriptions to new journals would be reduced to thirty and exchanges of unpublished reports to about thirty-five. It was anticipated that about thirty new books would be purchased per year. Surplus library material was distributed in the following ways: to other Harvard libraries went 813 books, 113 incomplete serials, and some pamphlets; to the American Meteorological Society seven books were transferred and fourteen sets of serials, which totaled 666 volumes; to the Weather Bureau, seventeen books, 157 sets of serials, and a large quantity of climatological data were given; the remainder was sold and the proceeds used to increase endowment.

When the library was first moved, it shared space with the Gordon McKay Library. Subsequently when Observatory personnel obtained their own premises in Pierce Hall, an independent library was created, although it was still cared for by the professional librarians of the Gordon McKay Library. Pictures of the four directors and other Blue Hill memorabilia decorate the passage leading to this library.

In October the last advance study group to visit Blue Hill attended a six-hour program in which lectures in weather radar, future use of satellites, local forecasting of snowstorms, clouds, and the future of the Observatory were presented.

Also in October, Wheatland, curator of historical instruments at Harvard, removed for preservation three loads of old instruments which had been used in earlier years. Soon afterward the remaining old and unused instruments were sold to friends, staff members, and Air Force personnel. Not long after this the Smithsonian Institution, having heard of the changes underway at the Observatory, inquired concerning the availability of the old instruments, but most of them had already been reassigned to the collection of historical scientific instruments at Harvard, thus preventing their transfer to Washington.

Extensive modernization of the Observatory began 26 November. Machines, tools, and supplies from the shop were moved to a temporary trailer in the back yard and the area made over into one well-lighted room.

The dining room table and corner cupboard were removed and a narrow counter was fastened to the east and south walls to serve as an eating place.

Personnel changes over the year were numerous. Sally Wollaston resigned on 4 September and Lorna Ridley on 5 September. Parties were held at Ponkapoag Pond and at the Observatory. Wollaston had served sixteen years at the Observatory. For years she hiked up and down the hill over trails on either the south or southwest side in all kinds of weather. She soon became a proficient observer and advanced to the position of chief observer. She not only maintained high quality meteorological records but assisted in the library, and in research projects that involved clouds, and performed some secretarial work. Ridley's work was mostly associated with the routine observations and special cloud photography. Both had taught mathematics before coming to Blue Hill and after leaving both went to Pennsylvania State University to obtain master's degrees in meteorology.

John Hamilton arrived in August and Fred Dawson in September to serve as observers. Observer Richard Royal left in September to attend MIT.

Lillie Hodges, librarian, transferred to Gordon McKay Library in October.

Air Force personnel remained. Kenneth Thompson was occupied with instrument work for the Air Force. He continued to operate Weather Log, and Conover and Newcomb stayed to complete the Cloud Photo project.

Summit personnel at this time are shown in figure 143. In November the nearby television station WGBH-TV doubled its power output to 100,000 watts, an unfortunate circumstance which was a factor in the ultimate demise of the Observatory by Harvard.

Two notes of interest were the sighting by Ken Thompson of a bobcat on the southwest promontory of the summit on 17 July, and on 28 November the occurrence of a severe southerly gale in which a maximum gust of 40 m sec.$^{-1}$ (89 mi. hr^{-1}) was recorded. No damage was caused.

In December, what became known as the Blue Hill Collection was shipped to the National Weather Records Center (NWRC) in Asheville, North Carolina. This collection included all of the autographic bound records made at Blue Hill for the period 1885-1953. It was thought that the last five years of records, 1954-58, should be kept at the Observatory for reference purposes by the Weather Bureau if they took over the observing program. Included in the shipment were the observation books and autographic records from Mount Washington Observatory for the

Figure 143. Summit personnel, 1958. Left to right, standing: Kenneth G. Thompson, Charles Cunniff (USWB), Ralph Newcomb, Richard M. Goody, Sally Wollaston, Lorna Ridley, Ralph Donaldson (GRD), Edwin Kessler (GRD), Albert Chemla (GRD), John Conover. Front row: Rudolph Loser (GRD), Lillie Hodges, John Hamilton, James Walker, David Atlas (GRD), and Roland Boucher. Courtesy S. H. Wollaston.

period 1932-58. The Blue Hill observation books were retained at the Observatory.

The Blue Hill observation books, or journals, and instrument books that contained calibration data and exposure notes through 2 December 1958 had been microfilmed at Harvard to form nineteen reels. A copy was retained at Harvard but it cannot be located; however, the original was sent to NWRC where it was received 2 May 1961. Content of the nineteen reels is given in appendix A, "Climatic and Other Records of the Observatory." Because the autographic records required so much time and money to microfilm, they were never filmed as originally planned.

In 1965 the Weather Bureau became part of the new Environmental Science Services Administration (ESSA). The NWRC was concurrently transferred from the Weather Bureau to the Environmental Data Service, a new element of ESSA which was designed to be the repository for all environmental information. In 1970 ESSA was renamed the National Oceanic and Atmospheric Administration (NOAA) and the Weather

Bureau became the National Weather Service. While these reorganizations took place, the records remained in Asheville.

In October 1975 the National Archives and Records Service (NARS) sent representatives to NWRC to inspect the facility and operating procedures. The autographic records from Blue Hill and Mount Washington were singled out as having little reference value, because they had never been used, and it was suggested they be shipped to Washington, D. C., for NARS appraisal and possible transfer to the National Archives.

During the spring of 1976 the autographic records were moved to the NOAA Records Holding Area in Rockville, Maryland, where they were cataloged for the first time by NOAA analysts, and in July 1976 they were offered to NARS.

Another collection of old meteorological journals from various locations was also moved from the Observatory. In January 1977 they were added to the collection at NARS.

In May 1977 NARS evaluated the entire collection and decided that only the very old journals should be retained. The remainder, some 125 cubic feet, was authorized for disposal. However, NOAA was reluctant to do away with the records. Accordingly, several scientists, including J. Murray Mitchell, Jr., and Norman L. Canfield, both of NOAA, and Helmut E. Landsberg, former chief of climatology for the Weather Bureau, were consulted as to whether the records should be saved. Their opinion was that they should be saved.

Meanwhile, Alexander McKenzie, now a trustee of the Mount Washington Observatory, learned of the situation. He and his wife went to Rockville and sorted through the material to rescue the autographic records and observation books that came from the mountain and its satellite substations. These records, along with their accompanying journals and observation books, are now on deposit in the White Mountains Collection at Dartmouth College, Hanover, New Hampshire.

To resolve the dilemma of the Blue Hill records, NOAA petitioned NARS to reevaluate its decision. Dr. Sharon Gibbs Thibodeau performed a detailed evaluation, which included favorable correspondence from Goody, and suggested that the records be permanently located at the Federal Archives Records Center in Waltham, Massachusetts. In view of her suggestion, the records were moved again to take up permanent residence, beginning 26 February 1980, in Waltham.

An inventory of the autographic and a few tabular records at the Waltham archives is given in tabular and graphic form in appendix A.

A copy of the observation and instrument book microfilms now accompanies the autographic records at Waltham. This greatly facilitates use of the data.

The five years of autographic charts, 1954-58, that were left at the Observatory, appear to have been lost. There is no record of their ever having been received at NWRC.

The very old meteorological journals were ultimately transferred to the National Archives where they now form part of Record Group 27, Records of the U.S. Weather Bureau.

A listing of these journals and a note on how they may be accessed is given in appendix A, part 4.

Correspondence between the Weather Bureau and Goody was initiated in an effort to have the government take over the climatological program. Landsberg, director of climatology at the bureau, was much in favor of continuing the climatological record, but the bureau found it very difficult to obtain funds for the undertaking. Goody set a rental fee, based on costs of operating the Observatory and the portion proposed for use by the bureau, of $3,000 per annum. The bureau simply could not meet this cost plus that of observers and equipment. It explored the possibility of telemetering data to Boston. This would require an initial outlay of $15,000 and estimated annual expense of $2,900. This was not accepted because of the high cost and fact that all the data that was desired would not be transmitted. Another scheme relied on Weather Bureau observers commuting to the hill from Boston or the airport and continuing use of the Observatory instrumentation. This idea was also abandoned.

1959

Early in the year Landsberg asked that a GS-5 position, having a salary of about $5,000, be created for the observational work and that the instrumentation be gradually replaced by standard Weather Bureau equipment. Meanwhile negotiations concerning the rental fee continued. Not until May was agreement reached, with the fee set at $2,400 per annum.

A twenty-five year lease was drawn up to provide the use of one floor of the tower, space on the tower roof for wind and solar instruments, and the enclosure area for the thermometer shelter and precipitation gages.

On 1 July the Weather Bureau officially took over operation of the climatological station. It had operated without a break, independent of the government, for seventy-four years and pioneered in many observational techniques.

James Walker, observer for Harvard, continued as a part time observer for the Weather Bureau. The new program consisted of one check observation per day at 0700 EST, at which time the charts were changed. An observer was on duty three or four hours per day weekdays and even less time on weekends.

The official record of average temperature was changed to one-half the midnight-to-midnight maximum plus minimum temperature to conform with Weather Bureau practice. However, for a while the average of the 0200, 0800, 1400, and 2000 EST hourly temperatures were computed. This average agrees closely with the true average of 24 hourly values. The Blue Hill observation books were used through December 1964.

On 27 July Robert Demaris became the official Weather Bureau observer; David Doe filled in as relief observer until September.

Over the period 17-28 August the station was moved from the first to the second floor of the tower. The standard barometers were raised 2.86 m (9.4 ft.) above their former elevation.

The Draper mercurial barograph, anemoscopes, and cinemograph were not transferred. For a while these instruments were used but they soon required attention and gradually fell out of service.

In October the Weather Bureau installed an "operations recorder" in a plywood panel set up in the northwest side of the second floor room. Eventually the Conover precipitation gauge, Foxboro temperature–dew point recorder, and the Aerovane recorder were added to this panel. The old masts on the tower were removed and a single mast near the center of the tower was erected to support the wind sensors. Their height above the tower was unchanged. Since Goody desired a clear view of the sun at all times, the university performed the work.

Meanwhile, in March, the records of diffuse radiation and radiation on a south vertical surface, which had been maintained by the Weather Bureau, were terminated. The occulting ring, which was used to obtain diffuse radiation by shadowing a pyrheliometer on the tower, was removed.

On 31 March the detailed cloud observations, which had been started by Brooks in 1932, were also terminated. Hourly cloud and visibility observations from 0700 to 2000 EST and a record of the beginning and ending of fog also ended. The Air Force radar operators were to note thunderstorms and hail.

The room over the shop, which had been occupied by the bureau's Solar Radiation Supervisory Station, for $700 per annum, was emptied on 3 April, and the station closed. Cunniff took a position with the Geophysical Research Directorate. The total radiation recorder was returned to the first floor observers' office and subsequently installed in the panel in the second floor office.

In May the upper bedroom was made into an office for Goody.

The tower roof was covered with another layer of felt paper and tar and a wooden slated floor was added in May.

A party was held at the Observatory for John Conover on 2 June, preceding his resignation, effective 30 June. He was presented a 35 mm camera, projector, and screen. About 28 persons attended. Conover first came to the hill in 1935 and was employed there, except for periods of "on loan," continuously from 1939 to mid-1959. Under the tutelage of Fergusson, Brooks, and others, and with the library at hand, he had acquired a tremendous amount of general and scientific knowledge without obtaining formal college degrees. He took a position in the Satellite Meteorology Branch of the Geophysical Research Directorate where his study of clouds could be continued on a grand scale.

In late June the contents of the temporary building in the back yard were removed and most of the equipment, which belonged to the Air Force, was transferred, for Conover's use, to the Geophysical Research Directorate. The building was disassembled shortly afterward.

A dinner party was given for Ralph Newcomb, at the Conovers, on 10 September. His wife, Newcomb, Wollaston, Ridley, and Winthrop and Carlson, and Mrs. Conover attended. Ralph Newcomb, who had previous meteorological experience with the Air Force, filled many needs at the Observatory. He was familiar with the weather codes, could plot weather data rapidly, and adapted quickly to new work. He helped in all phases of the Cloud Photo project and became an expert in the darkroom. He had resigned in July to take a position under Glaser at Allied Research Associates, Inc.

Kenneth Thompson resigned in October, but continued instrument work for the Air Force Weather Radar Group from his home in Maine.

Sterling Price Fergusson died 16 November at age 91. The writer drove him up the hill in May for the last time. Although some of the physical changes bothered him, they were accepted as "progress" toward the long-time purpose of the Observatory—to explore the upper atmosphere. Fergusson was best known for his mechanical ingenuity. He designed and constructed the meteorographs which were lifted by kites from Blue Hill. He is credited with the invention of the recording weighing precipitation gage, which is used in all parts of the world and at most stations in America. He experimented and tested new designs of anemometers and hygrometers and specialized in developing sensitive anemometers.

Fergusson was a member of several meteorological expeditions to observe meteorological effects of solar eclipses and the weather in Greenland. He was an honorary member of the Mount Washington Observatory, a member of the American Meteorological Society, an associate fellow of the Institute of the Aeronautical Sciences, and a fellow of the American Academy of Arts and Sciences. By all who knew him, he will be remembered for his quiet way and the polite manners of the perfect southern gentleman.

By year's end a photometer and amplifier had been assembled by Volz and Goody and measurements of twilight commenced at the Observatory.

Financial reserves continued to increase, mainly from contract overhead charges; a gift of $10,000 was received from the Higgins Fund and other gifts totaled $3,405.

1960

Cynthia Peterson was given a research assistantship to work with Goody on the measurement of nitrous oxide. The research staff was completed by the addition of John Noxon, who was appointed research fellow. His work involved the measurement of airglow, a third project to be undertaken at the Observatory.

Over the academic year 1959-60, three generous gifts were received. The National Science Foundation made a grant of $172,200, to be used over a period of three years, to support the research program. Other grants amounting to $122,000 were to support activities in Cambridge. The Storrow and Rotch families gave a total of $6,750 to the Meteorological Professorship Fund.

At this time the corporation approved the appointment of Goody to the newly created Abbott Lawrence Rotch Professorship of Dynamical Meteorology.

Volz published two interesting papers on rainbows in 1960 (see bibliography). One of these describes the rainbow in sea spray plus rain compared with that from rain alone. Due to the difference in refractive index between salt water and pure water, the sea spray bow was slightly displaced from the rainwater bow. The other paper explained the observation of a supernumerary rainbow high in the sky and not near the ground to undistorted small droplets at high levels and distorted larger drops near the ground. The larger droplets, which were distorted by their fall through the atmosphere, reflected the sunlight at different angles, thus preventing the formation of a supernumerary bow.

In March the old Draper mercurial barograph which had been in use since 1885 and the anemo-cinemograph or its duplicate, in use since 1891, and other instruments were taken to Harvard by Wheatland for historic purposes.

In August the Weather Bureau installed their standard bright sunshine sensor which was connected to an operations recorder. The sensor employed a photo cell which was elaborately baffled to allow for differing intensities of light as the sun changed elevation. The installation was for comparison with the Campbell–Stokes recorder, but results of the comparison are not known.

On 12 September Hurricane Donna passed northward to the west of Blue Hill. The maximum wind speed (over the shortest time for the passage of one mile of wind) was 41.1 m sec^{-1} (92 mi.hr^{-1}) from the south–southeast, and the peak gust reached 62.6 m sec^{-1} (140 mi. hr^{-1}). This was the second highest wind gust on record at the Observatory. Rainfall amounted to 7.09 cm (2.79 in.) with very heavy rates occurring between 1300 and 1500 EST.

In November George V. Fortin was made official-in-charge of the Weather Bureau Station.

1961

Work on the research projects at the Observatory progressed steadily. Dodd, a student, became interested in the airglow measurements and he attempted to measure turbulence in the emissions. Goody now spent about two-thirds of his time on the hill.

The workshop was re-equipped with a new lathe, milling machine, and a number of other essential tools. P. Venezia served as machinist.

In November the weather radar group of the Air Force left the hill. The equipment, including the tower in the center of the hilltop, was moved to property owned by the U.S. Army Quartermaster in Sudbury, Massachusetts. New quarters were erected at that site for the personnel.

1962

Renovations were finally drawing to an end. The library had been emptied and was in use as a laboratory, the kitchen and dining room had been modernized, and the varnished wood in the hall, stairs, and east wing was painted. An interior wall of wallboard had been added to the first floor of the tower, the radiators were enclosed, and new low cabinets were installed along the northwest and northeast walls. With exception of the third floor all the floors were covered with vinyl tiles. The furnace was enclosed and the exterior chimney was sheathed in copper. The basement of the west wing was made into a seminar room with a small darkroom to one side. A floor plan of the Observatory at this time is shown in figure 144. Upon completion of this work a restricted list of invitations was issued for 25 May, on which day thirty visitors came to inspect the building.

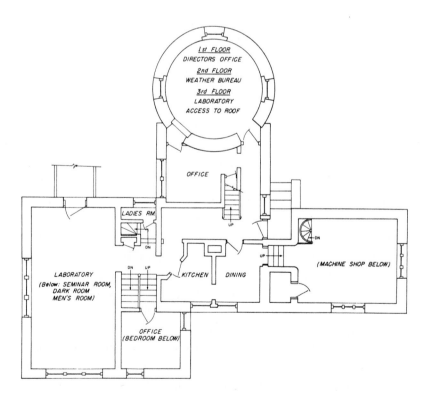

BLUE HILL   OBSERVATORY

Figure 144. Floor plan of the Observatory, 1962. Copyright President and Fellow of Harvard College. Buildings and grounds.

Figure 145. Photometer used to measure the twilight through five color filters. Courtesy F. Volz.

Volz completed his research on the twilight in February and returned to Germany (see bibliography, for 1962). Measurements were made with a photometer (figure 145) aimed 20° above the horizon in the direction of the sun at sunrise or sunset. Each period of observation began with the sun approximately 16° below the horizon and ended with it 3° above the horizon. Five color filters were used. Observations were restricted to clear sky in the direction of the sun from Blue Hill, but it was difficult to determine the amount of cloudiness along the relevant sunrise or sunset line which extended about 1000 km (621 mi.) over the horizon. However, techniques were found to allow for small amounts of cloudiness or signal bad data due to larger amounts of cloudiness along this line. Profiles of turbidity were derived up to 65 km (40 mi.) which showed significant amounts of dust at all levels. A wintertime maximum was revealed and the maximum found near the 20 km (12.4 mi.) level agreed with other data.

A theory of Bower's presumed that rainfall was enhanced as a result of meteor shower activity and subsequent increase in cloud nuclei. However, lack of correlation between meteor shower activity and the new dust content measurements dispelled the theory. The series of measurements which took place from September 1959 to November 1961 served well as a basic background distribution of dust, since over that period no volcanic aerosols had been ejected into the atmosphere. Whatever may be the ultimate consensus on the relationship between volcanic events and climate change, there is no doubt that it will be partly based on the extensive optical measurements of Volz, developed from the work performed at the Observatory.

Noxon and Goody reported the first observation of radiation by atomic oxygen at 6300 A and 6364 A in the day airglow (see bibliography). Figure 146 shows this historic measurement. The scan lines were made up of twenty independent recordings in the two regions of the spectrum. Detection of these small amounts of energy in the vastly more intense continuum due to scattered light was no small feat. The instrument was directed at an elevation angle of 45° and at right angles to the sun. The achievement then allowed the investigation of airglow from daytime, through twilight, and into the night. Similar observations were continued at Churchill, Manitoba, in November and December. There Noxon successfully measured oxygen at 5577 A and 6300 A during the day when the sun's elevation was 10°.

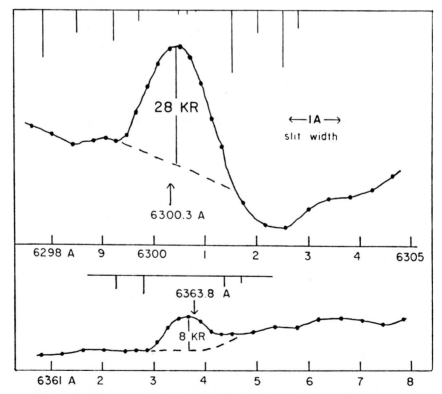

Figure 146. The first measurements of the dayglow at the 6300 and 6364 Angstrom 01 lines. J. Noxon and R.M. Goody, *J. Atmos. Sci.*, 19, p. 342, 1962.

Figure 147. Siderostat erected at the rear of the west wing in 1962. Entrance to the staircase was gained through a small door that had been placed in a window frame on the second floor. Tube into second floor has been removed. 1984. Courtesy J.H. Conover.

Within the year a siderostat, or solar tower, was completed for Peterson's work. This tower, shown behind the west wing of the Observatory in figure 147, collected the solar beam at its top, and by means of a motor-driven mirror directed the beam downward, and, with additional mirrors, into the upper floor of the west wing.

The three flat mirrors of thick glass were front surface coated at an expense of $1,500 each. On the upper floor of the west wing a large cement table about 20 cm (7-8 in.) thick was positioned on cement blocks to rigidly support instrumentation.

About this time machinist Venezia became ill and was forced to leave his job. William Albrecht, who had worked for Goody in Cambridge as a machinist, came to the hill in October to replace Venezia.

1963

Noxon continued the day airglow measurements of the 6300 A oxygen line with an improved scanning polarimeter. Large day-to-day and even hour-to-hour variations were found that did not seem possible to explain by photochemical processes alone. It was postulated that some of the variation arose from the excitation of oxygen atoms by electrons in the F region of the ionosphere. Night airglow measurements were made from an Air Force Cambridge Research Laboratory jet aircraft that flew along the 75th meridian from 55° to 85° north. Other measurements from a high-flying jet succeeded in recording the oxygen red lines during a solar eclipse.

Peterson's appointment terminated in September, but the program for measuring nitrous oxide was continued with the help of technical assistant Herbert McClees.

Adolph Johanssen was appointed Technical Assistant to Noxon. In July $18,720 was spent to restore the surface of the summit road and its drainage system. Forty percent of this expense was shared by WGBH. The Observatory had purchased a small vehicle and plow that was garaged in the back yard. The practice was for Albrecht or Noxon to plow the road, sometimes continuously during storms, to keep ahead of rapid accumulations of snow.

1964

Noxon was appointed senior research fellow without limit of time, and associate director of the Blue Hill Observatory. Goody wrote that the appointment "recognizes the need for more attention to be given to research at the Observatory."

During the year it was noted by the Weather Bureau that the Conover precipitation gage, the Weather Bureau sunshine sensor, and the Foxboro temperature and dew point recorder, were inoperative. With the exception of the sunshine sensor, these instruments were not in the bureau's inventory and therefore strange to their maintenance personnel. As a result they had ceased to work due to lack of care. Ultimately the Foxboro temperature and dew point recorder and the Foxboro recorder used in

Figure 148. Weather panel of instruments. Left to right: Aerovane wind speed and direction, total and normal incidence solar radiation, Foxboro temperature and dew point. Below, left to right: operations wind speed and direction recorder and atmospheric electricity. April, 1963. Courtesy F.W. Chapman.

the precipitation gauge were returned to the Foxboro Company. The precipitation gauge and its shield were taken by the writer. The bureau's panel of recorders at this time is shown in figure 148.

At the end of the year use of Brooks's forms on which the Bureau's observations were noted ceased.

1965

Beginning on 1 January weather observations were entered on Weather Bureau form number 610-10. One page of this form was used per day. On this form hourly and other weather data were tabulated.

Goody spent the fall term on sabbatical leave at the Kitt Peak National Observatory near Tucson, Arizona. There he revived his interest in planetary studies.

The addition of air conditioning in the summer greatly improved the working conditions in the laboratory at the Observatory.

About this time the spiral staircase that led from the upper to lower floors of the east wing was removed to provide additional space. The old staircase had posed problems for many years when instruments had to be carried from floor to floor. The stairs, shown in figure 149, were cast iron and the spindles between the steps and railing were beautifully cast in an

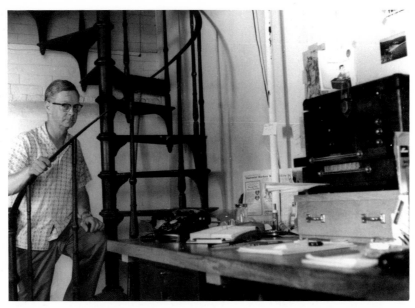

Figure 149. Mr. William Albrecht, machinist, and the spiral staircase leading from the shop to upper floor of the west wing. Courtesy W. Albrecht.

elliptical cross section. In its place the stairs that lead from the dining room
to the upper floor of the wing were divided to provide unsightly, narrow,
but straight, passages up and down.

It was learned recently that in 1974 a house for sale advertisement
made special note that it contained the Observatory spiral staircase.

1966

Noxon began observations of the emission lines in the solar corona.
His preliminary measurements indicated that data could be obtained out
to half a solar radius in the daytime. A more extensive program was
therefore started in collaboration with the Harvard College Observatory.

Measurements of the quantity of nitrous oxide in the atmosphere were
successful, but variations that might have led to conclusions about the
origin of the gas could not be established.

The method of measurement was spectroscopic. Light from the sun
that was brought into the laboratory through the siderostat was passed
through a spectrograph, separating a narrow range of wavelengths containing a fundamental absorption band of nitrous oxide. This band contains hundreds of lines overlain by hundreds of lines of other, unwanted
gases. A technique of pressure modulation was employed to automatically
compare the positions of the nitrous oxide lines with those in a laboratory

sample in a way that pays no attention to the unwanted lines of the other gases. The method was, therefore, selective for the gas contained in the laboratory sample.

The development of this pressure-modulated radiometer was an important byproduct of the work. Later, similar pressure-modulated radiometers were constructed at Oxford, England, and flown on satellites for high-altitude measurements of temperature and concentrations of a number of minor gases, both in the terrestrial stratosphere and on other planets.

Since the beginning of Goody's research at the hill the work was plagued by the intense electromagnetic radiation emanating from the WGBH-TV, FM, and state police antennas. Countless hours and considerable sums of money were expended in attempting to carry out the delicate measurements. The problem, illustrated by numerous field strength measurements, was discussed many times with WGBH personnel and finally a decision was reached to move the television station to an existing tower location in Needham. New measurements after the relocation showed that any improvement of conditions was offset by the FM station whose power was increased to 100,000 watts, and whose antenna was moved to the former TV tower. Discussion of the problem continued, since proper use of the Observatory remained in jeopardy.

1967

In the academic year 1967-1968 Goody was appointed Mallinckrodt Chair of Planetary Physics, thus allowing a new appointment to the Abbott Lawrence Rotch Chair of Dynamic Meteorology. His research interests by this time had shifted to planetary studies, which involved the Venus airglow and the meteorology of Mars. This work took him to the Mt. Palomar Observatory in California. In collaboration with the Goddard Space Flight Center, a probe to penetrate the Venus atmosphere was planned.

Noxon's analysis of the 6300 A twilight and night airglow now showed that a complete understanding could be obtained quantitatively in terms of direct solar effects combined with reactions on the ionosphere.

Weather observing equipment in use at Blue Hill in June was listed as follows:

Property of the Weather Bureau:

For solar　　　　　　　　Brown Electronic Multipoint Recorder

radiation:	Eppley total radiation sensor
	Normal incidence sensor with
	equatorial mount
Bright sunshine:	W. B. sunshine and precipitation
	indicator—not used as the official
	record.
	Campbell–Stokes
Wind equipment:	F104B 3-cup anemometer.
	F 005 direction transmitter,
	wired for four directions
	F 315 operations recorder,
	Esterline Angus Model AW

Not property of the Weather Bureau:

Aerovane system with
direction and speed
recorder

1968

Goody wrote in his report to the university's president in regard to the future of atmospheric sciences at Harvard (see bibliography for 1968):

The main achievement of the Physics of Atmospheres and Oceans group has been in the teaching and research carried out in Cambridge. Research at the Observatory has been of a frontline character but the limited nature of the site, the encroachment of greater Boston and the increasing interference of WGBH makes it increasingly difficult to do rewarding work there. No relief is expected in the future. On the other hand the work in Cambridge is slowly forming into a viable effort in the physics of the solar system. Joint plans between the Department of Geological Sciences and the Division of Engineering and Applied Physics have been formed to create a Center of Earth and Planetary Physics. Discussions are in progress with all interested parties to encourage these developments and to enable them to continue.

It was evident that research projects in which the Observatory was needed as a field station were coming to an end, especially in view of the interference from WGBH.

In July the Rotch monument, which was located a short distance from the Observatory, was cleaned and moved to a location inside the iron fence.

The move was necessary before vandals completely destroyed its engraving and embossed work. One of the most severe ice storms on record occurred on 13 March when 2.5 cm (1 in.) of ice accumulated on all surfaces.

As an ecological experiment the University of Vermont planted three Persian lilacs inside the enclosure in the spring. The flowers subsequently died, apparently due to extremes of cold, wind, and drought. Lilacs from the same clone were planted at about 150 sites in northeastern United States and 200 in Canada. Leaf and flowering dates were recorded and related to crop growth and climate.

1969

> On 7 April the corporation voted to authorize the opening of negotiations with a view to phasing out, over a period of approximately five years, Harvard's participation and interest in the operations carried on by the Blue Hill Observatory at the Great Blue Hill and further to authorize the use of endowment funds of the Blue Hill Observatory to recreate the Abbott Lawrence Rotch Chair in Atmospheric Science, to be administered by the Division of Engineering and Applied Physics, for use in connection with the Center for Earth and Planetary Physics.

A faculty committee was appointed to develop a program and raise funds for the Center for Earth and Planetary Physics.

Michael B. McElroy was appointed Abbott Lawrence Rotch Professor of Atmospheric Science. His previous research work was concerned with the physics and chemistry of the upper atmosphere of the earth and planets. McElroy's new work was to be performed in Cambridge.

When it came to the examination of papers pertaining to the Observatory property, it turned out that the structure, as well as the land, was a part of the 99-year lease which had been granted by the Commonwealth of Massachusetts after Rotch's death. This arrangement was identical to the original lease (see chapter 6). The Commonwealth was also not aware of this fact until it was pointed out to them. Termination of the lease, therefore, would return the entire property to the Metropolitan District Commission.

The Environmental Data Service, the government agency then responsible for the gathering of climatic data at Blue Hill, was duly notified of the Harvard plans. At the time about $13,000 per year was required to maintain the reference climatological station. This was composed primarily of rental to Harvard and the salary of a GS-5 observer. Immediate reaction in Washington was that the Observatory climatological record did not justify the expense of a salary plus maintenance of the building.

Meanwhile new snowfall records were broken in February. After an almost snowless season a storm on the 9th and 10th piled up 53 cm (21 in.) and on the 24th and 25th 71.6 cm (28.2 in.) fell in a twenty-four hour period to establish the all-time record for that time interval. Snow which fell from the 24th to the 28th amounted to 98.3 cm (38.7 in.), also the record for a single storm. Total snowfall for the month was 166 cm (65.4 in.) a new record for any month. Snow accumulated to 104 cm (41 in.) on the ground on the 27th, just 2.5 cm (1 in.) short of the 1948 record.

George Fortin suffered a coronary attack in June and the observing program was carried on by Fred Chapman, a former observer (1962-1964), Mary Earley from the Weather Bureau Airport Station in East Boston, and others. Fortin, after nine years of service at the station, retired on 31 October.

Just before Halloween, vandals split the back gate and scaled the tower, apparently from roof to roof. In the morning a flag was discovered flying from the anemometer mast. No other damage was sustained.

1970

On 5 January the Corporation voted upon the recommendation of the Dean of the Faculty of Arts and Sciences, to establish a Center for Earth and Planetary Physics. It was now expected that affairs at the Observatory would wind down rapidly and merge with those of the center.

Goody went on leave for the spring term at the Kitt Peak National Observatory where he continued his studies of motions of planetary atmospheres. The fourth Arizona Conference on Planetary Atmospheres was organized under his guidance.

Noxon continued airglow and auroral investigations at Blue Hill and from aircraft.

The National Weather Service, formerly the Weather Bureau, appointed William Cusick as official-in-charge of the station on 2 January.

On 17 May vandals demolished the standard rain gage, stole the Nipher shield surrounding it and the funnel from the weighing gage.

The gage and funnel were replaced but without a windshield. Vandals struck again on 2 September; in the process of scaling the tower they badly twisted the support for the Campbell–Stokes sunshine recorder. On this and on the previous occasion of climbing the tower, the vandals could have easily entered the building through the third floor trap door, but they did not.

On 22 May new forms, WB-B-16, were introduced for use by the station. Essentially they contained the same data as before on a single page for each day of the year. Monthly and annual summaries continued to be prepared for publication.

On 31 July this writer, who had heard rumors that Harvard no longer wanted the Observatory, wrote to White, an old friend and current administrator of the National Oceanographic and Atmospheric Administration. In that letter the poor maintenance of observing equipment at the Observatory was deplored, the monitoring of atmospheric pollution was suggested, and an answer to the rumor that the observational program might be terminated was sought. White replied that the rumor was indeed based on fact and budget austerity had forced a decision not to continue the observational program since the university could not pay for both maintenance of the building and the program. The establishment of a pollution measurement station was not seriously considered.

In a second exchange of letters Conover asked if NOAA would continue the observational program provided another Observatory landlord was found. White replied that they would be glad to continue under such an arrangement.

Correspondence then took place between Conover and J. Murray Mitchell, Jr., who, not knowing conditions of the lease, suggested that the takeover of the Observatory by a private organization might be the best solution. In a letter to the president of the American Meteorological Society, Mitchell urged the Society to take serious consideration of the problem. At this time there was concern over some of the endowment that had, for years, been used as the main source of funds for activities on the hilltop. By 18 September, through a discussion with Arthur Rotch, son of the founder and a great benefactor to an endowment fund, it was learned that the endowment problem had been addressed and settled some time earlier. Many years ago, before the Observatory had been founded, Harvard learned to prudently word its endowments, in effect, that allowed "the President and Fellows, in their opinion, should it be impractical or inexpedient, to continue as before, for funds to be used in another line of work as nearly as possible to the original." The problems of performing research on the hill were personally explained to Rotch and Storrow by Goody; furthermore, they understood that Harvard was not in the business of collecting climatological data. Reluctantly they agreed to allow the endowment income to be used in Cambridge, thus preventing any possible controversy.

It was then discovered that organization of a new foundation to carry on the Observatory would be a problem because the 99-year Metropolitan District Commission lease to Harvard was nontransferable. To obtain a new lease with a private organization would also be difficult; in fact, Abbott Lawrence Rotch arranged to have Harvard lease the land for him in 1896 because of that problem.

Prospects of continuing the observational program were bleak; nevertheless, a meeting was held on 8 October to discuss possible uses of the Observatory and to search for a new landlord. Those present were: Father Skehan, director of the Environmental Center at Boston College and assistant director of the Weston Seismological Observatory; Edward Brooks of Boston College; Wallace E. Howell of W. E. Howell Associates; James Mahoney of the Harvard School of Public Health; and Conover of the Air Force Cambridge Research Laboratory. Subsequently the Environmental Center and Harvard School of Health declined use of the Observatory. A consortium of schools had already been funded to study air pollution and they too declined. MIT and the American Meteorological Society were contacted again but little encouragement was found.

1971

In January Edward Brooks and Conover contacted the Director of Laboratories for the Division of Engineering and Applied Physics at Harvard to learn exactly when the Observatory would be given up. They learned that Harvard planned to vacate about October 1971. They also learned that NOAA was assuming it would cost $50,000 per year to maintain the Observatory, that Harvard would willingly transfer the building, probably for a $1 fee, and that Commissioner Sears of the Metropolitan District Commission would like to see an activity continued in meteorology or one having an historical aspect. It was also found that the Museum of Science was not interested in an acquisition.

Spengler of the American Meteorological Society, Brooks and Conover then planned to have David Ludlum, editor of *Weatherwise*, place an announcement in his journal in which the situation and need for a landlord were explained. An article was promised for the coming May issue. Finally, it was decided, as a last resort, to attempt to reverse NOAA's decision. A letter of solicitation, copied here in full, was sent to sixty-three meteorologists, twenty-two climatologists, and six friends of the Observatory. Nine of those solicited lived abroad.

15 Nobel Road
Dedham, MA 02026

Dear

The eighty-five year climatological record at the Blue Hill Observatory will end this coming fall if present plans are consummated.

Presently the National Oceanic and Atmospheric Administration (NOAA) rents space from Harvard University for personnel and equipment used to obtain the observations. Harvard plans to vacate the building in October 1971 and NOAA has stated that it cannot carry the expense of the building to maintain the record although it would continue observation-taking under a new landlord. Such a transfer of the building is feasible at token cost. The Metropolitan District Commission, upon whose land the Observatory stands, looks favorably upon a new lease for the continued use of the building.

Edward M. Brooks of Boston College; Kenneth C. Spengler, Executive Director of the American Meteorological Society, and I have been exploring various means of providing for the continuation of the station. We believe two possibilities exist; one to find an interested group, probably consisting of one or several universities, to pool together to keep the Observatory as a facility for their work and thus provide a new landlord for NOAA. The second possibility is to convince NOAA that termination of observations would be a grave mistake for the science of meteorology in view of the expenditure involved. This needs to be done even while pursuit of the first possibility is actively underway. We believe the second possibility is not unrealistic for two reasons: (1) NOAA's cost estimate for the future operation of the entire facility was reportedly $50,000/year while the actual cost is about $9,000 for building and road maintenance plus observer salaries of about $12,000/year and equipment maintenance. Further more, if NOAA were to acquire the facility they could then operate it for another $6,000/year, since their rental fee is $3,000/year, and continue the observations as well as new work that might be developing. (2) The scientific community did not have an opportunity to express its concern. Toward this end I ask you, as a leading scientist in this field, to write to NOAA and urge continuation of observations indicating what utilization you have made or what importance you place in the observations.

Some man-made modification of the climate at the Observatory may be taking place as the suburban sprawl spreads around the Blue Hills Reservation. However, the Observatory's location, 400 feet above the surrounding lowlands and in the midst of a permanent parkland covered by trees, is unique. Other sites having comparable long records and exposure are few or non-existent in the northeastern United States.

We also believe that the site should prove very useful for measuring atmospheric pollution at the northern end of the northeast megalopolis. In-city measurements could then be referenced to background or benchmark values obtained at the Observatory. In any event, your support is needed now, even if only to convince NOAA to continue without allowing a break in the records until a more permanent arrangement can be worked out.

Of course, if you or your colleagues might have use for the facility, this should be made known to us.

For the most effective impact from your letter, we believe it should be mailed to arrive in Washington, along with the others, over a relatively short interval of time. It is therefore suggested that your letter arrive before March 31, 1971.

Please address your letter to:
Dr. Robert M. White, Administrator
National Oceanic and Atmospheric Administration
Washington, D.C. 20230
A copy sent to me would be much appreciated.

Very truly yours,

John H. Conover

JHC:mp

Immediately after the letters were sent, White was informed of the activities that finally led to the world-wide appeal for continuing the climatological record. As noted, a solution by means other than a reversal of NOAA's plan would continue.

White received forty-one letters. All urged continuance of the climatological record; some were harshly critical of the government's decision. Others came from former professors that White had in school. Although White divested himself from the issue, because he held membership on the Harvard Visiting Committee to the School of Applied Physics and Engineering, a recommendation to continue the Blue Hill climatological record was soon made. The Observatory was to be administered, as before, by the Boston office of the National Weather Service. The decision was contingent upon the Metropolitan District Commission's taking over the building and negotiation of a reasonable lease.

At this point the announcement that was to appear in *Weatherwise* was postponed.

In May Brooks, Howell, Storrow, Mahoney and McKinny, of Harvard, Commissioner Sears and his staff of the Metropolitan District Commission and Conover, met at M.D.C., headquarters. Issues reviewed were: (1) Why the records should continue, (2) the concerns of the National Weather Service, (3) why the site should remain unchanged, (4) and the need for protection from vandals. The Commission agreed to cooperate and outlined a plan for making the Observatory a part of the Trailside Museum, which operated successfully at the base of the hill. They hoped to use a part of the building for lectures and demonstrations and to show the weather station. Plans were made to have a resident caretaker.

The Metropolitan District Commission accepted the Observatory from Harvard on 1 October. A 25-year lease was drawn up with the Weather Service in which the service paid $200 per month, or the same amount as it paid Harvard, for space in and on top of the tower as well as the small room adjacent to the tower. The rate was made negotiable every five years. On 27 September Richard Banks moved in as caretaker. The crisis was over and the climatological record continued without a break.

Until this time in 1971 Noxon had been winding down research activities and emptying the Observatory. Goody noted the following in his report to the President:

> It is a pleasure to pay special tribute at this time to the role which John Noxon, as Assistant Director of the Observatory, has played in the concluding years of the Observatory. Because of my duties in Cambridge, Noxon has borne much of the responsibility of administering the Observatory. In addition, he succeeded in these final years of the Observatory's operation at Blue Hill in gaining for the Observatory the reputation of one of the world's prominent observing stations for upper atmospheric phenomena.

Noxon was a pioneer; his work had made him a world authority on the airglow and chemistry of the upper atmosphere. This new knowledge was gaining importance in understanding changes in the upper atmosphere that were introduced by human activity, and in estimating the effects of these changes on our planet's surface. After leaving Blue Hill, Noxon continued his research at a NOAA laboratory in Boulder, Colorado. There he turned his attention to the issues of global pollution. He mapped the human impact on the atmosphere over a large area of the earth from the remotest regions of the Antarctic to the Arctic Circle. His untimely death came in 1985 at age fifty-seven.

On 2 May vandals struck again. The tower was scaled and the glass sphere of the Campbell–Stokes sunshine recorder was stolen. The instrument's frame was found undamaged on the roof. The instrument was recommissioned on 13 July.

It should be noted that over the past 13 years many activities listed administratively under Blue Hill Observatory took place in Cambridge. Goody maintained an office in Pierce Hall and from there he directed the teaching program while newly appointed professors, their assistants, and graduate students carried on their research in close proximity to other Harvard University facilities. The division of work between that which took place in Cambridge and at the Observatory is difficult to delineate, but an attempt has been made to set down only the activities that took place at the Observatory.

It is only fair to mention that Goody was successful in his effort to present atmospheric science at Harvard as an applied branch of traditional disciplines, and in that way he brought into the field mathematicians, physicists, engineers, chemists, and even some biologists. Important use was made of the Observatory, but the problem of electromagnetic interference made it all but impossible to continue experiments. Admittedly, new projects that might have been introduced there were becoming difficult to identify.

Compensation for loss of the Observatory came in the form of increased space at Pierce Hall. Also at that time Goody was appointed to a Mallinkrodt Chair in Planetary Physics, as already pointed out, and the Observatory endowment was applied to a new appointment in atmospheric science. At the same time a new entity was created to continue this new chair and thus, in a sense, to be the successor to the Blue Hill Observatory; the new entity was the Center for Earth and Planetary Physics.

Goody's success is shown by the fact that by 1984 the university supported five chairs in related areas of atmospheric and ocean science. Furthermore, fourteen senior appointments, including renewal of some that had previously expired, in atmospheric sciences and physical oceanography were made over the last twenty-five years.

Chapter XV

The Observatory and the Metropolitan District Commission: 1971–1985

Throughout this period numerous Metropolitan District Commission employees, either singly or as couples or families, served as caretakers at the Observatory.

For the next nine years few notes were logged but the climatological record was faithfully maintained and published, as before, by the Environmental Data and Information Service.

1972

The instruments were stolen from the thermometer shelter in April but they were replaced immediately.

1973

On 27 September a "wind accumulator" was installed. The equipment, which was not made operative until October of 1975, was provided for all reference climatological stations. It consisted of a cup anemometer and wind vane wired to a small computer and counters for the four cardinal points of the compass. Batteries ensured continuous operation. The counters were read daily at 0700 EST. From these data daily values of mean speed, resultant direction and speed, and other statistical parameters of variance could be obtained. The thermometer shelter was vandalized again in October. Conover removed his rain gage and the pole that supported the dewcel in the enclosure.

1974-1975

No significant events were logged during these years.

1976

On 16 February the tail of the Aerovane broke, apparently a result of corrosion, since winds were light at the time. This ended the Aerovane record of wind speed and direction.

Over the period 17-19 April, unusual early-season, record-breaking heat of 30.6°, 33.3° and 34.4°C (87°, 92°, and 94°F) was recorded.

A new system for recording wind gusts, the F-420-C, was installed in June. In this system output from a magneto attached to a three-cup anemometer was recorded on a strip chart.

In December an M-5130 integrator was installed. This device totalized both daily solar radiation on a horizontal surface and normal incidence solar radiation.

1977

On 10 May the heaviest late-season snowfall of record occurred. 20 cm (8.0 in.) fell on the summit.

In midsummer it was discovered that an additional correction of -0.3 mb had been applied to all station barometer readings instead of a +.33 mb correction since the barometers were moved to the second floor. Therefore, a correction of + 0.6 mb (0.33+0.3) was applied to readings taken between 27 August 1959 and 1 August 1977 to correct the data to the former first floor level.

By December it was noted that electromagnetic radiation from the TV station was causing many problems with the solar radiation integrators.

Relief observer Fred Chapman, resigned in October to join the Navy. He had served from January 1962 to October 1964 and from June 1969 to October 1977.

1978

The great "blizzard of '78" struck on 6-7 February. In thirty-three hours, beginning about 1030 EST on the 6th, 76.5 cm (30.1 in.) of snow fell, leaving 84 cm (33 in.) on the ground at 0700 EST on the 8th. Seven and one-half cm (3 in.) of this lay on the ground before the storm. The fastest one mile wind passage was 30 m sec^{-1} (67 mi. hr^{-1}). During the evening of the 6th, 35.8 m sec^{-1} (80 mi. hr^{-1}) gusts from the northeast were estimated. Roads were closed and the observer was unable to reach the Observatory on the 7th.

In May William Cusick retired and Robert Skilling, who had served as relief observer since October of 1960, was appointed official in charge. Richard Wilson filled in weekends.

In the fall, Conover sent a letter to George Cressman, director of the National Weather Service, outlining plans for a weather museum at the

Observatory. Conover, in this case, represented The Friends of the Blue Hills, an organization devoted to preserving the reservation and opening its facilities to the public. Cressman endorsed the plan; however, he was careful to note that changes on the summit that would affect the instrument exposures were unwanted.

Also in the fall much-needed repairs were made to the building; trees were cut southeast of the Observatory and around the enclosure and the old fence around the former Air Force plot was removed.

On 11 December an Alter wind shield was installed around the standard precipitation gage.

In the spring the enclosure thermometer shelter underwent extensive repairs. Rotted legs and braces were replaced to restore it to its original condition.

Tropical storm David passed to the west on 6 September. A peak wind gust of 34.9 m sec^{-1} (78 mi. hr^{-1}) was recorded at noon.

On 10 October 17.3 cm (6.8 in.) of snow fell on the summit making it the heaviest snowstorm on record so early in the season.

1980

The Campbell–Stokes sunshine mount was repaired and the normal incidence solar radiation sensor moved from the top of the south parapet railing to its original position at the side of the parapet. It had been shadowing the total radiometer at low sun in summer.

On 27 September the Observatory, thanks to efforts of the Friends of the Blue Hills, was made a National Historic Site. An open house was held at the Observatory to celebrate the occasion. Among those present were Roland Boucher, John Conover, Fred Lund, Edward Brooks, Rod Winslow, meteorologist-in-charge at Boston, and Robert Skilling.

The Metropolitan District Commission placed a new copper roof over the stairway leading to the second floor of the tower. Severe leaks had caused much of the ceiling to fall and some of the floor to rot.

Husar Rudolph (1981) of Washington University in St. Louis published an interesting study of visibility at Blue Hill. This record he found to be the best and longest in the country. He had studied visibility trends throughout the eastern United States; after locating the long record of the visibility of mountains from Blue Hill, a special analysis of the data were made. He found minimums of visibility in the decades 1910-20 and 1940-50. Although a definite cause–effect relationship was not established, he noted that during both of these periods the consumption of wood and coal peaked under the stimulus of wartime activities. From 1959 to 1980 these "mountain" observations were made only at 0700 EST. Beginning in 1981, at the suggestion of Mr. Rudolph, a second daily observation was resumed; first at 1100-1200 EST and later between 1300 and 1400 EST, corresponding to the earlier time.

During the year, William Minsinger, a young medical doctor, amateur meteorologist, and native of Milton, made several trips to the Observatory. He began to think seriously of restoration of the structure and forming a club where amateurs could get together and discuss the weather.

By December the solar radiation records were completely abandoned because of the great loss of records from electromagnetic interference. Unfortunately, the old strip chart records, which were satisfactory, were not reinstated.

A severe cold front Christmas Eve ushered in temperatures on Christmas Day that broke all records for that date. The maximum during the day was -19.8°C (-3.6°F) with a windchill factor as low as -54.4°C (-66°F).

1981

For a thirty-day period starting on 20 December 1980, the temperature averaged -9.6°C (14.8°F), one of the coldest thirty-day periods on record at the Observatory.

On 13 February the barometric pressure, reduced to sea level, reached 1052.4 mb (31.08 in.) a new record high at the station.

Early in April Minsinger installed a window shelter, which he had made from shutters, at the second floor north window of the tower. The shelter was much smaller than the old shelter that was removed about 1959, but served well to expose a few thermometers. At this time interest between temperatures measured at the window and enclosure shelters was re-kindled and a comparison was initiated. With the increasing height of hilltop vegetation, it was thought that differences between the two shelters might be greater than when less vegetation was present. It was also thought that the new temperature comparison might help to explain the fact that the Blue Hill average temperature, which is taken from the enclosure shelter, appeared to be increasing slightly more rapidly than at surrounding lowland stations.

In May a meeting was held at Minsinger's home to discuss the formation of a weather club. In addition to Minsinger, Robert Skilling, Richard

Wilson, a weekend observer, and Jack Anderson, a nearby cooperative weather observer, were present. The Blue Hill Observatory Weather Club and Museum was established the following month and the first meeting was held 21 September. The speaker at the first meeting was Don Kent, long-time radio and television local weather forecaster. Later nearly 3000 charts suitable for plotting the positions of hurricanes and showing the tracks of some of the most famous hurricanes in the Caribbean Sea and eastern North Atlantic Ocean were distributed with an announcement of the club's formation and its five-dollar membership fee. In response to the mailing, membership tripled to about 160.

In August a club member succeeded in having the Boston Edison Company donate and install a floodlight to illuminate the building throughout the night. The light was placed on an existing pole located west of the tower.

On 12 October Minsinger instituted live radio broadcasts over WJDA, Quincy, of the Blue Hill weather at 0855 and 1210 local time Monday through Saturday. For this ongoing program the brief talks were given by the observer on duty.

The first open house, sponsored by the club, was held on 14 November. Over 200 people climbed the hill to inspect the Observatory.

Objectives of the club were now defined in an ambitious list as follows:

1. Refurbish the Observatory.
2. Establish an amateur radio repeater station in the 145-147 MHz range. (This became impracticable due to radio interference.)
3. Establish an observer group that would send their observations to the Observatory.
4. Publish a newsletter.
5. Hold meetings at least quarterly and include guest speakers.
6. Establish a museum at the Observatory.
7. Investigate fund-raising techniques.
8. Publish a list of weather broadcast radio frequencies.
9. List publications that might be of interest to club members.
10. Seek group discounts for weather instruments.
11. Gather reports of acid rain.

1982

On 12 March the Weather Club and Museum was incorporated. Membership was classed as regular, student, contributing, patron, or corporate.

On 6 and 7 April, 35.6 cm (14.0 in.) of snow fell, making it the heaviest late-season storm on record. Record cold accompanied the storm with a maximum temperature of -6.1°C (21°F) and minimum of -10.6°C (13°F) on the 7th.

From 29 June until early September the club exhibited several instruments for obtaining meteorological data at Boston's Museum of Science; Weather Services Incorporated, a private organization, graciously supplied a live readout of the Chatham, Massachussetts, weather radar, a terminal showing weather reports and a facsimile machine that produced weather maps.

As a part of the fiftieth anniversary of the Mount Washington Observatory, a radio link on 6 meters was set up by David Doe, club member, between Mount Washington and Blue Hill. Alexander McKenzie and Al Oxton supervised activities on Mount Washington. Weather observations and reminiscences of the past were exchanged. From the time when the Weather Bureau took over the observations until November, no corrections were applied to the cup anemometer records. Beginning in November corrections, according to the following table, were applied to the "fastest mile" speeds:

1-25 mi. hr^{-1}	0
27-35	- 1
36-44	- 2
45-52	- 3
53-61	- 4
62-70	- 5
71-79	- 6
80-87	- 7

Average winds continued to be logged without corrections.

1983

The auxiliary power generator that was installed by Goody had been out of service for many years but much to the benefit of the caretaking family and observational program, it was completely restored for emergency use.

Repairs were made to the heavy iron gates in the fence and club members, on 29 October, caulked the flashing between the tower roof and parapet.

In November the Weather Club became associated with *The American Weather Observer*, a newly formed group of amateur weather observers.

During the year a representative from the Leeds and Northrup Company came to study the wind measuring systems and a student from the University of Massachusetts worked on methods of reducing barometric pressure to sea level.

In December a telephone answering service was installed. A recorded message included climatic data and the latest weather observation.

In July 1983 Minsinger and Conover went to the Metropolitan District Commission headquarters in Boston to explain the continuing deterioration of the building. In an effort to show why the structure should be preserved, a brief history of the Observatory was presented. Its role in American as well as international meteorology was explained, and the importance of the climatological record and need for its continuance was discussed. The Office of Community Affairs brought the matter to Commissioner William J. Geary, who discussed the need for immediate funds with the Office of Government and Community Affairs at Harvard. A lengthy letter, dated 3 October, followed in which it was proposed that Harvard financially participate with the Metropolitan District Commission the Blue Hill Observatory Weather Club and Friends of the Blue Hills in an effort to preserve and rehabilitate the Observatory. The history was reiterated including a description of the Observatory's reversion to the Metropolitan District Commission when the lease was resigned by Harvard. An estimate of $200,000 was given as the cost of complete restoration and $10,000 was asked of Harvard to provide immediate weather-proofing.

1984

Harvard recognized the worthy cause and responded to Commissioner Geary with a check for the full amount solicited.

Formal acceptance of the $10,000 check took place on 13 January at a press conference held at the Trailside Museum. Before TV cameras the Commissioner thanked Harvard University for making it possible to begin the work of renovation and expressed the hope that the Observatory could be completely restored in time for the 1985 centennial. The commissioner noted that three months earlier a task force had been formed to advise the commission on preserving, restoring, and managing the observatory's historically significant properties. Minsinger voiced his appreciation for the financial assistance and welcomed Harvard back to an active interest in the Observatory. He closed by saying, "Hopefully, hundreds of years from now, one could look back and see a standard chronology against which to measure how sunspots, industrialization, increased carbon dioxide in the atmosphere, and other factors have affected the weather." Commissioner Geary paid tribute to the volunteer efforts of the Blue Hill Observatory Weather Club, saying that with their help the weather station could begin its second century.

The check was transferred immediately to the Blue Hill Observatory Weather Club with the stipulation that it be used only for repairs to the Observatory.

In March the Metropolitan District Commission filed an application for a matching grant of $10,000 from the Massachusetts Historical Commission Survey, but this was not obtained.

Also in January WBZ-AM radio installed a second phone line to the observer's room. This line was to be used to relay weather reports, especially during storms, to the radio station for use in their forecasts.

An unusually severe snowstorm struck on 29 March. Wind gusts reached 48.3 m sec^{-1} (108. mi. hr^{-1}) at 1246 EST; 26.3 cm (14.3 in.) of snow fell. Below the summit the snow was wet and due to the heavy rate of fall accumulated rapidly, especially on pine trees. Dozens of the old Wolcott pines on the north side of the hill were broken or uprooted. In the valley the Dedham–Westwood–Needham area was severely damaged. In some areas power was not restored for a week.

In April the Maximum Inc. company installed their Maestro wind equipment. This equipment, a gift to the Observatory, retains peak gust speeds and indicates current speeds continuously. The sensor consisted of a small cup anemometer that turns a magneto to supply a voltage proportional to the wind speed. Other Maximum equipment at the Observatory indicated wind direction and speed on a dial. The panel of instruments, located on the second floor, at this time is shown in figure 150.

Wind sensors on the tower mast are shown in figure 151. In May the contracting firm of Gordon and Son worked on waterproofing the tower. It was first sealed with a silicone solution. The cracks were then plugged

Figure 150. Weather Service instrument panel. Left to right, top: Maximum wind speed and direction indicator, maximum digital wind speed, wind speed and direction indicators; directly below, wind component accumulator. Lower left: Operations recorder (wind passage by miles and direction), wind (gust) recorder. July, 1984. Courtesy J.H. Conover.

with a special plastic material, and finally the entire surface was coated with plastic. The final coat was gray but was supposed to oxidize to the former brown color of the tower. The company also supplied storm windows for the tower, for a small room behind the tower, and for the hallway, but all of them were not installed before winter. The waterproofing job, which cost $7,000, appeared good but leaks continued around the tower roof drains which run through the concrete walls. The structure at the end of this waterproofing job is shown in figure 152.

Figure 151. Wind sensors looking north. Left to right, top: Gust sensor, three-cup connected to operations recorder, direction indicator. Top, behind: maximum 3-cup. Below, left to right: Vane for accumulator, vane for operations, 3-cup for accumulator. July, 1984. Courtesy J.H. Conover.

Also in May, Richard Stimets of the University of Lowell supplied a tipping bucket precipitation gage made by Rainwise, Inc., of Bar Harbor, Maine. As each tip of the bucket, corresponding to one-hundredth of an inch of precipitation occurred, the precise time was printed on a calculator tape, thus allowing the computation of rates. The Blue Hill station is one of many in a network in eastern Massachusetts having the equipment. Preliminary analysis of the data show characteristic periodicities in the precipitation rates for different types of rain-producing storms. The gage was heated with electric lamps to melt the snow during winter storms.

In June Minsinger mailed 5,000 brochures announcing the Centennial Capitol Campaign. They were sent to residents of the Milton, Canton, Dedham, Westwood, Quincy, Randolph areas and to many business firms. This ambitious effort to raise more money for restoration and the museum proved unsuccessful, with less than ten replies and only $810 received.

In October the Observatory started to send its 0700 and 1000 EST observation to the Boston National Weather Service for inclusion in the continuous NOAA weather broadcast. The transmitter for this broadcast is located in the WGBH building atop the hill and the antenna is on the FM tower.

The end of the year was marked by a frantic rush to secure contractors and have the interior walls and ceilings of the tower, the small room behind

Figure 152. Blue Hill Observatory as seen from the northwest WGBH-FM tower on left. July, 1984. Courtesy J.H. Conover.

the tower, and the hallway repaired and painted. New floor tiles were placed in the small room and first and second floors of the tower. Although snow prevented the workmen from reaching the Observatory a few days, the work was completed in time for the hundredth anniversary.

Also completed at the last minute were the computation of mean 24-hour temperatures since 1959, thus making the entire 100-year record homogeneous. Former chief observer Sally Wollaston, and Conover corrected the bright sunshine record to make it homogeneous throughout its record.

Over the period 1971–1984 the basic climatological observations continued but maintenance and repairs by outside technicians of the wind recording and solar radiation recording systems were very poor. For this reason alone many breaks occurred in the records. The recording of radiation was eventually abandoned altogether and not returned to proven methods of the past. These problems clearly illustrate why systems that rely on electrical components having state-of-the-art status should not be installed at reference climatological stations as advocated by some. In spite of these problems, the observations were obtained as best possible and published. While the station was under Skilling's direction, it was commended year after year for having one of the lowest number of errors among the reference climatological stations.

Without the formation of the Weather Club under the capable guidance of Minsinger, the Observatory at this writing would literally have started to crumble. His objectives were being met. Three to five newsletters per year were issued, interesting speakers were engaged, and numerous other features enabled the amateurs to learn more about the weather and help them in observing the weather if they so desired. Through mailings, radio, and television the club objectives were publicized to increase membership to about 800 and collect about $14,750. This sum of money was used to cover expenses, repair the Observatory, and plan for the museum. Gifts in the form of instruments, waterproofing material, publication costs, outside lighting, and other items totaled approximately $7,000.

References

Randolph, H., J. M. Holloway, D. E. Patterson and W. E. Wilson, 1981: Spatial and temporal pattern of eastern U.S. haziness: a summary. *Atmos. Environment.* 15: 1919-1928.

Chapter XVI

The Centennial Anniversary: 1985

The centennial anniversary of the Observatory was toasted by observers and friends at the Observatory on the evening of 31 January. One hundred years earlier Observer Gerrish set maximum and minimum thermometers and read the dial of a cup anemometer at 2300 EST. These constituted "base" observations from which the official observations were made beginning at 0700 EST on 1 February 1885.

The evening event was reported live on WNEV-TV by meteorologist Harvey Leonard. Another camera crew from Adams Russell Cable in Braintree filmed the observation as well as the festivities for a special on Braintree Cablesystem.

Very early on 1 February a WCVB-TV 5 camera crew and meteorologist, Bob Copeland, arrived to broadcast weather segments live from 0545–0700 EST, Rob Gilman did his radio programs to stations in Salem, Marshfield, and Quincy while Ed Perry from WATD-FM in Marshfield broadcast between 0800 and 1000 EST. Don Kent also made his local radio broadcast across New England from the Observatory.

About 850 invitations to the anniversary (figure 153) were sent out. The program, which was arranged by President William E. Minsinger and members of the Blue Hill Observatory Weather Club and Museum, Inc., took place on 1 February. Over one hundred guests, some of whom are shown in figure 154, assembled at 1400 EST for the rededication of the Rotch Memorial monument. Among the distinguished guests were those associated directly with the Observatory in the past and present, friends of the Observatory, and Weather Club members. The Metropolitan District Commission, State Legislature, nearby town officials, the American Meteorological Society, Harvard University Archives, Milton and Canton historical societies, Mount Washington Observatory, National Weather Service, and National Oceanic and Atmospheric Administration were also represented. The back of the Rotch Memorial (figure 155) or the side facing the Observatory building had been inscribed, thanks to the generosity of the Rotch family and others, with climatological data obtained from the records of the past century.

Minsinger spoke concerning the significance of the climatological record and the aims of the Weather Club. The Reverend Donald E. Tatro

The Officers and the Board of Directors
of
The Blue Hill Observatory Weather Club and Museum, Inc.
Cordially Invite you
to the
One Hundredth Anniversary
Of The Blue Hill Meteorological Observatory
on February 1, 1985
at 2 p.m.
at the Observatory
Great Blue Hill
Milton, Massachusetts

Box 101
E. Milton, Mass. 02186
R.S.V.P.   For further details call: 617/698-5397
by Jan. 20, 1985   802/728-3318

Figure 153. Invitation to the 100th Anniversary of the Observatory.
Courtesy Blue Hill Observatory Weather Club and museum.

Figure 154. Rededication of the Rotch Memorial Monument. From left:
Dr. Kenneth C. Spengler, Executive Director, A.M.S.; Eleanor Vallier, Vice
President, Weather Club; an M.D.C. representative; Tom Mcquire, M.I.C.
N.W.S., E. Boston; William Minsinger, President, Weather Club. Courtesy
R.W. Green.

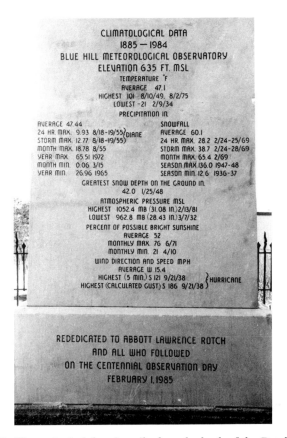

Figure 155. Climatological data inscribed on the back of the Rotch Memorial Monument. Courtesy J.H. Conover.

of the First Congregational Church in Milton gave the invocation. Thomas McGuire, newly appointed meteorologist-in-charge at the Boston Office of the National Weather Service, spoke briefly.

Although light freezing rain and sleet were falling, and the Observatory was enveloped in a dense cloud, the crowd took little heed of the weather by considering it not unusual for that location in mid winter.

The dedication ceremony was recorded by crews from WNEV-TV 7 and WLVI-TV 56. The latter carried out a brief interview with Minsinger.

Following the dedication the crowd returned indoors for refreshments and inspection of the newly redecorated three tower rooms, hall, stairway and small room beneath the stairs, figure 156. The remainder of the building, occupied by a caretaking couple, was closed.

Figure 156. Mr. and Mrs. Charles Pierce and meteorologist Jerry Brown, TV weather forecaster, in the Observatory tower. Courtesy J.H. Conover.

In addition to the government weather station instruments, a new Alden weather chart facsimile recorder, which had been presented to the Weather Club by the company, was on display. The machine, which was in operation, received its signals by radio from Norfolk, Virginia, or Halifax, Nova Scotia. The Maximum Company of Natick, Massachusetts, also presented the club with a precision aneroid barometer which was listed under the trade name "Proteus." Charles Orloff, a Weather Club member, demonstrated a computer that was temporarily connected to the Weather Services International (WSI) data bank at Bedford, Massachusetts. Current hourly weather observations were called up and listed on the screen. Two working model kites of the type used to lift meteorgraphs into the lower atmosphere were on display in the third floor tower room. The full size model (figure 157) of one of the smaller Hargrave kites used at Blue Hill was made by David Farrier. The other kite, made by Mark Meehl, was a model of one used by the Weather Bureau.

A bronze plaque, (figure 158) which indicated that the Observatory is now a national historic place, was presented to the Observatory by Ethel and John Conover. It was later mounted on the outside tower wall facing the northeast.

Figure 157. Working model of a small Hargrave kite on display at the Observatory. Courtesy J.H. Conover.

The crowd then slowly made its way down hill via private car, shuttle cars or on foot, to the auditorium of the Trailside Museum at the base for resumption of the program at 1600 EST (figure 159).

Figure 158. Bronze plaque presented to the Observatory by Ethel and John Conover. Courtesy J.H. Conover.

Figure 159. Audience that attended 100th Anniversary at the Trailside Museum. Courtesy J.H. Conover.

John Conover reviewed the importance of the Observatory in the developing science of meteorology in America. He then concentrated on the early cloud measurements and kite soundings at the observatory with brief comments about the Harvard radiosonde and the four Observatory directors.

J. Murray Mitchell, Senior Research Climatologist at NOAA, talked about the way his life had been shaped by his experiences and learning under Brooks at Blue Hill. He also emphasized the importance of the climatological record and the need for its continuance.

Edward M. Brooks, professor at Boston College and son of director Charles F. Brooks, introduced his mother, Eleanor S. Brooks, and other family members. He then presented an interesting biographical sketch of his father. He clearly illustrated the famous Brooks qualities of teaching and the ever-present weather observer.

Roland Boucher, research physicist at the Air Force Geophysics Laboratories, briefly related his experiences at Blue Hill while working with the weather radar data.

Last, before supper, Kenneth Spengler, executive director of the American Meteorological Society, discussed the society's relationship to Blue Hill, Brooks's founding of the society, and his staunch support of it. He then presented six books to Minsinger for use by the Weather Club and Museum. They were *Early American Winters, 1604-1820, Early American Winters II, 1821-1870, Early American Hurricanes, 1492-1870,* and *Early American Tornadoes, 1586-1870,* all by David Ludlum; *The History of Meteorology* by H. Howard Frisingner, and *The Thermal Theory of Cyclones* by Gisela Kutzbach.

At this time WBZ-TV 4 made its early-evening forecast by meteorologist Bruce Schwoegler from the observatory; Allan Bacon, Weather Club member, videotaped most of the proceedings.

For supper the Harvard Club supplied clam chowder, The Bents Company supplied crackers; the Parker House supplied rolls; and Hendries supplied ice cream. A giant cake decorated to depict the Observatory in 1885 and in 1985 was cut and served along with cookies and coffee.

After supper the lectures were resumed with an illustrated talk by Yeoman 1st class Fred Chapman, United States Navy, stationed at South Weymouth, Massachusetts. He related historical information and experiences as a Weather Bureau observer on the hill.

A certificate, signed by Milton selectmen James G. Mullenm Jr., Walter F. Timilty, and Walter A. Reilly, Jr., in which the day was proclaimed "Blue Hill Observatory Day in Milton" was read.

Frank Creedon, a master's candidate in historic preservation planning at Cornell University, then presented an illustrated talk about the Observatory and architectural changes over its lifetime.

A smaller crowd, this time, ascended the hill once again to watch fireworks which marked the end of the celebration. The summit remained in cloud with light freezing drizzle. At 2115 EST the fireworks began with spectacular bursts of muted colors, high above in the fog, to be followed by reverberating booms of the explosions. The noise was heard 6.5 km (4 mi.) to the west in Dedham. In addition to the radio and television broadcasts the significance of the Observatory and its past work was reported by the press.

Many invited guests who could not attend sent letters in which their memories in the hill were related. Other letters were of a congratulatory nature. A selection of these letters and a complete list of those who signed the guest book follows:

Barbara G. and Alexander A. McKenzie
Eaton, New Hampshire
January 15, 1985

W.E. Minsinger, M.D.
Box 101
East Milton, Massachusetts 02186

Dear Mr. Minsinger:

I am honored by your invitation to be present at the anniversary celebration of the Blue Hill Meteorological Observatory, and am planning to be there. If the weather prevents safe driving from New Hampshire, I'll regret being absent, and shall call the Observatory at the earliest possible opportunity.

With that possibility of being unable to attend, the following thoughts are offered for whatever use they may serve.

Generous donors of resources, thoughtful organization competent management, and reliable delivery of service or output to the eventual beneficiaries of an institution like Blue Hill are all necessary to its success. Sometimes, the people who perform the various tasks are masked by the organization, although sometimes they are very apparent. Too much individuality can be a problem, but devoted service can scarcely harm any organization. In other words, people are who get things done!

I was fortunate in being a small part of the Blue Hill Organization when growth of service to the community and innovations were exciting and beneficial. I tend to think of these changes in terms of the personnel I encountered in my rather brief stay with the Observatory during the period from mid-1935 to mid-1937.

Charles F. Brooks was a benevolent "boss" and potential good friend to all who wished to accept his selfless friendship. A very hard and enthusiastic worker himself, he made demands on his staff by setting a good example. His mind fairly bubbled with ideas, and if some of them proved impractical, more of them eventually found application.

During my tenure, Salvatore Pagliuca was Chief Observer, and provided the practical model of a good weather man at work. I had come to know him very well in the two years we served together at Mount Washington. As an "elder brother" full of practical wisdom, he kept me and other younger staff members at our tasks without being a taskmaster. I copied radio signals each morning during the worst weather and Sal and I plotted maps. His forecasts, required by the Blue Hill regimen, and especially needed by truckers and town road crews, vastly improved the accuracy of weather information needed in our region, where the Weather Bureau of that era shut down Friday afternoon and opened up for business again on Monday morning.

Mr. Sterling P. Fergusson, instrument maker, and kite maker of earlier times, was a pioneer in the technique that was soon to become the radiosonde. His humming of operatic airs, with an occasional brief solo (soto voce) together with his special brand of humor full of literary allusions both dated him and gave enjoyment during his part-time weeks on the Hill. Development of the radiometeorgraph work was in the hands of Arthur E. Bent.

Educated as a lawyer, Arthur Bent had been able, instead, to devote his life to other pursuits, of which amateur radio was followed as an engineering study. His quiet, generally deadpan comments could catch one unaware. His brand of humor was sophisticated, subtle, and often a little barbed. Besides the specialized radio work he did in assistance to Lange, Arthur provided improved equipment and ideas to the ongoing radio communications program, which included regular schedules with Mount Washington Observatory and hourly automatic tone transmissions for measurement of propagation variations at several distant points.

On one occasion, it was my unhappy duty, in Arthur's absence, to "drop" a complex antenna array on the tower that seemed—owing to loss of a guy wire in high wind—about to fall into one or two of Mr. Fergusson's delicate gust anemometers. After I cut another guy, the falling array narrowly missed the anemometers, but was, itself, bent out of shape and many of its fragile insulators shattered. Although Arthur had spent many hours designing and building, then erecting the antenna system, he accepted its loss without blaming others. After all the unwritten rule of Blue Hill was that meteorological observations and equipment had the utmost priority. Arthur's many gifts of expensive commercial as well as excellent handmade radio equipment made possible quicker results at no cost to the Observatory.

Often in touch by telephone or amateur radio, but an infrequent visitor, was Henry Southworth Shaw. He once explained his generosity to Blue Hill, and to many other organizations, with a self-deprecatory phrase that he was "fortunate in having access to funds that are available for scientific purposes." Much that was accomplished at Blue Hill would not have started without seed funds and continuing support from Mr. Shaw.

These were wonderful people with whom I have had the good fortune to be associated, and from whom I learned much. It would be impossible to list and catalog many other friends and valued associates, many of whom need no introduction to anniversary celebrants. They all contributed the blood, bone, and sinew of Blue Hill's success.

Of many of those outside the time of my own tenure, I think of the eccentric, and sometimes cankerous, Mr. Wells (in his celluloid collar) who retired, after many years of weather observations, just before my time. Charles B. Pear, an acquaintance, succeeded me as radioman/observer, and contributed vastly more to the ongoing radio work than I could have done.

My first trip up the Hill in September, 1932, was made memorable by a drenching shower, described with great delight as the effect of a cold-front passage, by Brooks, as raindrops dripped from his nose and hair!

I hope that I have not yet made my last trip.

> Sincerely,
>
> Alexander A. McKenzie

William E. Minsinger
President, The Blue Hill
 Meteorological Observatory
Box 101
East Milton, Massachussetts 02186

Dear Mr. Minsinger:

Thank you for your invitation to attend the Centennial Celebration. I regret that my schedule will not allow me to be there. The National Weather Service will be represented by Mr. Thomas McGuire, Acting Meteorologist in Charge of the Weather Service Forecast Office in Boston.

The National Weather Service and its early predecessor, the U.S. Weather Bureau, have a long history of involvement with the Blue Hill Meteorological Observatory. Our continuing interest is demonstrated by our provision of observational staff and instrumentation at the Observatory. We are very pleased with this association.

I extend best wishes to you and to the many friends of the Observatory who will celebrate this important occasion with you.

> Sincerely,

Richard E. Hallgren
Assistant Administrator
for Weather Services

Dear Minsinger:

I was very pleased to receive your invitation to the BHO Centennial, but I was away all of January and saw it too late to do anything about it.

I hope the celebration went well and I must congratulate you on the success of your efforts to maintain the Observatory as a historic site.

Sincerely,

Richard Goody

January 15, 1985
William Elliott Minsinger
President, The Blue Hill
 Meteorological Observatory
Box 101
East Milton, MA 02186

Dear Minsinger:

Your letter of January 1 informing me of the 100th Anniversary celebration of the Blue Hill Meteorological Observatory brings back many wonderful memories of my days at Harvard and the Observatory.

I was first introduced to meteorology by Charles Franklin Brooks, who was then in the Geography Department at Harvard while I was studying geology. He was a wonderful teacher and an enthusiast whose interests in the weather were contagious. He gave me my first part time job, which was changing the recording sheets on the wind and temperature recorders at the Geography Building at Harvard. He also gave me my first job in meteorology at the Blue Hill Observatory. Having been born in Dorchester, Massachussetts, and being familiar with the Milton area, I frequently went swimming at Houghton's Pond. I was not unfamiliar with the great Blue Hill, but always felt that the Observatory was a mysterious place.

When Charlie Brooks offered me a job at the observatory, I was overwhelmed. The job paid fifteen dollars per week, which was a fortune in those days. I lived at the observatory and would trudge up the hill with paperbags full of groceries. In the evenings, I would carry out my solitary observing functions. I remember having to go out and sketch the clouds in the sky, track them with a nephoscope, and take care of all the instruments at the observatory. It was a marvelous experience.

I made my first acquaintance with many of the famous names in meteorology while an observer there. I remember first becoming acquainted with Irving Hand of the Weather Bureau, whose office was at the Observatory ; and Fergusson, the creator of all those wonderful instruments. Above all, I remember the kindness and patience showed me by John Conover, who was the Chief Observer at that time.

I will always treasure these found memories. I am glad to see that the Observatory is still vital and there are individuals who care about its future. My best wishes to the Blue Hill Observatory Weather Club and Museum Incorporated for the wonderful work they are doing in connection with 100th Anniversary commemoration.

Best of luck for the second century!

Sincerely,

Robert M. White
President
National Academy of Engineering
Washington, D.C.

19 January 1985

Dear Mr. Minsinger:

Thank you for your kind invitation to be at the Blue Hill on 1 February. Unfortunately, I am already scheduled on that day and so will not be able to make it.

I worked at Blue Hill for Brooks and John Conover in April and May of 1943. I had finished my second semester at Dartmouth and was waiting to go into the Army. (I had spent the winter vacation from Dartmouth working at the Mt. Washington Observatory.) I performed observer duties and worked on completing climatological data at Blue Hill. I was the night watchman and sole occupant of the Observatory nights during that period. It was during the dim out and black out period and I remember it was very interesting to see how effective it was. I also remember some secret work that Dave Arenberg and some people from MIT were doing nights. When they were operating I was not allowed to watch them . . . secret defense work!

I left the observatory and entered the Army as a private on 21 June 1943. I spent the WWII year as a weather observer and upper air analyst and radiosonde operator in the U.S. and Greenland. I entered West Point in July of 1946 and retired from the Army in the grade of Lieutenant General on 31 August 1943, the last six years of which I served as the Inspector General of the Army. (Actually I was three months short of a full six years.)

I am enclosing a check for $25.00 toward the objectives of the Corporation which I hope will be helpful, however small.

Again, I appreciate your thoughtfulness in remembering me, and I wish I could be present. Please remember me to the Conovers.

As ever,

Richard G. Trefry
Lieutenant General (U.S. Army – Retired)

January 21, 1985

Dear Mr. Minsinger:

Thank you for your kind invitation to attend the 100th Anniversary Celebration of the Blue Hill Meteorological Observatory, but I am sorry that I cannot come at that time. I am afraid that the founders were not considering the convenience of future anniversary celebrants when they chose to start operations on Feb. 1 thus climatologically assuring the coldest weather of the year.

As for my recollections; they are very rosy. When I came to the Observatory as assistant observer in early 1936 it was to join a most congenial group of people led by Charles F. Brooks. Salvatore Pagluica was Chief Observer and we occupied the two bedrooms, Brooks, Robert Stone and S. P. Fergusson walked up the hill nearly every weekday. Later came H. H. Kimball with occasional visits by K. O. H. Lange, Arthur Bent and others often from the M.I.T. group working with Rossby. There was a high degree of cooperation among us; Blue Hill with its Mt. Washington satellite, M.I.T. and the U.S. Weather Bureau. Those were among the Observatory's best days, Brooks had brought together a group of young or young-minded energetic people mostly unfettered by marriage who were completely devoted to things they were doing. The pay was low but the rewards were great. Many other formal or informal students of meteorology came and went on. Brooks was ever ready to help young people and perhaps his greatest contribution was in the opportunities he offered and encouragement he gave to them. In my case, and I'm sure for many others, the Blue Hill experience set the pattern for my whole life.

To a degree, I suppose, the Observatory hastened its own fate by promoting the development of the radio-meteorograph which reduced the need for high elevation observations but I am sorry that the idyllic and monastery-like world that existed has not continued. However, I am sure that there will be many more great teachers who will inspire the young, even though they do not teach on top of a hill.

As for the future, I must commend the new organization and its officers for their efforts in carrying on and congratulate them on the fine celebration program that they have planned.

> Best wishes for the next 100 years
> and more!
>
> Charles B. Pear, Jr.

889 Indian River Drive   or   9 Second Street
Melbourne, Florida 32935   Mattapoisett, Mass. 02739

20 January 1985

William Minsinger
The Blue Hill Meteorological Observatory
East Milton, Massachussetts 02186

Minsinger,

Thank you for the kind invitation to the Centennial. Unfortunately I will not be able to attend because of activities associated with the next Space Shuttle Mission.

I have many fond memories of the time I spent at Blue Hill which are often recalled and will always be treasured—it was truly an unforgettable experience. It was my pleasure to be associated with C. F. Brooks, John Conover, Arnold Glaser, Dick Ashley, Lorna Ridley, Sally Wollaston, Murray Mitchell and many others too numerous to mention here.

While I was at the observatory in 1957 working on a weather satellite study, the first satellite was launched. Little did I realize then as we all watched Sputnik from the observatory tower, that my remaining working years would be spent on the space program. I left meteorology several years ago in favor of a career in computer science, but I have always retained a keen interest in the field. As I look forward to early retirement in about three years, I recall that my first job was at Blue Hill and I hope to be able to visit it again.

I trust the Centennial will be well attended and that the enclosed contribution will aid in the restoration.

Best Regards to all my friends who attend the festivities.

Very truly yours,

Ralph J. Newcomb

January 15, 1985

William E. Minsinger
President, The Blue Hill
 Meteorological Observatory
Box 101
East Milton, MA 02186

Dear Minsinger,

I regret very much that I will be unable to attend the celebration of the 100th anniversary of the Blue Hill Observatory. During my boyhood in Braintree, Great Blue was a familiar landmark on the western horizon. In later years I had many occasions to visit the Observatory, but never, as I recall, on official business. I have no serious stories to tell of my visits there.

However, I do recall one episode connected with the Observatory that I will always remember in amusement. Charles F. Brooks, the former Director, often appeared at meteorological meeting in the Boston area. On one occasion he arrived at an AMS meeting carrying an odd assortment of boards, clothing, papers and other objects. He had collected these on and about Blue Hill after the famous Worcester tornado of June 9, 1953. Odd items were found bearing Worcester labels at many locations in the Boston area, but nowhere was the strange precipitation from that storm better observed that at the Blue Hill Observatory (or so it seemed at the time).

Sincerely,

Richard J. Reed
Professor
University of Washington
Seattle, Washington

RJR:ca

Jan. 11, 1985

Dear Minsinger,

Thank you so much for inviting me to the 100th anniversary of the Blue Hill Observatory. I feel totally frustrated as there is nothing I would rather do than attend this outing. However, unfortunately, I must leave for the West Coast the night before the event.

I had many fond memories of the Observatory left over from my stay there. Memories of Mr. Wells, Furgy (Mr. Fergusson), Mr. Brooks, and other. Establishing regular 5 meter contact with Mt. Washington was, in those days, an exciting achievement, as was working with Pic, (G. M. Pickard) in the study of ultra shortwave propagation. I remember my frequent travels between Blue Hill, Mt. Washington, Exeter and Seabrook Beach and the many interesting friends I met along the way. These were wonderful people, real characters, each with his own strong personality. I cannot imagine any more unlike people than Furgy, Mr. Wells and myself. Yet even when working in close quarters, we always remained close friends, each respecting the others totally different life style. And Brooks, the serene master of it all, always encouraging each and every one of us.

I have not visited Blue Hill since the 30's. I had no real reason to. But now I am frustrated, as I long to revisit the old happy hunting grounds. Please remember me to any of the old gang that are left, and best of luck and good wishes to the new gang.

Al Sise
New England Masters Skiing, Inc.
Norwich, Vermont

January 11, 1985

William E. Minsinger
President, Blue Hill Observatory
Box 101
East Milton, MA 02186

Dear Minsinger,

Thank you very much for your kind invitation of January 1, 1985 to attend the 100th anniversary celebration of the founding of the Blue Hill Observatory. It is a matter of deep regret to me that circumstances will keep me here at that time and I will be unable to join you on this auspicious occasion.

Let me take the opportunity, however, to congratulate your organization for offering stability to this venerable institution. It is not only the historical interest that will be fostered, but also the educational opportunity for a younger generation. Above all the maintenance of the observational series is of greatest scientific importance. Without bench mark stations of this type we will never be able to pin down the climatic fluctuations and changes. There are too few reference points of this type and, I am sure, the value of the Blue Hill Observatory observations will increase with every coming year.

My very best wishes for your second century.

Sincerely,

H.E. Landsberg
Professor Emeritus
University of Maryland
College Park, Maryland

January 14, 1985

William E. Minsinger, M.D.
President, Blue Hill Observatory
Box 101
East Milton, MA 02186

RE: Centennial Celebration

Thank you for your kind letter of January 1st inviting me to the 100th Anniversary Celebration February 1, 1985. I regret I will be unable to attend, but do wish all there a very pleasant day.

I don't know whether there will be opportunity for those present to recount any instances of the time that they worked at Blue Hill, but I want to relate one. I was a weekend observer my senior year at Harvard. On Sunday December 7, 1941 at about 1:00 PM I was sitting in the kitchen eating a velveeta sandwich listening to the New York Philharmonic—I don't recall what was playing, when an announcer broke in and said Pearl Harbor had been bombed.

Shortly after that National Guard Troops came into guard Blue Hill and I remember eating greasy Army chow for a period of time while the observatory was being guarded.

Any way, best wishes.

Sincerely,

John S. McNayr
Lawton, Oklahoma

William Elliott Minsinger, President
The Blue Hill Meteorological Observatory
Box 101
East Milton, Massachussetts 02186

Dear Minsinger,

Thank you very much for your invitation to me to attend the 100th Anniversary celebration of BHMO. I surely wish that I could attend. But it would be very difficult for me to be in Massachussetts on the first of February, because I expect to be in Washington and New York during January 22-26, and my responsibilities here do not permit intensive travel.

I am deeply gratified to know that BHO is being restored and that the program is continuing. As you know, there were threats of various kinds over a substantial period of time, and from many quarters, some surprising. John Conover knows that history perfectly, and I have communicated my views to him separately. But all's well that ends well.

Please give my best wishes to John and to Ed Brooks, whom I have not seen in years, and also to Murray Mitchell, whom I have seen. I know it will be a delightful occasion and I will be with you in spirit!

Sincerely,

Edwin Kessler
Great Plains Apiaries
Purcell, Oklahoma

P.S. Please remember me also to Ken Spengler and to Bob Copeland, who was a student at MIT when I was there.

P.P.S. As for recollections, here is one. On snowy days it was not always possible to drive to the offices on top, so we parked our cars at the bottom, or drove as far as we could and pulled off the road where there was a space.

This was during the 1950s. Well, sometimes the road got pretty icy, and there was more than one occasion when I brought my sled, walked to the top with it, and then coasted down the road after work in the evening. Quite a ride! And rather dangerous—I don't think I would want my own son to do it, because sometimes one attained too high a speed. When that seemed imminent, the only prudent thing to do was to go off the road and come to a stop in the brush. Incidentally, are the American chestnuts still sprouting from the old stumps on Blue Hill? I always hoped that one of those trees would grow large again, but they always succumbed. I'll watch on TV.

January 14, 1985

William Minsinger, M.D., President
Officers and Board of Directors
Blue Hill Observatory

Dear Dr. Minsinger,

I greatly appreciate the invitation to the 100th Anniversary of Blue Hill Observatory and deeply regret that I will be unable to attend. Current family obligations make it impossible for me to be away overnight.

My work at Blue Hill was brief. In the summer of 1947, just after my graduation from Wellesley as an MA student in physics, I worked there two months. I am no student of language, but I spent most of that time translating captions in French. The memorable part of the experience was the lasting friendship I developed with Sarah Wollaston and Frances Wollaston Gates. In the fall of 1947 I went to work at Baird Associates. Two years later I married an MIT graduate student and we moved to Colorado. Boston days were a youthful adventure remembered with joy, but now I am much happier to be living on a remote hilltop in mid-Maine, far removed form everything urban.

My regards to John Conover, Sally and Fran and all good wishes to the Observatory "family."

Sincerely yours,
Ann Reiter Kendall
Dexter, Maine

October 24, 1984

Dr. William E. Minsinger
President
Blue Hill Observatory
Box 101
E. Milton, MA 02186

Dear Dr. Minsinger:

Thank you for your kind letter of October 17 inviting me to deliver a lecture at the February 100th Anniversary of Blue Hill Observatory. Much as I would like to do this I have to decline because of some unexpected physical problems which have forced me to be grounded in California for a while. Of course, the weather in February in your area is involved in this decision.

As you know, I feel that my early association with Blue Hill Observatory played a major role in my career and that I have nostalgic memories of the time in the 1980's there with Dr. Brooks, Dr. Haurwitz, Dr. Wexler and others. It is with great regret that I must decline your offer to participate in February.

Best of luck to you in your efforts and organizational plans.

Sincerely,

Jerome Namias
Research Meteorologist
Scripps Institute of Oceanography
La Jolla, California

JN/ch

January 17, 1985

William E. Minsinger, M.D.
President, Blue Hill Observatory
Box 101
East Milton, MA 02186

Dear Dr. Minsinger:

Thank you for sending me the program for the 100th Anniversary celebration of the Blue Hill Observatory and for inviting me to write some remembrances of my early connections with the Observatory.

It was late 1920's when I was in high school in Fall River that I became aware of Blue Hill and the pioneering work in meteorology that was going on at the Observatory. In fact I read a great deal of meteorology at that time and was thrilled at the excellent work that was done with the use of kites and various instruments. I read some of the works of Alexander McAdie and H.H. Clayton, and later on, Charles F. Brooks. I was also familiar to some extent with the pioneering research of A. Lawrence Rotch in earlier years. After high school, I attended a couple of meetings of the American Meteorological Society in Boston and was impressed and thrilled to see and hear Dr. McAdie and Dr. Brooks for the first time. Slightly afterward, I applied by letter to Professor McAdie for a job at the Observatory, but the funds were low and he said they had all they could do in supporting an observer.

A couple years later, I enrolled at M.I.T. after having taken a wonderful correspondence course in meteorology from Dr. Brooks, who was then at Clark University. It was then I got to meet Dr. Brooks in person, both at M.I.T. and at Blue Hill. He introduced me to Mr. Wells, the chief observer, and to a few other younger people with meteorological interests. Since I was existing on very limited funds, some from a grant given Mr. Henry Helm Clayton form the Smithsonian Institute, I asked Dr. Brooks for some part-time work, and with the recommendation of Dr. Rossby of Tech and my good record in the correspondence course under Brooks, he agreed to some limited research work for me and later for my best friend, Harry Wexler. It was a great pleasure and honor to be associated with the Observatory, and we worked hard to publish our research. It was then that I had published my work "Subsidence within the Atmosphere" as a Harvard Meteorological Monograph — a work that was to help my career materially and, I hope, enhance the image of Blue Hill. From time to time, Dr. Brooks sent me complimentary letters on it from his European and other colleagues. It was in these early years of my career that I got to know Mr. Sterling P. Fergusson, the first-rate instrument man who had sent up kites at Blue Hill, and developed many ingenious devices. Mr. Fergusson told me many tales of the early days at Blue Hill and it was no surprise to me when he returned to work there during his later years. He, H.H. Clayton, and Charles F. Brooks did a great deal to encourage me and to make possible a successful career.

It is with great regret that on this 100th Anniversary I cannot participate, but consider me there in full spirit and with best wishes for the years to come for the initial home of many of America's top meteorologists.

Sincerely,

Jerome Namias
Research Meteorologist
Scripps Institute of Oceanography
La Jolla, California

JN/mr

January 20, 1985

Dr. William E. Minsinger
President
Blue Hill Observatory
Box 101
E. Milton, MA 02186

Dear Dr. Minsinger:

It is with considerable regret that I inform you I will not be able to attend the 100th Anniversary celebration of the Blue Hill Observatory. I will at that time be on Mount Washington meeting with guests of our organization.

I have asked Mr. Alfred Oxford to represent the Mount Washington staff, and he has agreed to attend.

Please accept my congratulations on this important milestone for which much of the credit is yours.

Sincerely,

Guy Gosselin
Director
Mount Washington Observatory
Gorham, New Hampshire

5328 Wapakoneta Road
Bethesda, MD 20816

18 January 1985

Dr. William E. Minsinger
Box 101

East Milton, Massachusetts 02186

Dear Dr. Minsinger:

Thank you for your kind invitation to the Blue Hill Observatory 100th Anniversary Celebration. I sincerely regret that I will be unable to attend.

Although I never worked at Blue Hill I often visited during the years I was employed at the U.S. Army Natick Laboratories and also while studying for my Doctorate at Harvard with Dr. Brooks.

My very best wishes for a successful celebration and for continued operation of Blue Hill.

Sincerely,

Fernand de Percin

P.S. Please give my regards to John Conover when you see him. Thank you.

October 22, 1984

Dr. William Elliott Minsinger
President, Blue Hill Observatory
Box 101
E. Milton, Massachussetts 02186

Dear Dr. Minsinger:

Thank you very much for your kind letter, addressed to me at my home address in McLean, Virginia (which I suggest you continue to use in any future correspondence).

I am delighted to hear that progress is being made in the refurbishing of the observatory, and if my small contribution to the building fund has helped in any way to that end, it's a good feeling!

You are very thoughtful to invite me to join you for the 100th anniversary celebration, scheduled for February 1, and to give a brief presentation on a topic of my choice. I should be delighted to do both.

I look forward to hearing more about your plans for the celebration in due course, and to that opportunity to re-associate myself with the observatory which has meant so much to my career as a student of climatic change. I also look forward to meeting you and many other friends of Blue Hill on that signal occasion.

With kind regards,

Sincerely yours,

J. Murray Mitchell, Ph.D.
Senior Research Climatologist
National Oceanic and Atmospheric
Administration
Silver Spring, Maryland

January 23, 1985

Dr. William E. Minsinger, M.D.
President, Blue Hill Observatory
Box 101
East Milton, Mass., 02186

Dear Dr. Minsinger:

You are most kind and thoughtful to invite me to attend the 100th Anniversary Celebration of the Blue Hill Meteorological Observatory. I am proud to be included among those whom you thought would be interested. I am indeed. I have been holding back on this letter in hopes that I could accept but as the first of February approaches I foresee that I can not be with you, but I will certainly be grateful to learn about the ceremony and attendance and plans for prolonging the life of this very cornerstone of the science of meteorology in our beloved United States of America.

I am not a meteorologist; my efforts are in recording and preserving evidence of the history of aeronautics, I have a related interest in the first of all forms of man-made aircraft—kites. As a boy I made and flew them. As a Boy Scout I was proud to win the Aviation Merit Badge in which item 1 was "Have made a kite that will fly." Later when I had come to my position in this wonderful Smithsonian, I was asked to be examiner in that merit badge. I had learned that few boys knew how to make and fly kites. At the request of the Boy Scout Headquarters in New York City I wrote a booklet for their instruction-series: title—KITES. One item was red-penciled out. I had read about kite flying at Blue Hill and their use of piano wire for kite line, so I included that in the text. I had tried it and it was very good, having less drag than string, and stronger for its weight. But it can be dangerous, so that was crossed-out. Its date was 1931; Kite Lines Magazine obtained a copy and asked my permission to reprint it. I'll make a few changes, including "Don't use wire for kite line," and so that relic will bloom again. When I joined the staff of the Smithsonian and after some while was promoted so that I could use my brains as well as my hands one of the first accessions I sought for was an example of the Weather Bureau kite. It was duly received, and for years was displayed in the old Aircraft Building, now replaced by our magnificent building on the Mall. It is now in our Restoration Facility — a combined working place and associated Museum. From the enclosure you will note why I am proud to add that detail. During my service in the Navy, World War II, my interest in kites became useful as I invented a maneuverable kite that was a realistic ship-to-air target. More than 300,000 were made and sent to all ships and stations. When the Navy Test Center at Patuxent, MD learned of it I was asked to develop a kite-elevated system of lines including a horizontal line to be snatched by an airplane-flown hook, and a down-line to which a message container was attached. That system was very useful in expediting delivery of documents to headquarters. I received an official commendation for that. Currently, I maintain the Smithsonian's annual Kite Festival, in March. It is a happy occasion for hundreds of kite fliers and thousands of spectators. I have extended this report because I would never have become a capable kite flier had it not been for the inspiration I received from reading Dr. Rotch's book in which he described kite flying at Blue Hill. My Museum related interest led not only to obtaining an example of the huge Hargrave kite as used at Blue Hill, but also a treasured recollection of a visit from Mr. Fergusson when we discussed making and flying kites at Blue Hill. From the daughter of William A. Eddy, I received an original example of his deltoid kite. It is also displayed at our Facility. I am proud to add that this Institution assisted with his work at by allotting $500 from the Hodgkins fund for purchase of silk line.

I enclose a rehabilitation check to finance a "plank in the deck." Again my gratitude for your letter and folder.

Sincerely,

Paul E. Garber
Historian Emeritus
Smithsonian Institution
Washington, D.C.

18 January 1985

Dear Dr. Minsinger:

Thank you for your kind invitation to attend the hundredth anniversary celebration of the Blue Hill Observatory. I regret very much that I will be unable to attend it.

I am much interested to read about your plans for the future of the Observatory, and am enclosing my check for $100 as a modest contribution to the Restoration Fund.

Sincerely,

Bernhard Haurwitz

2523 Constitution Ave.
Fort Collins, CO 80526

Dr. William E. Minsinger
President, Blue Hill Observatory
Box 101
East Milton, Mass., 02186

Dear Dr. Minsinger:

Thanks for the invitation. I wish I could be present at the celebration of the 100th Anniversary of the Blue Hill Observatory. After courses in Harvard that included physics, meteorology, and climatology and an incomplete education elsewhere in theology, I had the privilege of being on the Blue Hill staff from 1936 to 1947 as what Director Charles Brooks called a "General Factotum," which included serving him as a secretary, librarian, observer, editor, and general assistant, even occasional coal-shoveler and coaxer of a sometimes recalcitrant hot-water furnace, as well as being his assistant as the Secretary of the American Meteorological Society, of which I later also became the Treasurer as the Society moved ahead in membership, activities, and publications.

During those years the Observatory had a staff (including part timers) ranging from less than a dozen to more than double that. Off the top of my head come some memories of a few of the contributions made by the Observatory in those years.

There was some considerable excitement about the development of new upper-air measurements by radiosondes, developed under Karl O. Lange and others, which sent back the temperature, humidity, and other meteorological data by radio from ascending balloons. Blue Hill's earlier upper-air explorations had been made by kites and instrumental recordings drawn back with the kite. I associate much of this work with Sterling P. Fergusson, who was still serving the Hill in the development and maintenance of instruments during my earlier years.

Cloud observation and analysis were being continued to help reveal and account for upper-air conditions, a field in which John Conover's Blue Hill experience looking from the hill top upward prepared him for his later work in the new possibilities of looking from space downward via receptions from an eye in a satellite. Blue Hill's century of history from hill-top, kite, and balloon observations has spanned the most astounding technical advancements in instrumentation for observing what goes on in the earth's atmosphere.

Considerable effort was involved with the operation of the Mount Washington Observatory, opened I think in the fall of 1931 under the leadership of Appalachian Club head of Pinkham Notch, Joe Dodge, and meteorological input by Blue Hill. Through radio communication this provided current weather as well as climatological data on conditions at a couple of thousand meters elevation and base stations.

Climatology, not only of Blue Hill and Base Stations, but climatological maps based on Weather Bureau and other observations of North America, drafted under Books, was a continuing significant contribution.

The effect of the sun on weather and climatology were being studied as well as the effect of the atmosphere on the sun to reveal atmospheric conditions. The Observatory provided quarters for and cooperation with I.F. Hand of the U.S. Weather Bureau for additional solar studies.

Through the publications of the American Meteorological Society, of which Brooks was a founder and central operator, certain documents, such as Jerome Namias' *Air Mass Analysis*, were important in the preparation and education of personnel for the meteorology of flying and other operations in the Second World War. Data from Blue Hill library's extensive collections from meteorological services and observatories around the world were used by military planners. The Observatory also cooperated in providing advice and data for the design of clothing, equipment, and operations in the war.

Please give my cordial greeting to those who may remember me at your Centennial Celebration. I wish I could be with you all.

I must add a footnote on my pleasant memories and appreciation of the Blue Hills Reservation itself, so wisely made a public park under the Metropolitan District Commission. I not only relished the daily ascents and descents mostly on the dirt or winter-snow-covered road to and from the

Observatory in all seasons, but I, my family, and friends, (from the Observatory and elsewhere) delighted in endless walks and picnics in various portions of the range at all times of the year, plus skiing on various back trails in the winter, and swimming at Ponkapoag in the summer, in this reservation where a Bostonian could quickly find a sample of beautiful and relatively unspoiled nature in a metropolis.

> Sincerely,
>
> Ralph Wendell Burhoe
> Center for Advanced Study in Religion and Science
> Chicago, Illinois

WALLACE E. HOWELL, SC.D.
SCIENCE CONSULTANT

> 36 S Mt Vernon Country Club Road
> Golden, CO 80401
> January 26, 1985

Dr. William E. Minsinger, President
Blue Hill Meteorological Observatory
P.O. Box 101
East Milton, MA 02186

Dear Dr. Minsinger,

Greetings to all the friends of Blue Hill, I regret that I live too far away to be with you tonight.

My first visits to Blue Hill were just over half a century ago during my freshman and sophomore years at Harvard. It was an added attraction, a dividend, for rock-climbing expeditions to the Quincy quarries. Soon, though, I was weighing possible fields of concentration, drawn both to science and to the outdoors. What better candidate than meteorology! The only trouble was that since the passing of Prof. Ward a few years earlier, the meteorology courses still in the catalogue were not being offered. I had to take my first weather courses at M.I.T.

So it was that during Blue Hill's fiftieth year I came to be walking up the road with Dr. Brooks. He asked me if I intended to continue my studies in this field. I did some very hasty mental calculation and decided I had nothing to lose by maintaining his interest in me, so I said yes. He said, what? I said yes. He said, what? I said YES! The upshot was I convinced myself.

Midterm of its first century, there was an infectious geniality, and extension to all the staff of Dr. Brooks' way of always looking on the hopeful side of things. Irving Hand, S. P. Fergusson, Charlie Pear, John Conover, all doing things they got a kick out of. I don't ever recall such a thing as a put-down. The Blue Hill attitude of that time has been for me a continually self-renewing source of strength through the half-century that has passed since then. Was it the spirit that Dr. Brooks cooked up, or the people who carry it abroad, that made the difference?

Now on the brink of its second century, I join in congratulations on a century of accomplishment and in trust that Blue Hill's tradition of open-hearted interest in the futures of the people who pass a portion of their careers there will continue and flourish.

Sincerely,

Wallace E. Howell.

A listing of those who signed the Guest Book at the Observatory in the afternoon follows:

W. E. Minsinger	President, BHOWC
Robert Skilling	Observer in Charge, BHO
David P. Hodgkin	President, Friends of the Blue Hills
Eleanor Vallier	V.P. Blue Hill Weather Club
Charles Orloff	Blue Hill Weather Club
Bob Copeland	WCVB-TV Ch. 5, Boston, MA
William Ralston	Rhode Island School of Design
Michael Skilling	10 Clifford Street, Hingham, MA
Allan Bacon	73 School Street, Hingham, MA
Al Oxton	Mount Washington Observatory
Bob Dixon	Adams Russell Cable
Paul F. Norton	Adams Russell Cable (Braintree)
Kevin Norton	U.S.A.F.
Mark Meehl	Boston, MA
I. Perry	19 Bola Road, Duxbury, MA
Rob Gilman	341 Highland Avenue, Wollaston, MA
Paul J. Clark	News Photog. TV-5, Boston
Larry Weisberg	News Photog. TV-5, Boston

William Curtin — Chestnut Street, Randolph, MA

Frank Bourbean — 21 Canton Street, Randolph, MA

Claire Broye — 110 Short Street, Easton, MA

Douglas Kent — 48 Emerson Street, E. Weymouth, MA

Don Kent — 48 Emerson Street, E. Weymouth, MA

Paul Vendetti — 224 Carlisle Road, Bedford, MA

H. F. Potts, Jr. — 55 Barnard Road, Granville, MA

Dean Talbut — 65 Winthrop Street, Fall River, MA

Eric Goldman — 63 Highland Avenue, Randolph, MA

Priscilla Ashton — 12 Neal Gate Street, Norwell, MA

George A. Martin — 1073 Riverside Drive, Methuen, MA

Glenn Villiard — 7 Evelyn Road, Holbrook, MA

Sam Perkins — 69 Glendale Road, Quincy, MA

Dr. Wen Tang — U. Lowell, 9 Oxbow Road, Lexington, MA

Arthur Adams, Jr. — 219 Dedham Street, Dover, MA

Joe Boccia — 29 Chapman Street, Beverly, MA

Arthur Johanneson — 15 Pleasant Street, Needham, MA

Ethan Mascoop — 232 Blue Hill Avenue, Milton, MA

Heidi Neipris — 17 Blake Road, Brookline, MA

Ruben Novack — 15 Acton Street, Maynard, MA

John Nicholson — 40 Winter Street, Weymouth, MA

J. Derek Seeley — Lee, NH

Joan C. Seeley — Lee, NH

Brian C. Seeley — Lee, NH

Charlie & Betty Pierce — Arlington, MA

Hurd C. Willett — Littleton, MA

Michael Cedrone — Stoughton, MA

Maggie Mcluett — Channel 7 News

Manton T. Spear — Lynnfield, MA

Sarah (Sally) H. Wallaston — Cummington, MA

Marion S. Philpot — Dedham, MA

Ethel Conover — Dedham, MA

David B. Clarke — Concord, MA

Marjorie Jeffries — Skied from 1268 Canton Avenue, Milton, MA

Joe Coyle	Quincy, MA
Selma Burgett	Quincy, MA
Malcolm Partridge	Braintree, MA
Mr. & Mrs. Warren Hirst	Duxbury, MA
Irwin Alpert	Lexington, MA
Danial Alpert	Lexington, MA
John H. Conover	Dedham, MA
Donald E. Tatro	Milton, MA
Fred Bates Lund, Jr.	Boston, MA
Ed Dunham	Radar Met., Raytheon Company
Evelyn Mazur	American Meteorological Society
Henry S. Howe	Canton, MA
John Wallace	Weather Service Corporation
Ken Spengler	American Meteorological Society
James G. Mullen, Jr.	Milton, MA, Selectman
Norman Lenison	Brookline, MA
Paul A. Blackford	Bridgewater State College
Phil Falconer	Schnectady, NY
John E. Hutchin	Milton, MA
Mike Coalombe	Norwood, MA
George F. Steinbacher	East Walpole, MA
Tom McGuire	National Weather Service, Boston
Jerry Brown	WLVI-TV Dorchester, MA
Bob Cunningham	Lincoln, MA & Kent Is., N. B. CANADA
Martha T. Curtis	Milton, MA
Clarke A. Elliott	Harvard University Archives
Rep. W. Joseph Manning	Milton, MA
Ray Falconer	Burnt Hills, NY
Ellen Anderson	M.D.C.
Stu Kofalis	Middleboro, MA
William Kofalis	Boston, MA
Tom Chishdon	Nashua, NH
Nicholas J. Trakas	Milton, MA
Sarah B. Brooks	Newton, MA
Phillip Erickson	Providence, RI
Laurence B. Stein, Jr.	Hingham, MA
Alan A. Smith	Mt. Washington Observatory
John L. Scott	

Russell & Jeanette Peuerey	Milton, MA
Joseph Kelly	Congressman Brian Donnelly's Office
Joseph M. Donavan	Jamaica Plain, MA
Garry Van Wart	Meteropolitan District Commission
E. O. Wendland	M.D.C.
William Moulton	M.D.C.
Nicholas A. Magendantz	Lincoln, RI
Elisu Magendantz	Lincoln, RI
Nancy Rotch Magendantz	Lincoln, RI
Mario Traficante	M.D.C. Community Affairs
Ann Prince	Metro Parks
Diane Keith	Metro Parks
Edward M. Brooks	Newton, MA
Claudia Altemus	Milton, MA
Debra Palmquist	Boston, MA
Merry Lang	Randolph, MA
Robert E. Lautzenheiser	Reading, MA
J. Murray Mitchell	N.O.A.A. Silver Springs, MD
Katherine Mitchell	McLean, VA
Frank Aicardi	Pine Street, Norwell, MA
Edward Adams	89 Harvard, Pembroke, MA
Richard N. Chase	47 Hutchins Street, Dorchester, MA
Frank L. Edmonds	6 Harrison Road, Salem, MA
Frank L. Creedon	16 Ravennah Road, West Roxbury, MA
Bob Smith	796 Jerusalem Road, Cohasset, MA

At the Trailside Museum:

Eric A. Walters	3 Hickory Lane, North Reading, NH
Scott A. Mandin	Partidge Lane, Boxford, MA
Ralph H. Lutts	Blue Hill Interpretive Center
Warren C. Scott	28 Fattler Road, Hingham, MA
Robert D. McArthur	Metro Park Blue Hill Reservation
Fred Chapman	54560 Underhill St., Otis AFB, MA
Russ & Tracy Ryths	P.O. Box 486, Hyde Park, MA
Zareen Poonen	7 Grove Street, Winchester, MA

Bjorn Poonen	7 Grove Street, Winchester, MA
James R. Comcau	23 Estes Street, Lynn, MA
David L. Emerson	1144 Burt Street, Taunton, MA
Korena Poonen	7 Grove Street, Winchester, MA
Allen R. Knowles	Friends of Blue Hill
Ross Vicksell	11 Foster Road, Burlington, MA
Pete Leavitt	35 Haynes Road, Newton Center, MA 02159
Jackie Boucher	Dutton Road, Sudbury, MA
Martha Robbins	715 Washington, Abington, MA
Donald R. Robbins	715 Washington, Abington, MA
Mr. & Mrs. M. F. Partridge	33 Nicholas Road, Braintree, MA
John & Sue Anderson	17 Ginley Road, Walpole, MA
Ken & Marita Lidman	Nashua, NH
Ralph & Denny Donaldson	Sudbury, MA
Nancy Sanders	Marblehead, MA
Frona Brooks Vicksell	Burlington, MA
Roland J. Boucher	Natick, MA
Elizabeth C. Carlson	Fryeburg, ME
Sue Koszalka	Dracut, MA, Univ. of Lowell - Meteorology
Eleanor S. Brooks	North Haven, CT
Edwin P. Nali, Jr.	Westborough, MA
Peter F. Blottman	North Andover, MA
Joseph Valade	Nashua, NH
Pamela Camara	Stoughton, MA
Roger Lawson	13 Juniper Brook Road, Northboro, MA
Joyce & Donald Duncan	Milton, MA
Fred Sanders	Marblehead, MA
Joe, Mary Lou, Rachel & Sarah Dow	Brookline, MA
Wallace E. Sisson	Milton, MA
Mr. & Mrs. DeCoursey Fales	Cambridge, MA
Margaret Plymire	
Ann Trakas	50 Meadowbrook Road, Milton, MA
Eleanor & Justine Alter	Brookline, MA

So ended the first hundred years of the Blue Hill Observatory where the first significant meteorological research in the western hemisphere was initiated. Among important achievements were: the first cloud climatology including heights and motions in America, the first soundings of the atmosphere anywhere in the world by means of kites, the first balloon-sondes and pilot balloon runs in America, the first practical radiosonde in the world, numerous studies in climatology and cloud physics as related to synoptic conditions, the most extensive measurements in the world of airglow, and basic turbidity values for the upper atmosphere, and the development of numerous instruments. Teaching in the atmospheric sciences and physical oceanography, both in Cambridge and on the hill, was a service provided by all four of our directors. As the Blue Hill climatological record continues into the second century, the scientific programs of the Observatory probably will never match former achievements. But the proud structure, now restored to withstand the pounding of future gales, cold, rain, and snow, remains as a monument to the science of meteorology in the United States.

Appendix A

Climatic and Other Records of the Observatory

Various climatic and other records of the Observatory are listed according to their present repositories:

1. At the Blue Hill Observatory, P.O. Box 101, Milton, MA 02186. Pertaining to the Blue Hill location unless otherwise noted:

Original Meteorological Observation Books:	1885–1984
[unbound (1960–1964)]	
Daily journals included after 1888	
Copies of U.S. Weather Bureau or	
National Weather Service daily	
observational forms.	1965–1984
Daily journal	Feb. 1885–Jan. 1886
Daily journal	Feb. 1886–Aug. 1886
Daily journal	Sept. 1886–Dec. 1886
Daily journal	1887
Daily journal	1888
Station log	1959–1984
Base station observations	Jun. 1886–Dec. 1887
Base station observations	Jun. 1888–Dec. 1888
Book of temperature and humidity	
differences, summit to valley	
during kite flights, diurnal, etc.	
Milton and Randolph observations	1887–1888
Book of weekly, ten day and annual	
means and extremes.	Jan. 1886–Dec. 1888
Working book, averages and extremes	1885–1910 to 1914
Kite flights	6 July 1911–6 July 1912
Book of wind direction frequencies	1886–1888
Miscellaneous summaries	1885–1895
Book of data tab of storm centers	
and their movement in relation	
to B.H.O.	
Record of Instruments, Vol. 1	31 Dec. 1884–1 Jan. 1932
Instrument Book (loose leaf)	1941–1959

Autographic records[1]:

Operations recorder—wind direction and movement (missing Nov. 1961, Aug.–Nov. 1964)	Dec. 1959–1984
Barograph (missing June 1960, Feb. 1970)	Jun. 1959–1984
Precipitation (missing Aug., Oct. 1962)	Nov. 1959–1984
Thermo–hygrograph (missing Sept. 1961, Jan.–Apr., Dec. 1963 Dec. 1964, Jan.–Apr., Dec. 1965, Jan.–Feb. 1966, Aug. 1974).	Jan. 1960–1984
Thermograph	Apr.–Nov. 1963 Dec. 1964–Apr. 1965 Jan., Feb., Apr. 1966
Ombroscope—duration of rain (missing Nov. 1960, Nov., Dec. 1961, Jan.–June 1967 1971–1977, Jan.–Apr. 1978.)	Jan. 1960–1984
Aerovane—wind direction and speed (missing Apr.–May, Nov. 1960, Feb.–Mar. 1964, Jan. 1965, Aug.–Dec. 1970, Jan.–May 1971, Nov. 1972)	Mar. 1960–Jan. 1976
Campbell—Stokes—duration bright sunshine (missing Mar. 1960, May 1961, July 1965, May–July 1 1971, Feb.–June 1976)	Jan. 1954–1986
Wind gust recorder	Oct. 1980–1984

2. At the National Archives—Boston Branch
      380 Trapelo Road, Waltham, MA 02154.
      Location specified
    Autographic records:

---

[1] These records through 1982 were transferred to the National Archives, Boston Branch 15 Oct. 1987.

Blue Hill Observatory Records Inventory

Box No.	Type of Data	Years	Location
G-1	Thermograph Disc. Dew Point Recording 32 Ft.	8/30/1946-1/2/1948	Blue Hill
	Thermograph Disc. Dew Point Recording 32 Ft.	12/31/1950-12/31/1952	Blue Hill
G-2	Thermograph Disc. Dew Point Recording 7 Ft.	7/2/1947-1/3/1950	Blue Hill
	Thermograph Disc. Dew Point Recording 32 Ft.	1/1/1949-1/7/1950	Blue Hill
G-3	Thermograph Disc. Dew Point Recording 42 Ft.	7/1/1947-11/1/1949	Blue Hill
	Thermograph Disc. Dew Point Recording 32 Ft.	1953	Blue Hill
	Thermograph Disc. Dew Point Recording 32 Ft.	1/7/1947-2/6/1947	Blue Hill
	Thermograph Disc. Dew Point Recording 42 Ft.	3/1/1947-8/11/1947	Blue Hill
	Thermograph Disc. Dew Point Recording 7 Ft.	1/2/1950-8/15/1951	Blue Hill
G-4	Thermograph Disc. Window Shelter Temp.	1/1/1940-12/31/1942	Blue Hill
	Thermograph Disc. Window Shelter Temp.	1/1943-12/1944	Blue Hill
G-5	Thermograph Disc. Window Shelter Temp.	1/1947-12/1948	Blue Hill
	Thermograph Disc. Window Shelter Temp.	1/1949-12/1950	Blue Hill
G-6	Thermograph Disc. Window Shelter Temp.	1/1951-12/1952	Blue Hill
	Thermograph Disc. Window Shelter Temp.	1953	Blue Hill
	Wind (Esterine–Angus Roll Charts	10/1945-3/1946	Blue Hill
	not indexed)		
	Duration of Sunshine		Blue Hill
	Duration of Sunshine		Blue Hill

	Element	Period	Station
G-7	Duration of Sunshine	1885-1900	Blue Hill
	Duration of Sunshine		Blue Hill
	Duration of Sunshine		Blue Hill
	Duration of Sunshine		Blue Hill
	Duration of Sunshine		Blue Hill
G-8	Duration of Sunshine	1901-1915	Blue Hill
	Duration of Sunshine		Blue Hill
	Duration of Sunshine		Blue Hill
	Duration of Sunshine		Blue Hill
	Duration of Sunshine		Blue Hill
G-9	Duration of Sunshine	1916-1930	Blue Hill
	Duration of Sunshine		Blue Hill
	Duration of Sunshine		Blue Hill
	Duration of Sunshine		Blue Hill
	Duration of Sunshine		Blue Hill
G-10	Duration of Sunshine	1931-1943	Blue Hill
	Duration of Sunshine		Blue Hill
	Duration of Sunshine		Blue Hill
G-11	Duration of Sunshine	1946-1947	Blue Hill
	Duration of Sunshine		Blue Hill
	Duration of Sunshine	1952-1953, Aug. 1956	Blue Hill
	Duration of Sunshine	1889-1905	Pickering Station

	Sunshine Records (type unknown) not indexed	1886-1888	Blue Hill
	Pole-Star Records	(dates uknown)	Blue Hill
	Moonlight Duration Record not indexed	1917, 1926	Blue Hill
G-12	Pole Star Records	1933-1941 (broken record)	Blue Hill
	Pole Star Records	1942-1948 (broken record)	Blue Hill
	Wind Record-Gust Experimental not indexed	1939-1941	Blue Hill
	Wind Record-Gust Experimental not indexed	1942-1950	Blue Hill
	Anemometer Charts	4/13/1919-10/31/1921	Blue Hill
G-13	Anemometer Charts	8/1/1918 to approx. mid-1925	Blue Hill
	Box 47 Weather Observations	1888-1910	Blue Hill & Valley Sta.
		1888 (incomplete)	
	Thermograph (duplicates)	1886-1892 (broken)	Blue Hill
	Barograph Anemoscope, Hydrograph		
	Daily Observation Sheets (long, narrow)	Jan.-June 1948, 1951 1957	Blue Hill
	Meteorological Observations	1911-1922 (incomplete)	Valley St.
		1911-1919 (incomplete)	Base St.
G-14	Meteorological Observations	Dates unknown	Blue Hill Base Sta.

ID	Description	Dates	Location
	not indexed		Blue Hill
	Wet/Dry 7.7	1914-1915	Blue Hill
	Daily Barograph	10/1920-10/1921, 1914	Mt. Wachusett, Mass.
	Monthly Meteorological Obs.	Approx. 1935-1941	
	Pressure, Humidity, Temp., Winds	1935-1936	Mt. Wachusett, Mass.
	Pressure, Humidity, Temp., winds not indexed		
	Pressure, Humidity, Temp., Winds	1937-1939	Mt. Wachusett, Mass.
	Pressure, Humidity, Temp., Winds	1936-1942	Mt. Wachusett, Mass.
B-1	Thermographs	1886-1893	Blue Hill
B-2	Thermographs	1894-1901	Blue Hill
B-3	Thermographs	1902-1909	Blue Hill
B-4	Thermographs	1810-1919	Blue Hill
B-5	Thermographs	1920-1929	Blue Hill
B-6	Thermographs	1930-1937	Blue Hill
	Thermographs	1891	Blue Hill Base Station
	Thermographs	1887-1888	Blue Hill Base Station
	Thermographs	7/88-12/88	Blue Hill Base Station
B-7	Thermographs	1890, 1893-1900	Blue Hill Valley Station
	Thermographs	1895-1897	Blue Hill Base Station
	Thermographs	1898-1904	Blue Hill Base Station
B-8	Thermographs	1889, 1901-1906	Blue Hill Valley Station

	Thermographs (1 volume)	9/30/1902-1/25/1912	Blue Hill Tower Station
B-9	Thermographs	1905-1909	Blue Hill Base Station
	Thermographs	1908-1914	Blue Hill Tower Station
B-10	Thermographs	1907, 1912, 1910-1914	Blue Hill Base Station
	Thermographs	1915-1921	Blue Hill Valley Station
	Thermographs	1913-1919	Blue Hill Base Station
B-11	Hygrometer/charts	1886-1893	Blue Hill
B-12	Hygrometer/charts	1894-1903	Blue Hill
B-13	Hygrometer/charts	1904-1913	Blue Hill
B-14	Hygrometer/charts	1914-1925	Blue Hill
B-15	Hygrometer/charts	03/30-12/31 & 1932-37 (Incomplete Volume)	Blue Hill
	Departures from normal temps (1 Volume)	1892-1907	Blue Hill
B-16	Hygrometercharts (1 Volume)	1900	Blue Hill Valley Station
	Hygrometercharts (1 Volume)	1899	Blue Hill Valley Station
	Anemometer Charts	1886-1890	Blue Hill
	Anemometer Charts (Broken Record—2 Vols.)	1926-1933	Blue Hill
	Anemometer Charts	1/1/1934-3/20/1934	Blue Hill
	Anemometer Charts	1932-1933	Blue Hill
	Anemometer Charts	1941	Blue Hill

B-17	Anemometer Charts	1930-1940	Blue Hill
	Hygrometer (2 volumes)	1901-1913	Blue Hill Valley Station
	Pole Start Records	1904-1912	Blue Hill
	Rain Gauge Chart	1886-1889	Blue Hill
	Fergusson Weighing Precipitation Gauge (2 Volumes)	1930-1938	Blue Hill
B-18	Fergusson Weighing Precipitation Gauge (5 Volumes)	1939-1952	Blue Hill
	Thermo-Hygrograph Charts (1 Volume)	1934-1938	Blue Hill Base Station
	Thermo-Hygrograph Charts (2 Volume)	1937-1946	Blue Hill Window Shelter
	Thermo-Hygrograph Charts	Enclosure B.H.O. Misc. Dates	Blue Hill
B-19	Thermo-Hygrograph	1937-1953	Blue Hill
	Cinemograph Charts	1891-1897	Blue Hill
B-22	Cinemograph Charts	1898-1905	Blue Hill
B-23	Cinemograph Charts	1906-1913	Blue Hill
B-24	Cinemograph Charts	1914-1921	Blue Hill
B-25	Cinemograph Charts	1922-1929	Blue Hill
B-26	Cinemograph Charts	1930-1937	Blue Hill
B-27	Cinemograph Charts	1938-1944	Blue Hill
B-28	Cinemograph Charts	1945-1950	Blue Hill
B-29	Cinemograph Charts	1951-1953	Blue Hill

B-30	Barometer (Monthly — 1 Volume)	1890-1929	Blue Hill
	Barometer (1 Volume)	1901-1906	Blue Hill
	Barograph (Weekly inep.)	3/1924-3/1930	Blue Hill
	Barograph (Weekly — 4 volumes)	1930-1955	Blue Hill
	Barograph (Monthly — 1 volume)	1930-1950	Blue Hill
	Barograph (Weekly — 1 volume)	1917-1924	Blue Hill
	Rain Gauge Charts (1 volume)	1894-1900	Blue Hill
B-31	Rain Gauge Charts (1 volume)	1890-1893	Blue Hill
	Rain Gauge Charts (1 volume)	1893-1894	Blue Hill
	Rain Gauge Charts (1 volume)	1895-1896	Blue Hill
	Rain Gauge Charts (1 volume)	1901-1911	Blue Hill
	Rain Gauge Charts (1 volume)	1893-1894	Blue Hill Valley Station
	Rain Gauge Charts (1 volume)	1895-1896	Blue Hill Base Station
B-32	Rain Gauge Charts	1897-1908	Blue Hill Base Station
	Rain Gauge Charts (1 volume)	1890-1892	Blue Hill Valley Station
	Rain Gauge (Precipitation) (Broken — 1 volume)	3/1920-7/1925	Blue Hill
B-33	Rain Gauge [Precipitation (1 volume)]	10/1911-12/1916	Blue Hill
	Rain Gauge [Precipitation (1 volume)]	1917-1919	Blue Hill
	Duration of Rain (1 Volume)	1904-1909	Blue Hill
	Anemograph	6-9/1926-1/12/1932	Blue Hill
	Anemograph	1-10/1933	Blue Hill
B-34	Duration of Rainfall (1 Volume)	10/23/1911-2/25/1913	Blue Hill

	Duration of Rainfall (1 Volume)	1914-1918	Blue Hill
	Duration of Rainfall (1 Volume)	1947-1948	Blue Hill
	Duration of Rainfall (1 Volume)	1951-1952	Blue Hill
B-35	Duration of Rainfall (1 Volume)	1910-1911	Blue Hill
	Duration of Rainfall (1 Volume)	1949-1950	Blue Hill
	Duration of Rainfall (1 Volume)	1946-1948	Blue Hill
B-36	Triple Register at 1416 Spruce St. Philadelphia, PA	1910 (complete)	Philadelphia, PA
		1913 (incomplete)	
		1913 (incomplete)	
	Anemoscope (3 Volumes)	1885-1887	Blue Hill
	Anemometers (3 Volumes)	1902-1905, 1906-1908 & Jan. & Feb. 1912	Blue Hill
B-37	Anemoscopes	1888-1892	Blue Hill
B-38	Anemoscopes	1893-1896	Blue Hill
B-39	Anemoscopes	1897-1902	Blue Hill
B-40	Anemoscopes	1903-1906	Blue Hill
B-41	Anemoscopes	1907-1910	Blue Hill
B-42	Anemoscopes	1911-1914	Blue Hill
B-43	Anemoscopes	1915-1918	Blue Hill
B-44	Anemoscopes	1919-1922	Blue Hill
B-45	Anemoscopes	1923-1926	Blue Hill
B-46	Anemoscopes	1927-1930	Blue Hill

B-47	Anemoscopes	1931-1934	Blue Hill
B-48	Anemoscopes	1935-1938	Blue Hill
B-49	Anemoscopes	1939-1942	Blue Hill
B-50	Anemoscopes	1943-1946	Blue Hill
B-51	Anemoscopes	1947-1950	Blue Hill
B-52	Anemoscopes	1951-1953	Blue Hill
	Anemometer	1901-1902	Blue Hill
B-53	Anemometer	1903-1906	Blue Hill
B-54	Anemometer	1907-1911	Blue Hill
B-55	Barometer	1895-1898	Blue Hill
B-56	Barometer	1890-1894	Blue Hill
B-57	Anemograph	1912-1918	Blue Hill
B-58	Anemograph	1934-1942	Blue Hill
B-59	Anemograph	1943-1950	Blue Hill
B-60	Anemograph	1951-1953	Blue Hill
B-60	Barometer	1901-1904	Blue Hill
B-61	Barometer	1905-1908	Blue Hill
B-62	Barometer	1909-1912	Blue Hill
B-63	Barograph	1913-1917	Blue Hill
B-64	Barograph	1918-1922	Blue Hill
B-65	Barograph	1923-1927	Blue Hill
B-66	Barograph	1928-1931	Blue Hill

B-67	Barograph	1932-1936	Blue Hill
B-68	Barograph	1937-1941	Blue Hill
B-69	Barograph	1942-1945	Blue Hill
B-70	Barograph	1946-1949	Blue Hill
B-71	Barograph	1950-1953	Blue Hill
B-72	Anemometer	1885-1887	Blue Hill
B-73	Anemometer	1888-1890	Blue Hill
B-74	Anemometer	1891-1893	Blue Hill
B-75	Anemometer	1894-1896	Blue Hill
A-76	Anemometer	1897-1900	Blue Hill
B-77	Barometer	1885-1888	Blue Hill
B-78	Barometer	1889-1890, 1892	Blue Hill
B-79	Thermograph	1933, 1934 (not bound)	Blue Hill Base Station
	Aneroid Barograph	1915-1924 weekly	Blue Hill
		(Incomplete, not bound)	
	Thermograph	1931-1933 (not bound)	Blue Hill Base Station
	Sea-Level Pressure—one bundle	1919-1924	
		(Incomplete, not bound)	
	Barograph (Monthly)	1951-1955	
		(Incomplete, not bound)	
	Force of Wind (1 Volume)	1887-1888	Blue Hill
	Force of Wind (1 Volume)	1889	Blue Hill

	Vertical Current Anemometer (French Forms)	1892	Blue Hill
	North Vertical Surface Solar Radiation	1949-1954	Blue Hill
	(Duplicate on file at NCC)		
	South Vertical Surface Solar Radiation	1948-1954	Blue Hill
	Diffuse Solar Radiation	1949-1954	Blue Hill
B-80	Pole-Star Records (glass slides)	1889-1890	Blue Hill
B-81	Pole-Star Records (glass slides)	1891-1892	Blue Hill
B-82	Pole-Star Records (glass slides)	1893-1894	Blue Hill
B-83	Pole-Star Records (glass slides)	1895-1896	Blue Hill
B-84	Pole-Star Records (glass slides)	1897-1898 (tray of 1903)	Blue Hill
B-85	Pole-Star Records (glass slides)	1899-1900 (tray of 1904)	Blue Hill
B-86	Pole-Star Records (glass slides)	1901-1902	Blue Hill

Time intervals for which the various records are available are shown in figures 160a and 160b.

Microfilm containing the following data, which are ancillary to the autographic records, pertain to the Observatory site unless otherwise noted.

Reel #1—Record of Instruments
> Volume I—December 31, 1884 to January 1, 1932
> Thermometers and Barometers
> Instrument Comparison and Testing of Instruments
> (Data of instruments in use at the observatory from the beginning of the station.)
> Volume II—1932
> Record of instruments in current use at the observatory and its cooperating stations, descriptions of standards and methods of standardizing, data of instruments tested.

Reel #2—Thermometer Comparisons
> Temperatures—Max., Min., and Precipitation, June 1, 1886 thru December 31, 1887. Map of Blue Hill, Mass.
> Temperatures and Precipitation at Randolph and Milton, Mass., January thru December 1887.
> Temperatures and Precipitation at Blue Hill, Mass., January thru December 1888.
> Blue Hill Daily Journal October 21, 1884–January 31, 1886 (includes a record of the beginning of the Blue Hill Observatory).
> Written descriptions of daily weather data, February 1, 1885 thru December 1888.
> Summary of Observations—February 1885 thru July 19, 1885.
> Record of Observations—Pressure, Air Temperature, Humidity, Wind, Weather, and Precipitation. Temperatures include Dry bulb and Wet bulb, February thru July 19, 1885.

Reel #3—Record of Observations (Con't of Reel #2)
> Pressure, Temperature, Humidity, Clouds, Precipitation and Wind, July 20, 1885 thru March 1887.

Reel #4—Record of Observations (Con't of Reel #3)
> Pressure, Temperature, Humidity, Clouds, Precipitation and Wind, April 1, 1887–December 1887; December 1888 and November, 1888; and January 1888–October 1888.

Reel #5—Record of Observations

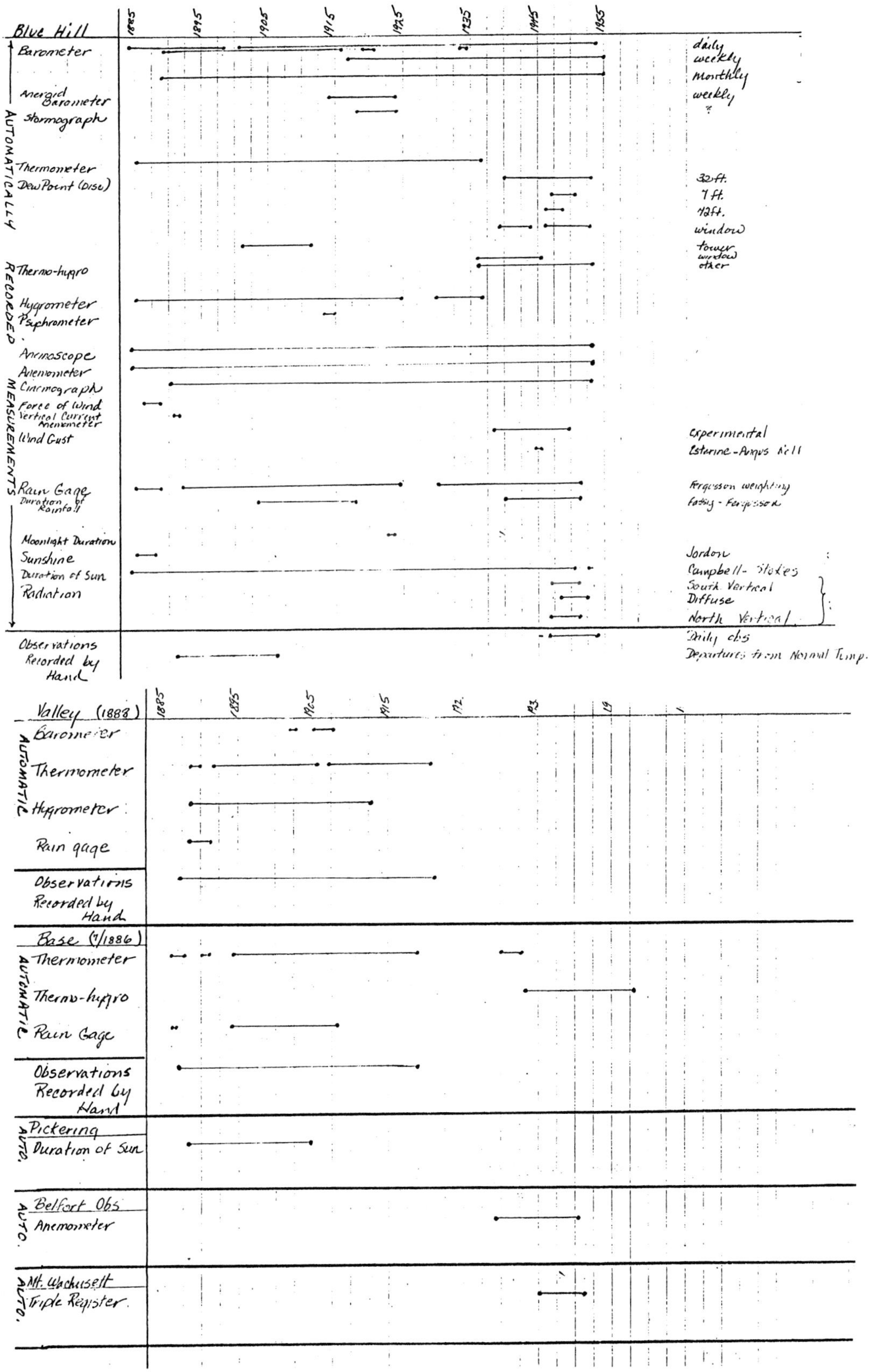

Figures 160 a & 160 b. Periods of available climatological records on file at the Federal Archives in Waltham, Massachusetts. Courtesy National Archives, Boston Branch, Waltham, MA.

Barometer, Thermometer, Dew Point, Humidity, Wind, Cloud Observation, Precipitation, Hydrometers and Visibility of Mountains, January 1889 thru December 1899.

Reel #6—Record of Observations

Barometer, Thermometer, Precipitation, Cloud Observations, Wind, Visibility of Mountains, January 1900 thru December 1912.

Reel #7—Record of Observations

Barometer, Thermometer, Precipitation, Clouds and Wind January 1913 thru December 1926.

Reel #8—Record of Observations

Barometer, Thermometer, Precipitation, Cloud Observations and Wind. January 1927 thru December 1937. (Changed type of forms August 1932.)

Daily Procedure of station on reel.

Reel #9—Record of Observations

Hydrometers used at station in 1938.

Pressure, Temperature, Humidity, Wind, Weather, Precipitation, January 1938 thru December 1943

Meteorological conditions of Hurricane of Sept. 21, 1938 at Blue Hill Observatory.

Phenological Observations of 1939 at Blue Hill Observatory.

Hourly Records: Air Temperature, Relative Humidity, Prevailing Wind Direction Mean Wind Speed, Amounts and Types of Precipitation, January 1943 thru December 1943.

Birds seen in vicinity of Blue Hill Observatory for 1943.

Charts on Station comparison for some dates

Reel #10—Record of Meteorological Observations

Pressure, Temperature, Humidity, Wind, Weather, Precipitation, Clouds, Visibility, January 1944 thru December 1947

Hourly Records: Air Temperature, Relative Humidity, Prevailing Wind Direction, Mean Wind Speed, Amounts and Types of Precipitation January 1944 thru December 1947.

Birds seen in vicinity of Blue Hill Observatory for 1944.

Hourly Cloud Codes and their amounts—effective Sept. 1, 1945.

Birds seen in vicinity of Blue Hill for 1945.

Reel #11—Record of Meteorological Observations

Pressure, Temperature, Humidity, Winds, Weather,

Precipitation January 1948 thru December 1952.

Snow Records for January 1948.

Hourly Records: Air Temperature, Relative Humidity. Prevailing Wind Direction, Mean Wind Speed, Amounts and Types of Precipitation January 1948 thru December 1952

Snowflakes, Types March 11, 1948.

Reel #12—Record of Meteorological Observations

Pressure, Temperature, Humidity, Wind, Weather, Precipitation January 1953 thru December 1957.

Notes on Hurricane "Carol", August 31, 1954.

Precipitation at Waterbury, Vermont June 23, 1957–July 7, 1957.

(Microfilmed with the June records.)

Reel #13—Record of Meteorological Observations

Pressure, Temperature, Humidity, Wind, Weather, Precipitation January 1958 thru December 1958.

Hourly Records: Air Temperature, Relative Humidity, Prevailing Wind Direction, Mean Wind Speed, Amounts and Types of Precipitation January 1958 thru December 1958

Meteorological Observations—Clouds in Detail.

January 1933 thru May 28, 1933.

August 1932 thru December 1932.

May 28, 1933 thru December 1933.

July 1934 thru December 1934.

January 1934 thru June 1934.

January 1935 thru September 23, 1935.

Reel #14—Record of Meteorological Observations

Meteorological Observations—Clouds in Detail, September 24, 1935 thru December 1938.

Reel #15—Record of Meteorological Observations

Meteorological Observations—Clouds in Detail January 1939 thru December 1942.

Reel #16—Record of Meteorological Observations

Meteorological Observations—Clouds in Detail January 1943 thru June 18, 1946 (Guide to coding clouds precedes January 1946).

Reel #17—Record of Meteorological Observations

Meteorological Observations—Clouds in Detail, June 19, 1946 thru December 1950.

Reel #18—Record of Meteorological Observations

Meteorological Observations—Clouds in Detail, January 1951 thru April 1956.

Reel #19—Record of Meteorological Observations

Meteorological Observations—Clouds in Detail, May 1956 thru November 1958, December 1 & 2, 1958, and for December 3, 1958.

3. The following records are housed at the National Climatic Data Center, Federal Building, Page Avenue, Asheville, N.C. 28801. Pertaining to the Blue Hill location unless noted otherwise. Microfilm (19 reels same as Fed. Archives—Waltham)

Climatic data by months, daily values. Feb. 1885–Dec. 1950
1976–1977
Climatic data by days, 610–10 or B–16
1965–1984
Climatic data by months, daily values
May 1885–Feb. 1901
at town of Milton, Mass.
(missing July and Aug. 1892)
Microfilm, 15 rolls of old meteorological records—same as the originals now in the National Archives in Washington, D.C.

4. At the National Archives: National Record Group 27, Records of the U. S. Weather Bureau. Scientific, Economic and Natural Resources Branch, National Archives and Records Administration, Washington, D. C. 20408

Requests for information about or access to these records should be directed to the National Scientific, Economic and Natural Resource Branch, National Archives and Records Administration, Washington, D.C., 20408.

Place	Period	Observer	Elements Observed
Boston, Mass.	1735	Rev. Samuel Checkley	Notes on weather
Boston, Mass.	12/15/1774 –9/20/1816 (incomplete)	John Jeffries	Daily entries of weather and a few wind directions and maximum & minimum temperatures
Boston, Mass. 51 Hancock St.	1821–1856	Jonathan P. Hall	Temperature Precipitation 1823–1856
Boston, Mass. Franklin Place	1851	Unknown	Various notes with a few temperatures

Cambridge, Mass. Harvard College	5/19/1780	Nathan Read	Notes on a dark day
Cambridge, Mass.	10/22/1785	Rev. Samuel Williams	Heavy rainfall
Cambridge, Mass.	12/11/1742	Prof. John Winthrop	Temperature, pressure, wind, precipitation (8/1749–1776) and lecture notes. See his "Meteorological Diary." This and the records from the following three observers are in the Harvard College Archive and discussed on pp. 140–144 and 286–290 of the *Monthly Weather Review* for 1908 by B.M. Varney.
Harvard College	12/31/1778		
	1790–1791	Prof. Samuel Williams	Temperature 3 times per day, wind direction, state of sky, remarks, & lecture notes. Tabulation of observations of extreme temperatures in various places in eastern Mass, 1/12/1752– 12/12/1876.
	1/12/1752 –12/12/1786	Prof. S. Webber	Pressure, temperature

	1800–1817	Prof. John Farrar	Temperature (1807–1812, 1813, 1816, 1817). humidity (1800–1806), storm of 9/23/1815
Cambridge, Mass. Botanic Garden	8/1865 2/1866	A. Fendler	Pressure, temperature, clouds and wind
Cambridge, Mass.	1840–1888	Harvard College Obs.	Original record books and pole–star slides in Cambridge. Published in Annals of the Observatory, v. 19, pp. 1–157.
Canton, Mass. Ponkapoag Vill–age	2/1/1813– 7/16/1819 12/30/1819 3/7/1827 1/15/1831– 1883 1884– 4/5/1886	Bazin Wallace Shaw (1886)	Temperature 2–4 times per day and comments on the weather.
Derryfield, NH	1/1809– 11/1813	Dr. Bentley	Temperature 4 times per day, weather, hourly cloudiness, pressure 2 times per day.
Epping, N.H. & other places such as Portsmouth & Dover, N.H.	1796–1823	William Plummer	Wind & weather notes
Hamden near New Haven, Conn.	25 yrs end- ing 3/31/1810	Jeremiah Alling	Register of weather in published form.
Kennebunk, ME	1/13/1809– 12/31/1823	W.B. Sewall	Summary of Temperature &

	(1842–1855, 1865–1868)		Precipitation record Very complete weather notes, indexed. (Temperature, pressure, precipitation, wind & sky.)
Leicester, Mass.	9/1889– 11/1895 (incomplete)	Leicester Academy	Pressure, temperature, wind direction & speed clouds and micellaneous.
Lunenburg, Mass.	1838–1875	Geo. A. Cunningham	Temperature, precipitation, wind, some phenological data, & misc. notes.
	1847–1875		Observations are given in detail. Summaries of data give longer period.
Milton, Mass.	3/11/1885– 12/31/1886	A.K. Teele	Maximum and minimum temperatures, wind & precipitation
Milton Center Mass. Reedsdale Road & Centre St., 42°16'N, 71°, 06'W, 60 ft. elev.	1/1849– 12/31/1888	Charles Breck	Temperature 2 or 3 times per day and comments on the weather
Nashua, NH	9/1886 & 7/1897 (incomplete)	Jackson Co.	Pressure, temperature, humidity, wind, precipitation, sunshine, weather & miscellaneous.

			Gives monthly temperature and precipitation back to Jan. 1886.
New Bedford, Mass.	Oct. 1812–1920	Samuel Rodman Thomas Rodman	Pressure, temperature (dry & wet), clouds, wind, precipitation. Original records of 1812–1905. Later reports from report of City Engineer.
New England	Before 1891	—	Book: "Historic Storms of New England"by Sidney Perley.
New Haven, Conn.	1873–1921	U.S.W.B.	Published table of monthly temperature & precipitation.
New York State	1816	—	Letter of Louis Krumbha quoting from article by Mrs. L.H. Hammond about crop failures due to cold weather.
Probably Newburyport, Mass.	March 1842	Unknown	Single sheet of mis–cellaneous data.
Newburyport, Mass.	1885–1890 (incomplete)	F.V. Pike	Pressure, temperature, humidity, precipitation, wind direction & speed, weather, & miscellaneous
Northeastern	Storm of	—	Precipitation data

US & Canada	10/4/1869		in published form.
Phila., Pa	1/1/1790	—	Book "A
	1/1/1847		Meteorological Account of the Weather in Phila.," by Charles Pierce.
Rutland, Vt.	1810	Rev. S. Williams	Letter on change of climate and notes on aurora, meteors, eclipses.
South Boston, Mass.	1806, 1807& 1808 (Boston)	Blake, probably	Notes
City Point	1841,1842, 9/1/1844– 12/31/1846 1/1/1849– 8/31/1851, July 1852– 1867, 1869, 1771–1892.	Samuel Blake	Weather with fairly complete record of temperatures.
South Boston, Mass.	1837–1838	Unknown	Temperature & notes.
Worcester, Mass.	June 1862	William N. Green	Temperature, pressure, wind direction, weather and remarks.

5. At Pierce Hall, Room G3C
    Harvard University
    Cambridge, MA.
    Pertaining to the Observatory:

Record of visitors	1889–1899
Record of visitors	1900–1902
Record of visitors	1910–1920
Record of visitors	1921–1938
Record of visitors	1938–1950
Record of visitors	1950–1957
Record of visitors	1957–1971

Author index and copies of letters	1891–1893
Author index and copies of letters	1893–1895
Author index and copies of letters	1896–1898
Author index and copies of letters	1898–1901
Author index and copies of letters	1902–1905
Author index and copies of letters	1905– 1909
Author index and copies of letters	1910–1913
Publications and letters received and sent	1886–1887
Publications and letters received and sent	1888
Publications and letters received and sent	1889–1892
Accession books to the library	1893–1900
Accession books to the library	1901–1908
Exchanges	1898–1900
Original ms. of reports in annals	1899–1900
Cloud observations, No. 1	1890–1891
Cloud observations, No. 2	1896–1897
Kite records	1895–1897
Kite records	1897–1903
Kite records	1903–1911

6. Evaluation of the meteorological data obtained at Blue Hill.

Mr. S.P. Fergusson prepared the following memoranda concerning the Observatory records:

Analysis of The Meteorological Record at Blue Hill Observatory 1885–1936. The work of the Blue Hill Observatory primarily has been research, particularly in aerology, and has been distinguished by the introduction of new and important methods of study, particularly that of soundings by means of kites and at the present time by valuable improvements in the technique of radio–sounding apparatus. It is most important that research should continue the major activity and that the maintenance of a continuous record of all elements be incidental to this end; there are many climatological stations, but only one Blue Hill Observatory where the opportunities for studies of meteorology are not subordinated to the needs of climatology, public service or routine or prescribed duties. But the record at Blue Hill is so valuable in one respect that a careful analysis of at least the chief elements is most desirable. Nearly all stations of observation in the United States are maintained for public service, chiefly forecasting, in cities where conditions of exposure are continually changing; consequently, although the records of the Weather Service began in 1870, and there are a few records of temperature and precipitation more than a century in length, there are almost none where conditions of exposure have been uniform a decade at a time. At Blue Hill, exposures of all instruments have been unchanged from the beginning of the record in 1885, and the data are exceptional for the study or detection of small or permanent changes of climate or unusual variations from a long average or normal. This is particularly true of data of the wind, which is the element most affected by changes of exposure.

Up to the time of Professor Rotch's death in 1912 the record was very complete, comprising automatic records of all elements excepting the kinds and motions of clouds and optical phenomena (hydrometeors); current means and normals of all elements were recorded in a "normal book" so designed that comparisons of any part of the record could be made almost at a glance. The condition of instruments, especially thermometers, barometers and anemometers likely to change as time passed was maintained adequately; thermometers were verified annually, barometers at frequent intervals, including two or more comparisons with standards at the Weather Bureau in Washington, while two anemometers were in continuous use; these last had been adjusted to agree with instruments tested by Dines in England and Marvin in this country.

After Rotch's death the income for maintenance of the Observatory at first was less than one–half that expended by him and only sufficient for the salaries of the Director, observer and janitor, instead of a meteorologist, observer and assistant with some temporary assistance, there was nothing except occasional gifts or grants for research and only sufficient income to maintain the more important (and the more durable instruments) and elements of the weather. Apparently, it seemed best for this limited staff of two to devote their time to keeping the building and some equipment in order and plan for resumption of the work in detail given up when Rotch died. No records of the condition of instruments in the current "record–book" or the special book containing the data of changes of instruments or technique, or on the original charts from the recording instruments, were made during the entire period, 1910 to 1931, and there is almost nothing to guide a student in an attempt to use the records. The twice–daily observations of the principal elements were maintained and these, probably are the sole dependence for data during the period 1914 to 1931.

When Dr. Brooks became Director in 1931 it was decided to summarize the entire record at Blue Hill, partly as a contribution to the International Polar Series of 1932 and chiefly because of the unusual value of a record maintained under invariable conditions for fifty years. Other work of immediate importance—the reconditioning and repair of instruments so that an accurate record could be resumed, and certain investigations, such as that of the eclipse of 1932, the equipment of the Mt. Washington Observatory, and other activities—prevented immediate attention to this task, and until 1936 the only work accomplished was a preliminary examination and classification of the record charts accumulated since 1912. (No binding of charts or forms was done during this period of twenty years, and in view of the cost of files for such a long period it seemed best to give ample time to the matter of selecting a suitable, inexpensive binder. The University Bindery devised a file satisfactory in every respect and in some details better than that in use earlier, and costing much less in quantity. The efficient handling of such a large accumulation of charts was a problem of some importance but, finally, a device for holding and drilling loose sheets was designed and constructed and at the present time nearly all the arrears of binding or filing have been brought up to date.)

The plan for summarizing the record of fifty years so far has included only the elements of pressure, temperature and wind, as the most important, leaving other items until time permitted the work. Of these, mean temperatures for the fifty years have been computed, and the wind record prepared for analysis.

The trend of temperature during the past 83 years in Milton (50 years at Blue Hill and 40 at Milton—see discussion, Annals, 1903) indicating a rise of about 1° Fahrenheit, deserves careful study. In all probability this is a long–period fluctuation, but its evaluation will depend partly upon the accuracy of thermometers, the zeros of which almost invariably rise, even after they are seasoned.

The values of atmospheric pressure also are dependent upon the condition of the barometers—constancy of vacuum, errors of thermometers, etc., changes of which can be detected only by comparisons of the three standard instruments with one another and with other standards, and occasional tests of the attached thermometers.

Humidity, sunshine, cloudiness and the frequency and amount of precipitation, all of which vary less regularly than other elements, may require more careful study to determine normals and departures therefrom than pressures and temperatures. The records of these elements have not been examined in detail as yet.

Repair and Reconditioning of Instruments in 1931. The first work of Dr. Brooks on assuming the Directorship of the Observatory was restoring the records to their original accuracy (1) by installing an accurate anemograph for comparison with the instrument in use during an indefinite period previously and known to be inaccurate, (2) testing all thermometers at the three stations, for zero and throughout the scale, (3) comparison of the standard barometers and testing their attached thermometers, and in addition, comparing these barometers with new instruments recently compared with the British standard at Kew, the standards at Washington and the Canadian Meteorological Office; this last work was completed early in 1932. Other restorations since 1931, including the addition of new and modern equipment and the standardization of apparatus long in use have brought the record up to a standard higher than has been attained during any earlier period.

Condition of the Records during the Period 1913–1931. The absence of notes or other data of changes in instruments or technique already referred to has necessitated careful examination of all record charts and books containing daily observations so that some basis might be found for the discussion of the records during this period and the preparation of notes for the information of others wishing to use these data. As stated, the records of temperature, humidity, pressure and precipitation have not been examined in detail, but, inspection during the process of assembling and binding revealed the condition described below, which, with other suspected deficiencies, must be considered in any adequate use.

Deterioration of equipment evidently began about the time of Mr. Rotch's death and continued throughout the period; there are long periods between the failure of an instrument and its replacement and in some instances there was no replacement.

Thermographs. The instruments at the Base and Valley were in operation a few years after 1913, but when they failed (probably due to failure of clocks) the records at these stations were not resumed.

At the Observatory, there are periods of months (perhaps totaling more than a year) when the file consists of blank charts dated but with no notes or explanation. The remaining records seem to be fragmentary and some appear to be simply personal estimates as are the records of wind velocity and direction, through most of the period 1913–1931.

Hygrographs. The record of humidity is probably more uncertain than that of temperature because the instruments (both psychrometer and hygrograph) are more difficult for an amateur or student to keep in good condition. Apparently the hygrograph was not in use between 1920 and 1925 for no record–charts have been found covering this period.

Precipitation gauges. The ombroscope (duration of rainfall) disappeared after a few years and only the clock–cylinder remains. The recording rain and snow gauge evidently failed about the same time for there are no records through a considerable period; during a visit about 1925 the instrument was, and had been for some time, open and fully exposed to the weather. (Notwithstanding this—perhaps years of this exposure—a test of the spring–balance showed, in 1932, that the instrument was accurate within about 4 percent. After some 30 years of use and perhaps five of exposure to the weather, and without previous cleaning or adjustment.)

A new recording gauge of the same pattern was installed about 1928, but, unfortunately, with the standard gauge, was placed on the roof of the tower, apparently in accordance with the custom of the Weather Bureau, but in ignorance of the comparisons of gauges on the roof and on the ground during the first year of observations (1885) which indicated a deficiency of rainfall and (according to memory) almost no snowfall.

Barographs. The Draper barograph—the most accurate of those in use—became unserviceable, apparently through failure of its clock, during this period and although the clock appears to have been running, it was not connected with the board carrying the chart and the observer moved the board by hand apparently several times daily in an attempt to give at least a personal estimate of the changes of pressure. Fortunately, there was a duplicate record by the Richard weekly mercurial barograph which may cover at least a large part of the period when the Draper instrument was unserviceable.

Anemographs. The record of the wind has been examined in detail. Because of the severe conditions at Blue Hill two anemographs have been maintained since the beginning in 1885, and from 1891 to the present time the data have been obtained chiefly from the Richard anemo–cinemograph, one of the very few instruments recording velocities direct without the necessity of calculating them from a chronogram, as is customary at practically all stations in this country. This instrument, however, is complex, although very durably built, and requires competent care by one who understands it. The transmitter in use was designed at Blue Hill and provided with sealed bearings so that it could function for months (at one time more than one year) without attention; but the rotor—a delicate wind mill—was easily deformed by ice and a duplicate had been kept in reserve in case of damage.

A duplicate record was maintained by a modified Draper anemograph—having a rather heavy, substantial four–cup Robinson cup–wheel about twice the dimensions of the Weather Bureau instrument but adjusted to agree with it. This instrument also recorded mechanically the movement of the wind. It was very durable and required little attention.

There has been but one record of the direction of the wind—that from a Draper anemoscope of simple, durable construction and seldom in need of adjustment.

The deterioration already mentioned was very conspicuous in these wind–records, and first became evident to others than the Observatory staff by the extraordinary velocities published in the years 1922 and thereafter. Velocities are usually very constant and change very little through a long period of years although the mean for an occasional month may be abnormal. Although the velocities, as will be explained hereinafter, are approximately "true" velocities and about 83 percent of those indicated by the Robinson anemometers having a factor of 3, some of these published values for summer months were larger than the indicated means for February 1886 and one other winter month, and about twice as high as the normal for the month. Evidently the observers were entirely unaware of the condition of the instruments and unable to make or plan suitable adjustments or corrections.

The bearings of the Draper anemograph were worn out after a few years and this instrument was discontinued about 1917. (The lower bearing—a disk of hardened steel—was worn or drilled through by the spindle.) It was replaced by a pressure–tube instrument (a Dines–Baxendell "anemo–biagraph)" but this apparently was never standardized for the records, covering several months, have no explanatory notes and some are not timed or dated.

The cinemograph rotor became deformed (perhaps by ice) probably before 1920, and since high winds tend to alter the pitch of the sails so that it over–registers, two new rotors made at this time evidently were based on the deformed one, the sails of which were set at about 12° instead of the most efficient angle or pitch of 45°. Tested in the wind–tunnel of Mass. Institute of Technology, this flat–pitched rotor gave velocities much too high, as should have been expected: at a V of 15 m s^{-1} it registered 30 m sec^{-1}. No data of the period of use of this rotor are on record but its performance indicates that it could easily have yielded the extraordinary high velocities referred to. The recording apparatus of this instrument evidently was in bad condition during many years, for the record of velocity does not agree with that of movement at the upper edge of the chart and only during the daylight period of about 6 or 7 a.m. does it appear to have been actually made by the instrument actuated by the wind. During the period from 8 p.m. to 7 a.m. the observer drew a free–hand line apparently indicating his opinion or estimate of the changes of velocity and sometimes another intended to be a record of the movement of the wind, though, as stated, this did not always agree with the estimate of velocity. Possibly, something might be made of these records by carefully comparing them with the record at Boston, but it is believed, after carefully comparing the Blue Hill and Boston records since 1885, that a straight interpolation of the adjusted Boston record will be sufficient to maintain the normal for the entire period of the Blue Hill record. As will be seen from the discussion of available data, the agreement of adjusted means is reasonably close. All extremes, however, are very uncertain; perhaps entirely lost.

Fragmentary records of velocity during the period 1918–1931 were obtained from the anemo–biagraphs referred to, and in 1926 to 1929 from a new three–cup anemometer in use during a few months of 1926, 1927 and 1928. The records of the latter instrument were tabulated together with others from the older four–cup instrument for possible use, but, because of defective exposure (shielding of one anemometer by another) appear to be of very little use. (This shielding was mentioned by Mr. McAdie in notes on this comparison.)

The direction of the wind reported by the Draper anemoscope appears to have been very uncertain from about 1916. It has been said by U.S. Navy students at Blue Hill during the war, that the wind–vane was loose on its spindle which could be set to indicate any point of the compass; that these students purposely changed the position of the cylinder at times so that it indicated anything but the true direction until the observer discovered its condition and reset, without securing the cylinder, in a position estimated to be correct. The record has not been examined in detail, but during many years the charts of the anemoscope appear to bear the same kind of "estimated" records that characterize the cinemograph; perhaps a few brief records by the instrument during the day and the remainder drawn in by hand to resemble that normally made by instrument, but unmistakably the work of the observer. In many cases the record begins at 8 p.m. but always ends at 4 p.m. daily.

The direction of the wind can not be interpolated as accurately from the Boston record as can that of velocity and it may be that this should be used only during periods of strong, steady winds from certain directions; this use will require careful examination in detail of both records. During much of the period 1916–1931 the Blue Hill record will probably have to be designated as missing.

Description of Records, Particularly Those Obtained between 1913 and 1931, for the Information of All Wishing to Use Them. Because of the entire absence of all descriptions or notes in record–books and files and charts, it is very necessary that an adequate description indicating the character and accuracy of all automatic records be pasted in every file and every record–book for the guidance of those making use of the records. Probably a single sheet printed in small pica will suffice for all items and an edition of about 500 should be enough. The wording of the text doubtless will have to be adapted to a more thorough examination of the records than has been possible so far, but it is suggested that the preliminary draft herewith may cover the essentials without injustice to the observers during the doubtful period of 1913–1931, and be useful to the present staff and all who come after us.

(Accompanying this memorandum:

"Draft of Instructions to be attached permanently to record–books and files of charts. Analysis of Record of Wind–velocity at Blue Hill Observatory and Boston.")

Blue Hill Observatory,

17th August, 1937.

S. P. Fergusson.

His instructions "to be attached to the record books and chart files," which was never done, follows:

Description Of The Records in This File, an Estimate of Their Accuracy and Suggestions for Their Use. (Edited by their author.)

Results of Examination of Records at the Observatory.

Barographs. Record appears to be nearly complete until the clock failed (apparently about 1924). Later, the clock became disconnected from the chart–board which was moved by the observer at irregular intervals during a day but there are no time–marks on the charts.

The record nearest continuous is that of the Richard mercurial barograph whose scale is 2 x 1 and in "kilobars" (millibars). The "monthly" barograph was also maintained during part of the period and gaps in the other records may perhaps be filled from that of this instrument. The record at Boston may safely be used in interpolations.

Thermographs. These were maintained at the Observatory, Base and Valley stations for several years after Rotch's death, but afterward the latter two were discontinued; for some time before discontinuance the records were fragmentary and very uncertain. The instrument at the Observatory was given better care but there is evidence that missing traces were interpolated from some source not indicated or are simple estimates of the observer. About 1923 to 1930 a large proportion of the charts in the files are blank, containing nothing but a date without explanation of any kind. Interpolations from the record at Boston are safe only during the prevalence of strong, westerly or southerly winds and easterly storms—not sea–breezes—or during off–sea winds the temperatures at the Observatory and Boston differ considerably.

Hygrographs. These instruments, much more difficult for unskilled observers to maintain in good condition, became unserviceable after a few years and during at least one period of five years (about 1920 to 1925 or later) there was no hygrograph in use; also, due to uncertainties as to accuracy of thermometers and doubtful condition of wet–bulb coverings, the observations at regular hours must be regarded as uncertain and the entire record of humidity during the period 1913–1931 of doubtful value. Interpolation from Boston seems inadvisable at any time, although this may be of some use when all modifying conditions are allowed for.

Anemographs. Because of severe conditions and the strain on anemometers at Blue Hill two velocity anemometers have been maintained from the beginning, all adjusted to the Weather Bureau standard (designed and manufactured by James Green after 1855). Until 1898 data of velocity or movement of wind were recorded as if the cup–wheel revolved with one–third of the velocity of the wind—i.e., or that the "factor" of Robinson anemometers was 3, regardless of dimensions or proportions. It had been known for many years (since Stow's comparisons in 1872) that this assumption was incorrect but not until the studies of Dines and Marvin (1888–1896) with whirling apparatus was it known with reasonable assurance what changes should be made in standards of velocity, and not until about 1902 or later (chiefly because of the desirability of keeping the records comparable) was the standard changed in England; in this country "true" velocities were not officially authorized until 1928, but, in order that the records might be comparable with others from kite–meteorographs and measurements of clouds, the Blue Hill anemometers, beginning with 1898, were altered to record velocities nearly correct at the most frequent velocity at the Observatory. The most satisfactory method of accomplishing this was to change the wheel–work of the instruments so that their rate would be 83 percent of that obtaining before the change; this was accomplished by using a gear with 60 teeth instead of the one having 50 teeth in gear with the spindle. (The use of 61 or 59 teeth would be caused, respectively, under– or overregistration.) The Richard anemo–cinemograph, having a rate nearer constant than that of the Robinson anemometers, was adopted as the standard anemometer about 1891 and has been in use ever since. A Draper–Robinson instrument has provided a duplicate record.

But one anemoscope (a Draper instrument) has been in use since the beginning; this is a simple, rugged direct–recording instrument, seldom out of order and easily kept in condition.

The cinemograph is one of the very few instruments recording velocities direct on a scale without calculation being necessary, and therefore is a most desirable and useful anemograph. It is complex but very durably built, and requires competent care by one who understands it. Its durability is indicated by continuous operation with only minor repairs since 1890.

The records from all the anemographs have been examined and adjusted to agree with standard or "true" velocities according to the scale adopted in 1898. Monthly means of velocities recorded by all instruments in use during the period 1913–1931, whether of one or several months, have been compared with that of the cinemograph, and missing or doubtful data have been obtained from the Boston record; details of this work will be found in another note filed with the records. It was found possible to interpolate monthly means from the Boston record after allowing for a slow decrease of the mean velocity at Boston since 1890, and the mean difference between that and the mean at Blue Hill which is 1.63 times higher. During some months in almost every year there are differences of about 1 mi. hr^{-1} between these adjusted means, possibly due to different directions (sea–breezes) but the annual means are in very good agreement. This adjusted record of means, however, should not be used to adjust extremes or hourly or daily means, for often there is little resemblance between the hour–to–hour records at the two stations.

Examination shows that the record from the cinemograph probably began to fail shortly after 1912, although the mean velocities determined and published appeared to be at least approximately correct for several years afterward—until 1914 to 1916. Because of the ease with which data can be evaluated from the charts there was evidently an effort made to trace upon them some kind of estimate of the changes of velocity long after the instrument became unserviceable. This will be apparent on inspection of charts during many years beginning about 1920 or earlier; referring to an example, the chart dated............ bears two kinds of tracings; that marked A (from about 8 p.m. to 7 a.m. next day was evidently drawn by hand—the instrument's pen could not move backward and forward in time or in regular loops, nor is it likely that the color of the ink would change abruptly every day at 6 a.m. and 8 p.m. with the exception of a few storms these "estimated" tracings could not be connected with wind–data from any other instrument or source; nor does the record of movement of the wind at the top of the chart (mostly "estimated" as was that of velocity) have any relation to the tracing supposed to indicate velocity. The actual record made by the instrument itself from 7 a.m. to 8 p.m."

Signed S. P. Fergusson

Probably August 1937.

When Dr. Brooks became director in 1931 his first priority was to restore accuracy in the observational program. The instruments were rebuilt, thermometer errors determined, and entries in the observational books were carefully checked. Thermometer corrections were not routinely applied until about 1940. From that time until about 1959, when the Weather Bureau took over the program, calibrations, which were tied into the National Bureau of Standards, were performed about once a year; corrections were applied and checks meticulously made. As the Weather Bureau instruments replaced those in former use, corrections to thermometers and anemometers ended. An apparent decrease in speed of 0.5 m sec^{-1} (1 mi. hr^{-1}) occurred at about this time. Investigation is underway by the author.

Hicks barometer number 872, which had been used prior to 1959, continued to serve as the standard barometer for the Weather Bureau and National Weather Service.

Beginning in 1959 the bureau published average temperatures based on one–half the daily maximum plus minimum, midnight to midnight. Since the previous record had been reduced to means of 24 h, or their equivalent, a discontinuity was introduced. However, for those interested, 24 h means are now available upon request to form a homogeneous record.

A discontinuity was also introduced in the Campbell–Stokes sunshine record, beginning in July 1975, when corrections were applied for the time interval from sunrise until the time the card could show a burn; corrections were similarly applied for sunset. Wollaston and Conover homogenized the record through 1984 in January 1985 by applying monthly corrections, the cards were not reread. This record of unadjusted monthly totals and its continuation is also available.

Under the bureau and Weather Service, data entered on the observational forms are considered accurate although computational checks are not made and some errors have been found in the chart notations. Most serious has been the loss of significant numbers of original charts between 1959 and 1978.

Appendix B

Historical Instruments of the Observatory

Many of the instruments that have been discussed in the foregoing chapters are now at the Collection of Historical Scientific Instruments, Science Center, Harvard University, Cambridge, Massachusetts 02138. The instruments have been tagged with a circled number corresponding to the numbers in the following list:

1. Radio-meteorograph. Intermediate Model #31272. (cardboard box)
2. Radio-meteorograph. Early model, single tube, balsa box.
3. Balloon meteorograph. In wicker basket, pressure bellows missing.
4. Assmann meteorograph 1902. Pressure, temperature and relative humidity. Strip chart. Made by Kgt. Aëronaut Observatorium Reinickendorf-W. No. 17.
5. Dines helicoid air meter, 1887. Made by R.W. Munro, London.
6. Negretti & Zambra clock for inverting thermometers at specific times.
7. Thermograph complete with clock and 24-hour drum. Probably first thermograph at Blue Hill. First chart 22 March 1886. Made by Richard Freres, Paris. No. 2612.
8. Hair hygrometer (single hair).
9. Radio-meteorograph. Single tube, balsa box. Made by Feiber Instrument Company, Cambridge, MA. No. 241.
10. Radio-meteorograph. Two tubes, balsa box. Made by Feiber Instrument Company, Cambridge, MA. No. 181.
11. Radio-meteorograph. Final model, cardboard box.
12. Radio-meteorograph. Pilot model of Simmonds–Lange type. No. TPC 23.
13. Radio-meteorograph. First commercial model Simmonds–Lange type. No. TPH I6.
14. Balloon meteorograph—Jaumotte. Pressure bourdon tube removed. Air scope appears to have been added later.
15. EarIy model "L" radio-meteorograph in bakelite frame. Pressure change takes the place of a clock. No humidity. Possibly used on aircraft for tests.

16. Early model, radio-meteorograph mechanism. Temperature and pressure. RMF 98.

17. Radio-meteorograph, single tube, balsa box. Made by Feiber Instrument Company, Cambridge, Massachusetts. No. 952.

18. Archibald's Kite. Cloth, employed in experiments with anemometers, 1882-84.

19. Kite meteorograph. Pressure, temperature, relative humidity, and wind speed. Made by S. P. Fergusson.

20. Aspirated temperature, relative humidity, and dew point recorder. Designed by A. McAdie.

21. Hygrograph, 24-hour drum and clock. Probably the first used at Blue Hill Observatory. Made by Richard Freres, No. 20888. Shield around hairs added later.

22. Hair hygrometer. Hottinger and Cie.

23. Hair hygrometer. Hottinger and Cie.

24. Samples of coverings for kites. 1894–1905. (In large envelope.)

25. Ventilated case holding dry and wet bulb thermometers graduated in Kelvin Kilograd. Relative humidity and absolute humidity are obtained from a nomogram attached to the inside. First experimental model of McAdie's.

26. Owens dust counter. Made by Casella.

27. Maximum, minimum, dry and wet bulb thermometers in cases. For use in a portable exposure frame. Made by Baudin.

28. Device for determining the true wind while at sea on a moving vessel. Designed by A. L. Rotch 1903. Made by Casella & Company, London.

29. Fineman standard nephoscope, 1882.

30. Owens dust counter.

31. Nephoscope, c. 1890.

32. Nephoscope, Fineman Model 1880.

33. Parts of nephoscope, McAdie Model 1920.

34. Anemometer, c.1900. Fan type. Scale in meters. Compass on base. Made by Richard Freres, Paris.

35. Dines meteorograph. c. 1908. Pressure and temperature.

36. Nephoscope 1895. Designed by S. P. Fergusson for Clayton.

37. Nephoscope. Made by S. P. Fergusson in use at Blue Hill Observatory about 1932 - 1950. Eyepiece on a stand that could be moved around on the floor.

38. Parachute, red, for balloon meteorographs.

39. Dines kite meteorograph, 1902. Pressure and temperature. Wooden case, cloth cover.
40. Dines anemometer head.
41. Thermometer in vacuum.
42. Thermometer, kilograd and Fahrenheit scales.
43. Thermometer, kilograd and Fahrenheit scales.
44. Wind direction indicator attached to wind vane shaft inside Observatory.
46. Anemo-cinemograph, spring driven. Made by Richard Freres.
47. Anemoscope. Made by Draper.
48. Vertical wind component recorder.
49. Mercurial barometer temperature correction scale. Designed by S. P. Fergusson.
50. Ultraviolet meter, used about 1938.
51. Pendulum timer switch for nephoscope.
52. Spectrometer, A. Hilger, London.
53. Two curved sails from the first cinemograph mill used in 1891.
54. Pressure-tube anemometer. Made by W. H. Dines of England for use in comparisons of anemometers At Blue Hill Observatory, 1892 - 1896.
55. Flask.
56. Model pressure tube anemometer with direction scale after Lind Scale, in miles.
57. Regnault's dewpoint measuring device. Made by L. Goloz et Fils. Const RS, Paris 1884. No. 135. (Chrome plated in later years.)
58. Marvin bright sunshine sensor. Made by J. P. Friez, Baltimore, Maryland.
59. Jordan sunshine recorder. Uses light-sensitive paper or film. Made by Negretti and Zambra, London. In use 1886.
60. Pickering model sunshine recorder. Made for use on five successive days. (For noon-sunset only, other half-item 60b.)
61. Plotting machine for determining cloud heights by triangulation. Designed by Clayton, constructed by Fergusson. First used 1890.
62. Automatic camera used to take pictures from kites. Made for A.L. Rotch by L. Gaumont and Cie, Paris. May be set to take pictures from 80 min to 2 h after start of timer.
65. First thermograph lifted by a kite to sound the atmosphere. Made by Fergusson. Tracing of first sounding with explanations on clock drum. 4 August 1894.

66. Parachute for Blue Hill Observatory radio-meteorograph.

67. Windmill anemometer. 8 blades. Stopwatch and movement dial in meters. Made by Jules Richard, Paris.

68. Pyrheliometer—normal incidence. Made by E. Ducretet, Paris. No. 50.

69. Frame for Clayton nephoscope. 1886. First nephoscope used at Blue Hill Observatory, and probably in the United States.

70. Nephoscope for use on shipboard.

71. Black bulb thermometer in vacuum. H. J. Green. No. 3680.

72. Black bulb thermometer in vacuum. H. J. Green. No. 2455.

73. Thermometer. °F with Kilograd back. H. J. Green. No. 38130.

74. Evapometer.

75. Glass pressure tube anemometer. Presented to Blue Hill Observatory by W. Dines for tests with other anemometers.

76. Portable wet and dry thermometer. (One thermometer broken.)

77. Turnover thermometers by Negretti and Zambra. Nos. 64353 and 62040.

78. Multiscale thermometer. Negretti and Zambra, London. No. 45871.

79. Maximum thermometer, Negretti and Zambra.

80. Thermometers (4) Negretti and Zambra. Kilograd scale. (One broken.)

81. Thermometers (3) H. J. Green. Kilograd scale. One has what was probably McAdie's first experimental multiscale back.

82. Multiscale thermometer backs by H. J. Green.

83. Thermometers (3). Kilograd scale. H. J. Green.

84. Thermometers (2). H. J. Green. Dating back to 1894.

85. Old station maximum thermometer.

86. Thermometer #1250, °C

87. Thermometer, -100 to 30°C. Baudin No. 16398.

88. Thermometer. Noncylindrical tube rectified by Baudin (1885).

89. Minimum thermometer. J. Hicks, London. No. 181510. Milk glass back, oak case.

90. Maximum thermometer. J. Hicks, No. 45981, London. Milk glass back, oak case.

91. Thermometer, °F. Hicks 252161 on homemade back. Used as the standard at Blue Hill Observatory, 1885 - 1959.

92. Maximum thermometer -39° to 0°F. J. A. Seitz. Boston, Massachusetts.

93. Minimum thermometer. H. J. Green. No. 21935.

94. Dry and wet thermometers. °K. Milk glass backs. By H. J. Green. Nos. 27307 and 27308—1915. Originally property of S. P. Fergusson.

95. Thermometer -30° to 124°F. Made by Huddleston, Boston. Thermometer used by Reverend Charles Brech, Milton, Massachusetts, beginning in 1849.

96. Standard thermometer. Hicks No. 766523. Broken. (Not the Blue Mill Observatory standard.)

97. Black bulb in vacuum. Hicks, London. No. 34874.

98. Portable barograph. Enregistreurs Richard, Paris. Used by A. L. Rotch in his travels. Bears a trace up Mount Washington and return, 7 November 1937.

99. Campbell–Stokes sunshine recorder, possibly from Geographic Institute, Harvard University.

100. Graves spectroscope. By John Browning, London.

101. Mohn cloud theodolite. (Cost, $100.)

102. Draper's self-recording raingage. Clockwork and collector missing. (Capacity, 5 in. Cost, $175.)

103. Same as item 28. Assembled device for determining true wind from a moving vessel.

104. Wooden split-tail wind vane. Weather Bureau model by J. P. Friez. (Use on Blue Hill?)

105. Draper cup anemometer.

106. Carrying case, conical shape.

107 and 108. Anemometer. Dines portable. By Casella, London, No. 156

109. Miscellaneous chart drums, some with clocks.

110. Kite meteorograph. Pressure temperature, relative humidity, and wind speed. Made by S. P. Fergusson, 1898.

111. Frame, with some cloth of six-sided kite. Label illegible.

112. Crova actinometer. Black and gold balls with compensating ambient temperature elements. Made by Richard Freres, Paris.

113. Kite (unassembled in long wooden box including cloth made by L. Hargrave in 1896 for prize offered by Boston Aeronautical Society. Awarded prize as best kite for use in high winds. S. Cabot, H. H. Clayton, S. P. Fergusson, Committee on Tests.

114. Self-recording pressure plate anemometer. In use June 1887–1889. Designed by A. L. Rotch.

115. Split-tail wind vane, used until 1941 at Blue Hill Observatory.

116. Maximum gust anemometer. Pressure plate on front unattached. See item 134. Split-tail. Designed by A. L. Rotch. Used 1886–1889.
117. Smithsonian silver disk pyrheliometer.
118. Remote weather station recorder.
119. Damping vanes for Draper anemoscope.
120. Clock drive for Draper anemoscope.
121. Three-cup wheel. BHO No. 2, from Draper anemometer, 1931 (?)–1960.
122. Radio-meteorograph recorder—Lange model.
123. Split tail wind vane. Connected to anemoscope. In use 1941–1960. Made by S. P. Fergusson.
124. Meteorograph case. Made by Fergusson.
125. Sunshine recorder by Usteri-Reinochan, Zurich. No. 63. Designed by Dr. Maurer.
126. Wind vane and cup anemometer connected to Draper recorder, 1931 - 1960. Three-cup anemometer. No. 1 (broken).
127. Remote weather station transmitter.
128. Heated pressure tube anemometer. Made by John H. Conover.
129. Case for weighing precipitation gage. Designed by Rotch. Made by Richard, Paris. Put in use 1886. First gage to successfully record water equivalent of snowfall.
130. Screen to go around precipitation gage of item 129.
131. Remote weather station sensors and encoder.
132. Base, weighing and recording mechanism to Rotch precipitation gage (part of item 129).
133. Vertical wind component anemometer.
134. Plates of pressure plate anemometers.
135. Snow gage designed by Rotch.
136. Part of anemometer to fit on a kite meteorograph.
137. Wicker case for early balloonsonde.
138. Weekly and monthly mercurial barograph. ("J" tube with float.)
139. Draper daily mercurial barograph.
140. Draper anemograph, daily. Recorded 1938 hurricane wind.
141. Hargrave kite frame.
142. Hargrave kite frame and cloth.

The following instruments at the Blue Hill Observatory, P.O. Box 101, Milton, MA 02186. Tagged with circled numbers as described below: BHO 126. Raingauge. Small conical orifice atop a triangular collector. Collector sinks with accumulation of water, against three springs. At full capacity, probably 20 cm, the collector dumps and the cycle is repeated. Vertical motions of the collector are graphed on a 24-hour chart. Clock is below chart drum. An enameled time scale below the drum shows the time. No. 36. Made by Th. Usteri-Reinacher, Zurich.

BHO 127. Recorder for pressure tube anemometer. Serial No. 104-38, Cat. No. 454. Made by Julian P. Friez and Sons, Inc. In use 1943-1959, connected to Conover heated anemometer.

BHO 128. Time-lapse camera mount with azimuth scale. Made to fit mounts on the tower parapet. In use 1956-1959.

BHO 129. Micromax temperature-recorder potentiometer. 3 mv full scale. Made by Leeds and Northrup Co. Serial No. 251755M. Used to record normal incidence solar radiation.

BHO 130. Rangefinder. 1.5 m base coincidence type. Model 4, No. 150, 1918. Mark XI. Made by Bausch and Lomb Co. Used by Brooks to measure balloon and cloud heights, 1943-58.

BHO 131. Metal tripod to support rangefinder.

BHO 132. Three feet to hold legs of tripod.

BHO 133. Wooden tripod to support telescope, telescope missing.

BHO 135. Esterline recorder, used by Weather Service.

BHO 136. Bright sunshine sensor, U.S. Weather Bureau. No. 156409 WBZ.

The following is a list of miscellaneous paper materials.

1. Original records of miscellaneous kite soundings, 21 December 1895–15 September 1911.
2. Reprints—old and books—the old reprint file.
3. Pictures.
4. Miscellaneous correspondence.
5. Newspaper clippings (Rotch period).

Appendix C

Bibliography

This bibliography lists work fully or partially performed by Blue Hill personnel at the Observatory. Work by staff members "on loan" to other institutions and funded from other sources is not included.

Although the Mount Washington Observatory was closely affiliated with Blue Hill during Brooks's directorship, work done by personnel assigned to that station is not included. An exception occurred in the late 1940's and early 1950's when closely related work of staff members who occupied space in Cambridge was included.

Before the days of abstract listings it was the practice to place reviews of work published elsewhere, or abstracts of lectures, in journals. Rotch, Clayton, McAdie, Brooks, and Stone authored many of these reviews but the policy here has been to exclude most of them and attempt to include only references to work that appears original in scope.

1884

Rotch, A. L. Establishment of a meteorological station on Blue Hill. *Amer. Meteor. J.* 1:304-305.

—— The Blue Hill Observatory. *The Tech.* 3 December 1884, 46.

1885

Rotch, A. L. Inversion of the wind's diurnal period at elevated stations. *Amer. Meteor. J.* 2:29-33.

1886

Clayton, H. H. A brilliant aurora. *Sci.* 8:124.

—— Anemometer exposure. *Sci.* 8:458.

—— An experiment in long range prediction. *Amer. Meteor. J.* 2:457-463.

—— Barometer exposure (effect of wind on the barometer). *Sci.* 7:484.

—— Cause of a recent period of cold weather in New England. *Sci.* 8:233, 281.

—— Loomis' "Contributions to meteorology" — The origin and development of storms. *Amer. Meteor. J.* 3:212-220, 270-280, 356-367, 417-426, 459-467 and 4:34-46.

—— Thunder storms moving from east. *Amer. Meteor. J.* 3:387.

—— Visibility of Bishop's Ring. *Amer. Meteor. J.* 3:94.

McAdie, A. Atmospheric etricity at high altitudes. *Amer. Meteor. J.* 2:415-421.

Rotch, A. L. The Connecticut tornado. *Amer. Meteor. J.* 3:310-316.

—— The mountain meteorological stations of Europe. *Amer. Meteor. J.* 2:445-457, 500-510, 538-548 and 3:15-24.

—— The Sonnblick Observatory, the highest meteorological station in Europe. *Amer. Meteor. J.* 3:385-386.

1887

Clayton, H. H. A sensitive wind-vane. *Sci.* 9:342.

—— Barometer exposure. *Sci.* 9:316.

—— The barometer during thunderstorms. *Sci.* 9:392, 418.

—— The distribution of the weather in storms and anti-cyclones as affected by local influences. *Amer. Meteor. J.* 4:74-82.

—— Tracings from the self-recording instruments to illustrate certain meteorological phenomena. *Results of the meteorological observations made at the Blue Hill Meteorological Observatory in the Year 1886*, Boston, MA: Alfred Mudge and Son, 44-45.

Rotch, A. L. *An account of the foundation and work of the Blue Hill Meteorological Observatory.* Boston: Alfred Mudge and Son. 29 p.

—— An experiment in local weather predictions. Amer. Meteor. J. 3:454-458.

—— Hourly readings of wind velocity and atmospheric pressure in the United States. *Amer. Meteor. J.* 4:54-55.

—— *Results of the meteorological observations made at the Blue Hill Meteorological Observatory in the year 1886.* Boston: Alfred Mudge and Son. 42 p.

—— Some results derived from the hourly observations of atmospheric pressure at the Blue Hill Observatory during 1886. *Amer. Meteor. J.* 4:260-262.

—— Some results of wind observations made in 1886 at the Blue Hill Meteorological Observatory. *Amer. Meteor. J.* 4:17-21.

1888

Clayton, H. H. A thirty-day period of thunderstorms, the moon and the weather. *Amer. Meteor. J.* 4:407-409.

—— An unusual auroral bow. *Sci.* 11:289.

—— Diurnal cloud and wind periods at Blue Hill Observatory during 1887. *Amer. Meteor. J.* 5:321-332.

—— Instructions for observing clouds (a review of Abercromby's instructions with comments by H. H. C.). *Amer. Meteor. J.* 5:379.

—— Local weather predictions. *Amer. Meteor. J.* 4:409-417, 482-484, 5:38-43.

—— Weather predicting. *Sci.* 11:22-23, 56-57.

Fergusson, S. P. A new self-recording rain gauge. *Amer. Meteor. J.* 5:321.

Rotch, A. L. The Austrian meteorological station on the Sonnblick (10,170 feet high). *Amer. Meteor. J.* 5:13-19.

—— The closing of the U.S. Signal Service stations on Pike's Peak and Mt. Washington. *Amer. Meteor. J.* 5:284-286.

—— The new Swiss Meteorological Observatory on the Santis (8,200 feet high). *Amer. Meteor. J.* 5:99-105.

—— The organization of the meteorological service in some of the principal countries of Europe. *Amer. Meteor. J.* 5:49-62, 241-259, 393-409, 481-492. Reprinted by Register Print & Publishing Co., Ann Arbor, Michigan. 115 p.

Upton, W., and A. L. Rotch. Meteorological observations during the solar eclipse of August 19, 1887, made at Chlamostino, Russia. *Amer. Meteor. J.* 4:356-369, 450-459.

1889

Clayton, H. H. A device for facilitating the reading of electrical anemograph records. *Amer. Meteor. J.* 6:330.

—— Cloud observations; Introduction. *Ann. Astron. Obs. of Harvard College.* 20 Part 1:50-57.

—— Comparison of the thermometer shelters. *Ann. Astron. Obs. of Harvard College.* 20 Part 1, Appendix A:115-120.

—— Formation des nuages. *Mem. du Cong. Meteor.* Paris. 80 pp.

—— Investigation of the normal difference of temperature between the base and the summit. *Ann. Astron. Obs. Harvard College.* 20 Part 1, Appendix B:120-130.

—— Some diurnal and annual oscillations of the barometer. *Amer. Meteor. J.* 6:151-153.

—— Tracings from the self-recording instruments illustrating certain meteorological phenomena. *Ann. Astron. Obs. of Harvard College.* 20 Part 1, Appendix D:137-141.

—— Verification of weather. *Amer. Meteor. J.* 6:211-219.

Fergusson, S. P. Investigations of the marked inversions of temperature between the base and the summit. *Ann. Astron. Obs. of Harvard College.* 20 Part 1, Appendix C:131-137.

Rotch, A. L. *Meteorology at the Paris Exposition*. 46 p.

—— Observations made at the Blue Hill Meteorological Observatory in the year 1887 with a description of the Observatory and its work. *Ann. Astron. Obs. of Harvard College*. 20 Part 1:1-114.

—— Observations made at the Blue Hill Meteorological Observatory in the year 1888 with a statement of the local weather predictions. *Ann. Astron. Obs. of Harvard College*. 20 Part 2:147-267.

—— Sur les appareils employes a l'observatoire de Blue Hill pour measurer les mouvements de nuages et pour enregistrer la nebulosite. *Mem. du Cong. Meteor.* Paris. 77 p.

1890

Clayton, H. H. Cloud observations in 1889. *Ann. Astron. Obs. of Harvard College*. 30 Part 1:39-75.

—— Notes on cirrus formation. *Quart. J. Royal Meteor. Soc.* 16:16-20.

—— Sonnblick meteorology (Review of book by J. Hann). *Amer. Meteor. J.* 7:327-329.

Fergusson, S. P. A new recording rain and snow gauge. *Amer. Meteor. J.* 7:231-233.

—— Observations made at Blue Hill Meteorological Observatory in the year 1889 with a statement of the local weather predictions. *Ann. Astron. Obs. of Harvard College*. 30 Part 1:1-38.

Rotch, A. L. Observations made at the Blue Hill Meteorological Observatory in the year 1890. And appendicies: 1. Summary of U.S. Sig. Service records, Boston, MA 1890. 2. Summary of meteor. records at BHO 1886 and humidities 1887 and 1889. 3. Summary of meteor. records at BHO during lustrum, 1886-1890. Also ref. to descriptions of the work at the BHO from 1885-90. *Ann. Astron. Obs. of Harvard College*. 30 Part 2:81-201.

—— *Quelques resultats des observations au Mont Blanc*. Limoges.

1891

Clayton, H. H. Cloud heights and velocities at Blue Hill Meteorological Observatory. *Amer. Meteor. J.* 8:108-122.

—— Long range weather predictions (relation of cloud observations to). *Amer. Meteor. J.* 8:62-68.

—— Progress in meteorology (correspondence). *Amer. Meteor. J.* 8:91-92.

—— The chief features of the diurnal and annual periods at Blue Hill, as shown by the (meteorological) tables (for lustrum 1886-1890). *Ann. Astron. Obs. of Harvard College*. 30 Part 2, Appendix 2:191-198.

—— Verification of weather forecasts (made at Blue Hill). *Amer. Meteor. J.* 8:369-375.

Fergusson, S. P. An ink recorder for the electrical anemograph. *Amer. Meteor. J.* 8:18-19.

—— Vertical anemograph at Blue Hill Observatory. *Amer. Meteor. J.* 8:382-383.

McAdie, A. Experiments on atmospheric electricity at Blue Hill Meteorological Observatory. *Amer. Meteor. J.* 8:233-235.

—— Franklin's kite experiment (repeated at Blue Hill). *Amer. Meteor. J.* 8:97-108.

Rotch, A. L. A geographical exhibition in Boston. *Amer. Meteor. J.* 8:93-94.

—— Meteorology at the French Association. *Amer. Meteor. J.* 8:301-303.

—— Mountain meteorology. *Amer. Meteor. J.* 8:145-158, 193-211.

—— Reports on the following questions were presented to the International Meteorological Conf. at Munich in 1891. Question 3. The epoch for reading extreme thermometers. Question 12. The appointment of a definite zone round the zenith for the estimation of the amount of cloud. *Rept. of the Conf., London Meteor. Off.* Official no. 103, Appendix 8, 9.

—— Sur la mésure des hauteurs et des vitesses de nuages í l'observatoire de Blue Hill (Etats Unis). *Assoc. Francaise pour l'Advancement des Sciences, Cong. de Marseille, 1891, notes et extraits.* 346.

—— The high-level meteorological observatories in France. *Amer. Meteor. J.* 8:316-325.

—— The highest meteorological station. *Amer. Meteor. J.*, 8:92-93.

—— The local weather predictions at Blue Hill Meteorological Observatory. *Amer. Meteor. J.* 8:58-61.

—— The meteorological observatory recently established on Mont Blanc. *Amer. Meteor. J.* 7:443-446.

1892

Clayton, H. H. Cloud observations. Includes tabular data for 1891. *Ann. Astron. Obs. of Harvard College.* 40 Part 1:25-49.

—— Measurements of cloud heights and velocities. *Ann. Astron. Obs. of Harvard College.* 30 Part 3:207-268.

—— Recent efforts toward the improvement of the daily weather forecasts (containing chart of average cloudiness at Blue Hill in cyclones). *Amer. Meteor. J.* 9:128-134.

Fergusson, S. P. A discussion of the vertical component of the wind during November and December, 1891. *Ann. Astron. Obs. of Harvard College.* 40 Part 1, Appendix B:59-62.

—— The anemograph for vertical currents at Blue Hill Meteorological Observatory. *Amer. Meteor. J.* 8:481-483.

McAdie, A. Abstract of a report upon some experiments on atmospheric electricity. *Ann. Astron. Obs. of Harvard College.* 40 Part 1, Appendix A:53-58.

—— Experiments with kites during thunderstorms. *Electrical World.* Sept. 1892.

Rotch, A. L. A meteorological balloon ascent at Berlin, October 24, 1891. *Amer. Meteor. J.* 9:245-251.

—— Observations made at the Blue Hill Meteorological Observatory in the year 1891. *Ann. Astron. Obs. of Harvard College.* 40 Part 1:1-24. Also list of instruments at the Blue Hill Observatory. 40, Part 1: 50-52.

—— Sur la componante verticale du vent à Blue Hill. *Annuaire de la Soc. Meteor. de France.* 40:103-104.

—— The international meteorological conference at Munich. *Amer. Meteor. J.* 8:433-458.

—— The mountain meteorological stations of the United States. *Amer. Meteor. J.* 8:396-405.

Upton, W., and A. L. Rotch. Meteorological and other observations made in connection with the total solar eclipse of January 1, 1889, at Willows, California. Ann. Astron. Obs. of Harvard College. 29 Part 1:1-34.

1893

Clayton, H. H. Six and seven day weather periodicities. *Amer. Meteor. J.* 10:35-44, 322-325.

—— The effect of high winds on the barometer (at Blue Hill). *Amer. Meteor. J.* 9:563.

—— The movement of the air at all heights in cyclones and anticyclones, as shown by the cloud and wind records at Blue Hill. *Amer. Meteor. J.* 10:170-178.

—— and W. H. Fergusson. Sudden temperature changes (at Blue Hill). *Ann. Astron. Obs. of Harvard College.* 40 Part 2, Appendix D:125-138.

Fergusson, S. P. Anemometer comparisons. *Amer. Meteor. J.* 9:421-429.

—— Anemometry. *Proc. Chicago Conf. on Aerial Navig.* 1893:104-113.

McAdie, A. Experiments on atmospheric electricity made by direction of the chief of the Weather Bureau at Blue Hill Meteorological Observatory, July 12–August 12, 1892. *Ann. Astron. Obs. of Harvard College.* 40 Part 2, Appendix C:120-124.

—— The electrification of the lower air (at Blue Hill) during aurora displays. *Amer. Meteor. J.* 9:443-448.

Rotch, A. L. Observations at the Blue Hill Meteorological Observatory for the year 1892. *Ann. Astron. Obs. of Harvard College.* 40 Part 2:67-120.

—— The highest meteorological station in the world. *Amer. Meteor. J.* 9:282-287.

—— The meteorological stations on Mont Blanc. *Amer. Meteor. J.* 9:411-414.

Ward, R. DeC. Thunderstorms in New England during the years 1886 and 1887. Reviewed by H. H. Clayton and the thunderstorm movements in cyclones compared with the cloud movements at Blue Hill in cyclones. *Amer. Meteor. J.* 10:365.

1894

Clayton, H. H. A study of the short wave-like oscillations shown by the barograph at the Blue Hill Meteorological Observatory. *Ann. Astron. Obs. of Harvard College.* 40 Part 3, Appendix E:195-202.

—— Meteorological records obtained in the upper air by means of kites. *Amer. Meteor. J.* 11:297-303.

—— Rhythm in the weather. *Boston Commonwealth.* (17 Nov. *Amer. Meteor. J.* 11:376-380; *Z. Meteor.* 1895:22-25.

—— Six and seven day weather periods. *Amer. J. Sci.* 147:223-231.

Eddy, W. A., and H. H. Clayton. The Eddy malay tailless kite (with meteorological applications). *Sci.* Amer. 71:169-170.

Fergusson, S. P. The pole-star recorder. *Amer. Meteor. J.* 11:62-64.

Fergusson, W. H. The average weather conditions in a period of 26.68 days. *Ann. Astron. Obs. of Harvard College.* 40 Part 3, Appendix F:203-205.

Rotch, A. L. Meeting of the International Meteorological Committee. *Amer. Meteor. J.* 11:303-310.

—— Observations at the Blue Hill Meteorological Observatory for the year 1893. *Ann. Astron. Obs. of Harvard College.* 40 Part 3:143-194.

—— The meteorological services of South America. *Amer. Meteor. J.* 11:201-211.

—— The Upsala meeting of the International Meteorological Committee. *Nature.* 51:185-186.

1895

Clayton, H. H. A cyclonic indraught at the top of an anticyclone. *Nature.* 52:243-244.

—— Barometric undulations. *Symon's Mon. Meteor. Mag.* 30:29-30.

—— Relation of clouds to rainfall (at Blue Hill). *Amer. Meteor. J.* 12:110-116.

—— Velocity of air currents. *Amer. Eng. and R. R. J.* Aug.:385-386.

Fergusson, S. P. The meteorgraph for the Harvard Observatory on El Misti, Peru. *Amer. Meteor. J.* 12:116-119.

Rotch, A. L. *Mountain Observatories. The Happy Thought.*

—— Observations at the Blue Hill Meteorological Observatory for the year 1894. *Ann. Astron. Obs. of Harvard College.* 40 Part 4:211-264.

—— The meteorological observatory on Mount Cimone, Italy. *Amer. Meteor. J.* 12:219-221.

—— The physiological effects of high altitudes. *Amer. Meteor. J.* 12:221-223.

—— Studies of the upper air. Boston *Commonwealth*, Apr. (Summary in *Amer. Meteor. J.* 12:90-93.)

1896

Clayton, H. H. Cyclones and anticyclones (cirrus movements in). *Sci.* 3:325.

—— Das Einströmen im ob even Theil Einer Antickleno. *Z. Meteor.* 13:176-178

—— Discussion of the annual and diurnal periods, as shown in the preceding tables. (Summaries for the lustrum and decade ending 1895.) *Ann. Astron. Obs. of Harvard College.* 40 Part 5, Appendix H:363-379.

—— Discussion of the cloud observations made at the Blue Hill Observatory. *Ann. Astron. Obs. of Harvard College.* 30 Part 4:273-500.

—— Gebrauch von Drachen bei der Messung von Wolkenhöhen. *Z. Meteor.* 13:140.

—— The origin of stratus cloud and some suggested changes in the international methods of cloud measurements. *Nature.* 55:197-198.

—— The seven day weather period Nature. 54:285. (Abstract *Amer. J. Sci.*) *Nature* 54:285

—— The seven day weather period. *Amer. J. Sci.* 152:7-16.

—— The use of kites for meteorological observations in the upper air. *Nature.* 55:150.

Fergusson, S. P. A high kite ascension at Blue Hill. *Mon. Wea. Rev.* 24:327-328.

—— Anemometer comparisons. *Ann. Astron. Obs. of Harvard College.* 40 Part 4, Appendix G:265-299.

—— Kite experiments at the Blue Hill Meteorological Observatory. *Mon. Wea. Rev.* 24:323-328.

—— Materials used in kite experiments at Blue Hill Meteorological Observatory. *Means's Aero. Ann.* 138-140.

Rotch, A. L. Emploi de cerfs-volants pour enlever des instruments meteorologiques enregistreurs à l'Observatoire de Blue Hill. *Rapt. Conf. Meteor. Int. Reunion de Paris,* 1895. Appendix 8:85-87.

—— Etude des conditions météorologicques des couches supérieures de l'atmosphere par des cerf-volants. *Archiv. des Sci. Phys.et Nature.* Geneve, Quatrieme périod 15 October 2:371-373.

—— Observations made at the Blue Hill Meteorological Observatory in 1895. *Ann. Astron. Obs. of Harvard College.* 40 Part 5:305-362.

—— The chance of observing the total solar eclipse in Norway. *Sci.* 3:356-357.

—— The exploration of the upper air by means of kites; *Rept. Brit. Assoc. Adv. Sci.* Liverpool meeting, 1896, Trans. Sect. A:728. (Abstract.)

—— The international hydrological, climatological and geological congress at Clermont-Ferrand. *Mon. Wea. Rev.* 24:367.

—— The International Meteorological Conference at Paris, September 1986. *Mon. Wea. Rev.* 24:365-367.

—— The meteorological use of kites at Blue Hill. *Amer. Meteor. J.* 12:393.

—— The new meteorological observatory on the Broken. *Amer. Meteor. J.* 12:1-3.

—— The relation of the wind to aeronautics. *Mean's Aero. Ann.* 105-110.

1897

Clayton, H. H. Cloud measurements at Blue Hill. *Mon. Wea. Rev.* 24:135-136.

—— The velocity of a flight of ducks obtained by triangulation. *Sci.* 5:26.

Fergusson, S. P. The early use of wire in kite flying. *Mon. Wea. Rev.* 25:135.

—— The highest kite ascension at Blue Hill. *Mon. Wea. Rev.* 25:392.

——, and H. H. Clayton. Exploration of the air by means of kites. *Ann. Astron. Obs. of Harvard College.* 42 Part 1. Appendix B.: 41-128. Reviewed By R. Suring, *Meteor. Z.*, Apr.: 25-27 and by D. LeBois, *La Nature*, 10 Juin: 27-30.

McAdie, A. Franklin's kite experiment with modern apparatus (repeated at Blue Hill). *Pop. Sci. Mon.* Oct. 739-747.

Rotch, A. L. Cloud observations and measurements at the Blue Hill Meteorological Observatory, Milton, Mass. *Mon. Wea. Rev.* 25:12-13. Reprinted in *Nature* 55:614.

—— Exploration of the air. *Appalachia.* 8:179-189.

—— La météorologie et les cerfs volants. *L'Aérophile.* Mar. 46-47.

—— Meteorological investigation in the free air at the Blue Hill Meteorological Observatory. *J. Assoc. Eng. Soc.* Boston. July, 38-44.

—— Observations at the Blue Hill Meteorological Observatory for the year 1896. *Ann. Astron. Obs. of Harvard College.* 42 Part 1:1-27.

—— On obtaining meteorological records in the upper air by means of kites and balloons. *Proc. Amer. Acad. Arts and Sci.* 32:245-251. (Reprinted in Nature 56:602-603).

—— Progress of the exploration of the air with kites at Blue Hill Meteorological Observatory, Mass., U.S.A. Rept. *Brit. Assoc. Adv. Sci.*, Toronto meeting, 1897, Trans. Sect. A 569. (Abstract).

—— Results from the highest kite flight. *Sci.* 6:561-562, *Nature* 56:540.

—— The exploration of the air by means of kites. *Nature.* 57:53.

—— The international and hydrological meetings. *Sci.* 5:17-19.

—— The use of kites to obtain meteorological records in the upper air at Blue Hill Meteorological Observatory, U.S.A. *Quart. J. Royal Meteor. Soc.* 23:251-253.

Sweetland, A. E. A study of special cloud forms. *Ann. Astron. Obs. of Harvard College.* 42 Part 1, Appendix A:28-40.

1898

Clayton, H. H. Examples of the diurnal and cyclonic changes in temperature and relative humidity at different heights in the free air. *Blue Hill Meteor. Obs.* Bull. 2. 4 pp. (Reviewed in *Nature*, 58:59.)

—— Weather harmonics. *Sci.* 7:243-245.

Fergusson, S. P. The highest kite ascensions in 1897. *Blue Hill Meteorol. Obs.* Bull. 1:2 pp. Reprinted in *Nature* 57:372. *Mon. Wea. Rev.*, Sept:1897: 392 Translated in *Bull. de la Soc. Belg d'Astronomie*, Mars 1898:163-164.

—— The storm of January 31-February 1, 1898. *Blue Hill Meteor. Obs.* Bull. 3, 5 pp.

Rotch, A. L. Drachen und Fesselballons für Meteorologische Zwecke. *Illustrirte Aeronautische Mettheilungen.* April, 51.

—— Les cerfs-volants et les ballons dans la météorologie. *L'Aérophile.* Avril-Mai, 64-65.

—— L'usage des cerfs-volants à l'Observatiore de Blue Hill pour obtenir les observations météoroloques. *Bull. Soc. Astron. de France.* Sept., 377-381

—— Progress in the exploration of the air with kites at the Blue Hill Meteorological Observatory, Mass. (Read before Boston meeting August 1898.) Abstract printed in *Proc. A.A.A.S.* 47, Meeting 127. Printed in full with notes in *Mon. Wea. Rev.* 26:355-356 and *Aero J.* Jan. 1899: 17-19.)

—— Progress in the exploration of the air by means of kites at the Blue Hill Meteorological Observatory, Mass U.S.A. *Rept. Brit. Assoc. Adv. Sci.* Bristol Meeting 1898. Trans. Sect. A 797. (Abstract.)

—— Rapport sur les moyens employés au Blue Hill Météorological Observatory pour obtenir les observations meteorologiques avec des cerf-volants. *Protokoll Int. Aeron. Komm. Strassburg 1898.* Appendix 18:119-120.

—— The eighth general meeting of the German Meteorological Society. *Mon. Wea. Rev.* 26:160.

—— The exploration of the air. *Appalachia.* 8:179-189.

—— The exploration of the free air by means of kites at Blue Hill Meteorological Observatory, Mass, U.S.A. *Quart. J. Royal Meteor. Soc.* 24:250-261.

—— The international aeronautical conference at Strassburg. *Sci.* 8:846-848.

Sweetland, A. E. Tornado at Hampton Beach, N.H. July 4, 1899. *Mon. Wea. Rev.* 26:308-309.

1899

Clayton, H. H. On a recent recurrence in weather—a lunar or 30-day period. *Symon's Meteor. Mag.* 34:68-70; 35:88-89.

—— Studies of cyclonic and anticyclonic phenomena with kites. *Blue Hill Meteor. Obs. Bull. 1.* 19 pp. Translated in *Das Wetter* 16:85-92, 114-116, 139-144. Reviewed in *Ciel et Terre*, 16 Mai 1899:146-147, and *Petermann's Mitteilungen*, 9.

—— The thermometric scales for meteorological use. *Nature.* 60:491.

—— Weather periodicities. *Proc. Amer. Acad. Arts and Sci.* 34 No. 22. (Reviewed by Ward, R. DeC. *Sci.* 10:537-538. Abstract by LeBois, D. *La Nature* 24 Mars. 1900, 275-278.)

Fergusson, S. P. Formation of cumulus clouds over a fire. *Sci.* 10:86.

—— Progress of experiments with kites in 1897-98 at Blue Hill Meteorological Observatory. *Sci. Amer. Supp. 1209*, 4 March 1899, 19375-19377. also (*Blue Hill Meteor. Obs. Bull. 3.* 1899, 8 p.)

Rotch, A. L. Bericht uber die Erforschung der Atmosphare durch Drachen am Blue Hill Observatoriums und an verschiedenen anderen Stationen Amerikas. *Bericht Des Int. Meteor. Komitee.* St. Petersburg, 1899, Anhang 6.

—— Progress in exploring the air with kites, *Rept. Brit. Assoc. Adv. Sci.* Dover meeting, 1899, Trans. Sect.A, 655-656. (Abstract.)

—— Rapport sur l'exploration de l'air par les cerf-volants à l'Observatoire de Blue Hill et á differents stations en Amérique. *Rapport du Comite Meteor. Int.* Reunion de St. Petersburg, 1899, Appendex 6.

—— The highest kite ascent (August 26, 1898). *Illustrite. Aero. Mittheilungen.* 17 January.

Sweetland, A. E. Two remarkable snowstorms. *Blue Hill Meteor. Obs. Bull.2.* 1899: 8 pp.

1900

Clayton, H. H. Kite-flying. *The Universal Cyclopedia.* 6:639-642.

—— Measurements of cloud heights, velocities and directions. *Ann. Astron. Obs. of Harvard College*. 42 Part 2 Appendix C:193-280.

—— Recent exploration in the upper air and its bearing on the theory of cyclones. Nature. 61:611-612. (Translated in *Ciel et Terre*, 1 Sept. 1900, 323-325.)

—— Studies of cyclonic and anti-cyclonic phenomena with kites. *Blue Hill Meteor. Obs. Bull. 1*. 19 p.

Fergusson, S. P. Progress in meteorological kite-flying. *Sci.* 12:521-523.

Rotch, A. L. Fifth report on the use of kites to obtain meteorological observations at Blue Hill Meteorological Observatory, Mass. U.S.A. *Rept. Brit. Assoc. Adv. Sci.* Bradford Meeting, 1900, Trans. Sect. A, 650-651. (Abstract.)

—— International congress of aeronauticals (and meteorology). *Extra Rept. to Int. Univ. Expos.* Paris 8:846-852.

—— International congress of meteorology and aeronautics at Paris. *Sci.* 12:796-799.

—— Kites vs. balloon. (Kite Flight of July 19,1900). *Sci.* 12:193. Translated in *Z. Meteor.* 1900: 524.

—— Observations made at the Blue Hill Meteorological Observatory in the years 1897 and 1898. *Ann. Astron. Obs. of Harvard College*. 42 Part 2:133-192.

—— Physical observations during the total solar eclipse. *Sci.* 11:752-753.

—— *Sounding the ocean of air.* London and New York 184 pp. Soc. for Promoting Christian Knowledge.

—— The eclipse wind. *Nature.* 61:589.

—— The use of kites to obtain meteorological observations. Tech. Quart. 89-99. Reprinted with revisions in Appendix. *Smithsonian Rept.*, 1900, 223-231.

1901

Blue Hill Observatory. *Repts. of the Dir. Harvard College Obs.* 9-10.

Clayton, H. H. The eclipse cyclone and the diurnal cyclones. *Proc. Amer. Acad. Arts. and Sci.* 36:307-318; also *Ann. Astron. Obs. of Harvard College*. 43 Part 1:33.

—— The eclipse cyclone, the diurnal cyclones, and the cyclones and anti-cyclones of temperate latitudes. *Quart. J. Royal Meteor. Soc.* 27:269-292.

—— The effect of diminished air pressure on the pulse. *Sci.* 14:696.

—— The influence of rainfall on commerce and politics. *Popular Sci. Mon.* Dec. 1901, 158-165.

Rotch, A. L. A meteorological balloon ascension at Strassburg, Germany. *Mon. Wea. Rev.* 29:298-300.

—— A method for the scientific exploration of the atmosphere by means of kites. Second conv. of Weather Bureau officials, Milwaukee, WI 1901. *Bull. No. 31, U.S. Wea. Bur.* 66-67.

—— A new field for kites in meteorology. *Sci.* 14:412-413.

—— Kite flying at sea. *Symon's Meteor. Mag.* 36:164-165.

—— Meteorological observations with kites at sea. *Sci.* 14:896-897.

—— Meteorological observations with kites in the United States. Verhandlung des 7. *Int. Geog. Cong.* 1899, Berlin, 1901, Part 2, 399-401.

—— On the systematic exploration of the atmosphere at sea by means of kites. *Rept. Brit. Assoc. Adv. Sci.* Glasgow Meeting, 1901, 724. also *Engineering* 11 Oct. 1901, 514.

—— The chief scientific uses of kites. *Aero. J.* 5:56-59.

—— The exploration of the atmosphere over the ocean. *Nature.* 65:4.

—— The use of kites to obtain meteorological observations. *Report of the Smithsonian Institute,* for 1900:223-231.

1902

Blue Hill Observatory *Reports of the Dir. Harvard College Obs.*, 12.

Clayton, H. H. The daily barometric wave. *Sci.* 15:232.

—— The volcanic eruption in Martinique and possible coming brilliant sky glows. *Sci.* 15:791-792.

Rotch, A. L. Die Erforschung der Atmosphúre Über dem Ozean. *Protokoll Int. Kommission für Luftshiffahrt.* Berlin 1902 Beilage 11.

—— Kites and wireless telegraphy. *Nature.* 65:198.

—— Reviewer Meteoroloische Optic. (Neudrucke con Schriften und Karten Über Meteorologie und Erdmagnetismus herausgegeben von Prof. Dr. Hellman.) *Sci.* 16:352-353.

—— Observations and investigations made at the Blue Hill Meteorological Observatory in 1899 and 1900. *Ann. Astron. Obs. of Harvard College.* 43 Part 2:39-83; and published references to the work of the Blue Hill Meteorological Observatory from 1896-1900, 106-109; and additional errata in the Blue Hill Meteorological Observatory observations 1890-98, 110.

—— Sondages atmosphériques executés à l'aide de cerf-volants à Blue Hill Meteorological Observatory. *Ann. de la soc. Météor. de France.* 50th year, 1902, 78-80.

—— The circulation of the atmosphere in the tropical and equatorial regions. *Mon. Wea. Rev.* 30: 181-183.

—— The exploration of the atmosphere at sea by kites. *Quart. J. Royal Meteor.* Soc. 28:1-8.

—— The international aeronautical congress. *Sci.* 16:296-301.

—— The international aeronautical congress at Berlin. *Mon. Wea. Rev.* 30:356-362. (Abridged in *Nature.* 68:137-141.)

—— The measurement of the wind at sea. *Sci.* 15:72-73.

Sweetland, A. E. A discussion of the temperature during fifty years at Milton, Mass. *Ann. Astron. Obs. of Harvard College.* 43 Part 2, Appendix B:91-105.

—— A study of the visibility of distant objects during the lustrum 1896-1900. *Ann. Astron. Obs. of Harvard College.* 43 Part 2, Appendix A:84-90.

1903

Blue Hill Observatory. *Reports of the Dir. Harvard College Obs.* 9.

Clayton, H. H. A second Bishop's Ring around the sun and the recent unusual sunset glows. *Sci.* 17:150-152.

—— Professor Alexander Graham Bell on kite construction. *Sci.* 18:204-208, and *Sci. Amer. Suppl.*, June 1903, 22975.

—— Results from the kite-meteorographs and simultaneous readings at the earth's surface 1897-1902. *Ann. Astron. Obs. of Harvard College.* 43 Part 3, Appendix C:164-214.

—— The diminishing size of the new Bishop's Ring around the sun. *Nature,* 69:270-271.

—— The eclipse cyclone of 1900. *Quart. J. Royal Meteor.* Soc. 29:47-53.

—— The 27-day period in auroras and its connection with sunspots. *Sci.* 18:632.

Fergusson, S. P. Kites and instruments employed in the exploration of the air at Blue Hill Meteorological Observatory, 1897-1902. *Ann. Astron. Obs. of Harvard College.* 43 Part 3, Appendix D:215-239.

Rotch, A. L. Audibility and meteorological conditions. *Rept. Brit. Assoc. Adv. Sci.* Southport meeting, p. 581. also Engineering. 9 Oct. 1903, 490.

—— Kite observations at the Blue Hill Observatory and the use of this method on the Tropical Oceans. *Rept. Brit. Assoc. Adv. Sci.*, Southport meeting 1903:565-566. also Engineering. 9 Oct. 489.

—— Les sondages de l'atmosphire au-dessus des oceans équatoriaux. *Memoires, Annuaire de la Soc. Meteor. de France.* 51st year, 201-202.

—— Meteorological observations with kites at sea. *Sci.* 18:113-114. (Abstracted in *Nature* 69:65-66. and in *Aeronautical J.* 8:63.)

—— Meteorology at the British Association. *Sci.* 18:657-661; correction in 19:239.

—— Obituary of Arthur E. Sweetland. *Sci.* 17:799.

—— Observations and investigations made at the Blue Hill Meteorological Observatory in 1901 and 1902. *Ann. Astron. Obs. of Harvard College.* 43 Part 3:115-155.

—— Progrès dans l'exploration de l'air à l'Observatoire, de Bleu Hill, et un projet de sonder l'atmosphère au-dessus des mers equatoriaux. *Rept. Int. Meteoro. Committee.* Southport Appendix 5.

—— Progress in exploring the air at Blue Hill Meteorological Observatory, and a project in making atmospheric soundings over equatorial oceans. *Bericht. des int. meteor. Komitees,* 1900 and 1903, Anjang 5.

—— The effect of meteorological conditions upon audibility. *Ann. Astron. Obs. of Harvard College.* 43 Part 3:156-163.

—— The first use of the word "barometer." *Sci.* 17:708. (Reprinted in *Mon. Wea. Rev.* 31:142. Translated in *Z. Meteor.* 38:368.)

—— The investigation of the atmospheric circulation in the tropics. *Sci.* 17:178-179.

—— The new Bishop's Ring. *Nature.* 68:623.

—— The unusual sky colours and the atmospheric circulation. *Nature.* 69:173-174; correction in 69:304. *Sci.* 19:209-210.

—— Un curieux éclair en boule dans la Tour Eiffel. *Bull. de la Soc. Astron. de France.* 17:483. (*Annuaire de la Soc. Meteor. de France,* 51st year, 205-206. Reprinted in *Ciel et Terre.* 24th year, 544.)

1904

Blue Hill Observatory. *Repts. of the Dir. Harvard College Obs.* 9-10.

Clayton, H. H. A study of some errors of kite-meteorographs and observations on mountains. *Mon. Wea. Rev.* 32:121-124.

—— The diurnal and annual periods of temperature, humidity and wind velocity up to 4 kilometers in the free air and the average vertical gradients of these elements at Blue Hill. *Ann. Astron. Obs. of Harvard College.* 58 Part 1:5-62.

—— The study of sunspot cycles. *Symon's Meteor. Mag.* 39:8.

—— Various researches on the temperature in cyclones and anti-cyclones in temperate latitudes. *Beitrage zur Phy. der freien Atmos.* 1: Part 3, 93-107.

Rotch, A. L. An instrument for determining the true direction and velocity of the wind at sea. *Quart. J. Royal Meteor. Soc.* 30:313-316.

—— Five ascents to the observatories on Mont Blanc. *Appalachia.* 10:361-373.

—— Rotch, A. L. Hann's handbook of climatology, rev. translated by R. DeC. Ward, *Amer. Naturalist,* Nov.-Dec., 1904, 899-900.

—— Kite work by the Blue Hill Meteorological Observatory and the United States Weather Bureau. *Mon. Wea. Rev.* 32:567-568.

—— La Temperature de l'air dans les cyclones et anti-cyclones d'après les observations à l'aide de cerfs-volants à Blue Hill Observatoire. *Rapport Conf. Int. d'Aerostation Sci.* St. Petersburg. Supp. 8. (Translated by R. deC. Ward in *Sci.* 23:274.)

—— Present problems of meteorology. *Sci.* 20:872-878. also address to Section of Cosmical Physics of Int. Cong. of Arts and Sci. at St. Louis, 1904.

—— Project for the exploration of the atmosphere over the tropical oceans. (Abstract.) *Rept. 8th Int. Geog. Cong. in the U.S. in 1904.* 322. also *Nat. Geog. Mag.* 15:430.

—— Registration balloons. (Proposing their use in the United States.) *Sci. Amer.* 91:95.

—— The temperature of the air in cyclones and anti-cyclones as shown by kite-flights at the Blue Hill Meteorological Observatory, U.S.A. *Rept. Brit. Assoc. Adv. Sci.* Cambridge Meeting 468-469 and *Symon's Meteor. Mag.* 39:172.

—— The use of kites for meteorological observations at sea. *Sci. Amer.* 91:479

1905

Blue Hill Observatory. *Repts. of the Dir. Harvard College Obs.* 7-8.

Clayton, H. H. The lifting power of ascending currents of air. *Mon. Wea. Rev.* 33:390-391. Reprinted in *Quart. J. Royal Meteor. Soc.* 41:70-71.

Fergusson, S. P. Two new meteorological instruments: 1. The automatic polar star light recorder: 2. The ombroscope. *Quart. J. Royal Meteor. Soc.* 31:309-316.

Miller, E. R., H. H. Clayton and S. P. Fergusson. International definitions and symbols. *Mon. Wea. Rev.* 33:524-527.

Rotch, A. L. Denkmäler Mittelalterlicher Meteorologie. (Neudrucke von Schriften und Karten Über Meteorologie und Erdmagnetismus herausgegeben von Prof. Dr. G. Hellman). *Sci.* 22:116.

—— Die Bergkrankheit in den Alpen und im Himalaya. *Die Umschau.*, Frankfurt, 24 Juni, 1905.

—— Eclipse shadow bands. (Instructions for observing them.) *Nature.* 72:307-308. Abst. in *Bull. Soc Astron. de France*, 19th year, 1905, 368-370; *Ciel et Terre*, 26th year, 1905, 280-281.

—— Extended exploration of the atmosphere by the Blue Hill Meteorological Observatory. *Sci.* 22:57-58.

—— Inversions of temperature and humidity in anti-cyclones. *Nature.* 71:510-511.

—— On the first observations with registration balloons in America. *Proc. Amer. Acad. of Arts and Sci.* 41:347-350.

—— Optical refraction in the lower atmospheric strata as affected by meteorological conditions. Abstract read before *Amer. Assoc. Adv. Sci.* at Philadelphia. *Sci.* 21:335.

—— The exploration of the atmosphere above the Atlantic. *Nature.* 72:244. (Reprinted in *Mon. Wea. Rev.* 33:209. Summary of results in *Nature*, 72:538, *Sci.* 22:414.)

—— The first observations with balloon-sondes in America. *Sci.* 22:76-77. Reprinted in *Sci. Amer.* 27 May 92:419.

—— The Saint Petersburg Conference on the exploration of the atmosphere. *Sci.* 21:461-465.

—— The temperature and the drift of the air at great heights above the American Continent obtained by means of registration balloons. *Sci.* 21:335. (Translated in *Ciel et Terre.* 26th year, 1905, 72, and *III. Aeron. Mitteilungen*, 9:88.)

—— and L. T. de Bort. Sur les preuves directes de l'existence du contre-alize. *Comptes rendus de l'academie des Sciences.* 141:605-608.

——, and ——. The exploration of the atmosphere over the tropical oceans. *Nature.* 73:54-56. (Abstracted in Nation, 15 March 1906, 223-224.)

1906

Fergusson, S. P. The errors of absorbtion hygrometers. *Ann. Astron. Obs. of Harvard College.* 58 Part 2, Appendix:126-141, plus plates.

Rotch, A. L. Conference de M. Rotch à l'Observatoire de Meudon le 18 Septembre, 1900, sur l'emploi des cerfs-volants pour les observations meterologiques. *Procis-verbaux de Congris Int. d'Aéronautique.* Paris, 1906, 22-23.

—— Observations and investigations made at the Blue Hill Meteorological Observatory in 1903 and 1904. *Ann. Astron. Obs. of Harvard College.* 58 Part 2:67-118.

—— Proof of the existence of the upper anti-trade and the meteorological conditions at lesser heights in the Northern Tropics. *Bull. Amer. Geog. Soc.* 38:128-130.

—— Résultats des sondages aériens dans la région des Alizes. *Comptes Rendus de l'academie des Sci.* 142:918-921

—— Results of the Franco-American expedition to explore the atmosphere in the tropics. *Proc. Amer. Acad. of Arts and Sci.*, 42:263-272.

—— The Franco-American Expedition to explore the atmosphere in the tropics. *Sci.* 24:603.

—— The lightning rod coincident with Franklin's kite experiment. *Sci.* 24:780.

—— The international meteorological conference of Innsbruck. *Sci.* 23:975-977.

—— The meteorological conditions above the Tropical North Atlantic. *Hann Band der Meteor.*, 41:270-275.

—— Unusual sky-colors. *Ann. Astron. Obs. of Harvard College.* 58 Part 2:119-121.

—— When did Franklin invent the lightning rod? *Sci.* 24:374-376.

——, and L. T. de Bort. The vertical distribution of the meteorological elements above the Atlantic. *Nature.* 73:449-450.

Wells, L. A. The effect of meteorological conditions upon optical refraction in the lower atmospheric strata. *Ann. Astron. Obs. of Harvard College.* 58 Part 2:122-126.

1907

Rotch, A. L. An autumn passage of the Col du Géant. *Appalachia.* 11:211-221.

—— Benjamin Franklin and the first balloons. *Proc. Amer. Antiquarian Soc.* 18:259-274.

—— Benjamin Franklin's original letters about balloons. *Proc. Amer. Antiquarian Soc.* 19:100-102.

—— Did Benjamin Franklin fly his electrical kite before he invented the lightning rod? Worcester. (MA) The David Press. 8p.

—— Die meteorologischen Verhaltnisse Über St. Louis. *Illustrierte aeronautische Mitteilungen.* June 1907, 193-194.

—— Les conditions météorologiques au-dessus de Saint Louis et la coupe Gorden-Bennett. *L'Aerophile,* 15:216-217.

—— Résultats des premiers balloons-sondes en Amerique. *Fifth Conf. de la Comm. Int. pour l'Aérostation Sci.* a Milan, 29-30.

—— Résultats principaux des campagnes de l'Otaria. *Fifth Conf. de la Comm. Int. pour l'Aérostation Sci. a Milan.* 1907, 105-109.

—— The balloon in science and sport. *Navigating the Air, Aero Club of America.* 117-126.

—— The exploration of the atmosphere at sea. *Sci. Amer.* 96:70.

—— The highest ascent by man. *Amer. Mag. of Aeronautics.* August, 1907, 123.

—— The international aeronautical conference at Milan. *Sci.* 25:841-845.

—— The longest balloon voyage. *American Mag. of Aeronautics.* October 1907, 35.

—— The meteorological conditions above St. Louis. *Amer. Mag. of Aeronautics,* July 1907: 14-15 and *Sci. Amer.,* 96:271.

—— The use of registration balloons in obtaining meteorological conditions at great heights. *The Amer. Aeronaut.* November 1907,17-18.

—— Un rapport inédit du deuxième voyage en balloon. *L'Aérophile.* 15:303.

1908

Bibliography 1901-1905. (Blue Hill Observatory work). *Ann. Astron. Obs. of Harvard College.* 58 Part 3:223-228.

Clayton, H. H. The meteorology of total solar eclipses, including the eclipse of 1905. *Ann. Astron. Obs. of Harvard College.* 58 Part 3:192-216, plus plates.

Rotch, A. L. Atmospheric explorations conducted by the Blue Hill Meteorological Observatory. *Aeronautics.* 2:22-23.

—— Discussion on the theory of wave motion. *Report of Brit. Assoc. Adv. Sci.* Dublin meeting, 609-610.

—— Observations and investigations made at the Blue Hill Meteorological Observatory in 1905. *Ann. Astron. Obs. of Harvard College.* 58 Part 3:147-228.

—— On the first observations with sounding balloons in America. *Aeronautics.* 2:20-22.

—— The balloon vs. the aeroplane. *Aeronautics.* 3:7.

—— The balloon-sondes at St. Louis. *Sci.* 27:135

—— The eclipse shadow-bands. *Ann. Astron. Obs. of Harvard College.* 58 Part 3:217-222

—— The isothermal layer of the atmosphere. *Nature.* 78:550-552.

—— The warm stratum in the atmosphere at heights exceeding eight miles in the United States. *Rept. of Brit Assoc. Adv. Sci.* Dublin meeting. Sept. 591.

—— The warm stratum existing at a great height in the atmosphere. *Amer. Assoc. Adv. Sci.* Hanover meeting, 1908, Sect. E, 28:381-382.

—— The warm stratum in the atmosphere. *Nature.* 78:7.

—— Very high cumulus clouds. *Sci.* 27:783.

Wells, L. A. Features of the twenty years observations (at Blue Hill Meteorological Observatory). *Ann. Astron. Obs. of Harvard College.* 58 Part 3:189-191.

1909

Clayton, H. H. The distribution of the meteorological elements around cyclones and anti-cyclones up to three kilometers at Blue Hill Meteorological Observatory. *Ann. Astron. Obs. of Harvard College.* 58 Part 1:85-92, 11 plates.

——, and S. P. Fergusson. Exploration of the air with balloon-sondes, at St. Louis and with kites at Blue Hill Meteorological Observatory. *Ann. Astron. Obs. of Harvard College.* 58 Part 1:1-84.

Rotch, A. L. Aerological observations during international week. *Sci.* 30:914-915.

—— Die obere Inversion im Osten der vereinigten Staaten. *Z. Meteor.* 44:554-555.

—— Die warme Schicht der Atmosphäre oberhalb 12 km in Amerika. *Z. Meteor.* 44:22-23.

—— Rain with low temperature. *Mon. Wea. Rev.* 37:21-22. Reprinted from *Boston Transcript,* 20 January 1909.

—— *The Conquest of the Air.* New York: Moffat, Yard and Co. 192 pp.

—— The aerological congress at Monaco. *Sci.* 30:193-199.

—— The highest balloon ascension in America. *Sci.* 30:302-303

—— The highest human ascent. *Sci. Amer.* 101:295.

—— The highest meteorological observations in America. *Rept. to Brit. Assoc. Adv. Sci.* Winnipeg meeting, 415.

——. General results of the meteorological cruises of the "Otaria" on the Atlantic in 1905, 1906 and 1907. *Nature.* 80:219-221.

——, and L. T. de Bort. *Etude de l'atmosphére marine par sondages aériens, Atlantique moyan et région intertropicals.* Travaux Scientifiques de l'Observatoire Dynamique de Trappes. Gauthier-Villars: Paris 1909.

1910

Rotch, A. L. A brief account of the Blue Hill Meteorological Observatory. *Pubs. of the Astron. and Astrophy. Soc. of Amer.* 1:56-57.

—— Aerial navigation of to-day. *Sci.* 31:507-508.

—— The atmospheric ocean. *Sci. Amer.* 103:324-325.

—— The coldest region of the atmosphere. *Sci. Amer.* 103:9.

—— The relation of the wind to aerial navigation. *Aircraft.* March, 19-20.

—— Twenty-five year's work at Blue Hill Meteorological Observatory. *Tech. Reviewed April* 227-229.

1911

Clayton, H. H. A study of clouds with data from kites. *Ann. Astron. Obs. of Harvard College.* 68 Part 2:171-192.

Palmer, A. H. Pressure oscillations of short wave-length. *Ann. Astron. Obs. of Harvard College.* 68 Part 2:210-229.

—— Wind velocity and directions in the free air. *Ann. Astron. Obs. of Harvard College.* 68 Part 2:193-209.

Rotch, A. L. Drachen und Fallschirme. *Moedebeck's Taschenbuch zum praktischen Geburach für Flugtechniker und Luftdchiffer.* 1911:149-184.

—— Observations and investigations made at the Blue Hill Meteorological Observatory in the years 1906, 1907 and 1908. *Ann. Astron. Obs. of Harvard College.* 68 Part 2:97-169.

——, and A. H. Palmer. *Charts of the atmosphere for aeronauts and aviators.* New York: John Wiley and Sons. 96 p.

1912

Rotch, A. L. Aerial engineering. *Sci.* 35:41-46.

—— Book review of Aerial Navigation. A popular treatise on the growth of aircraft and aeronautical meteorology. by A. F. Zahm.

1913

McAdie, A. Air drainage of a great city. *Trans. of the Commonwealth Club.* 8:480-486.

1914

Brooks, C. F. The ice storms of New England. *Ann. Astron. Obs. of Harvard College.* 78 Part 1:77-84 plus plates.

McAdie, A. Blue Hill Observatory, *Rept. of the President of Harvard College, 1913-1914*: 219-221.

—— Introduction of new units at Blue Hill Meteorological Observatory. *Ann. Astron. Obs. of Harvard College.* 73 Part 1:85-90.

—— Recording moonlight and night cloudiness. *Sci. Amer. Suppl.* 78:109.

—— Standard units in aerology. *Sci.* 39:391-392.

—— The founder of the Observatory: a review of the scientific work of Abbott Lawrence Rotch. *Ann. Astron. Obs. of Harvard College.* 73 Part 1:60-73.

—— The paradox of the east wind. *Pop. Sci. Mon.* 85:292-295.

—— The rainfall of California. *Univ. of Calif. Publs in Geor.* 1:127-240.

—— The storm of March 1, 1914, measured in new units. *Sci. Amer.* 110:238.

—— Observations and investigations made at the Blue Hill Meteorological Observatory in the years 1909 and 1910. *Ann. Astron. Obs. of Harvard College.* 73 Part 1:3-59.

Wells, L. A. Features of the twenty-five years observations. *Ann. Astron. Obs. of Harvard College.* 73 Part 1:74-76.

1915

McAdie, A. Advances in the study of storms. *Sci. Conspectus.* 4 No. 3 August.

—— Blue Hill Observatory, *Rept. of the President of Harvard College, 1914-1915*: 237-239.

—— Get the units right. *Sci.* 41:647-648.

—— Observations and investigations made at the Blue Hill Meteorological Observatory in the years 1911, 1912, 1913 and 1914. *Ann. Astron. Obs. of Harvard College.* 73 Part 2:95-167.

—— Temperature inversions in relation to frosts. *Ann. Astron. Obs. of Harvard College.* 73 Part 2:168-177 plus plates.

—— The theory and practice of frost fighting. *Sci. Mon.* 1:292-301.

—— Weather conditions on the Pacific Coast. Nature and Science on the Pacific Coast. *Amer. Assoc. for the Adv. of Sci..* San Francisco. 62p.

1916

McAdie, A. Aviation and Aerography. *Aviation and Aeronaut Eng.* 1:8-11.

McAdie, Blue Hill Observatory. *Rept. of the President of Harvard College. 1915-16*:236-239.

—— Battles and rainfall. *Sci. Mon.* 2:170-173.

—— Has war affected the weather? *Atl. Mon.* 392-395.

—— Muir of the mountains. *Sierra Club Bull.* 10:20-22.

—— Observations and investigations made at the Blue Hill Meteorological Observatory in the year 1915. *Ann. Astron. Obs. of Harvard College.* 73 Part 3:183-210.

—— The Blue Hill Meteorological Observatory. *Harvard Grad. Mag.* 24:605-610.

—— Thermometer scales. *Sci.* 43:854-855.

—— The new thermometer scale. *Proc. of the Nat. Acad.* 2:670-672.

—— The winds of Boston and vicinity. *Ann. Astron. Obs. of Harvard College.* 73 Part 3:211-231 plus plates.

1917

Forbes, W. Ice saints. *Ann. Astron. Obs. of Harvard College.* 83 Part 1:51-59.

McAdie, A. A new thermometer scale. *Sci. Amer. Suppl.* 83:75.

—— Blue Hill Observatory, *Rept. of the President of Harvard College, 1916-1917:* 224-228.

—— Forecasting the seasons. A subject of great importance in connection with planting and growth of crops. *Sci. Amer. Suppl.* 84:50-51.

—— International units and symbols in aerography. *Sci.* 46:360.

—— Observations and investigations made at the Blue Hill Meteorological Observatory in the year 1916. *Ann. Astron. Obs. of Harvard College.* 83 Part 1:1-27.

—— Protection from frost. *J. Royal Hort. Soc.*

—— Saving the crops from injury by frost. *Geogr. Rev.* 4:315-358.

—— The desirability of scientific units in aerography. *Ann. Astron. Obs. of Harvard College.* 83 Part 1:47-50.

—— The passing of the Fahrenheit scale. *Geog. Rev.* 4:214-216.

—— *The principals of aerology.* Chicago: Rand, McNally and Co. London: George C. Harrap. 318 p.

—— The unit of pressure. *Sci.* 45:385.

—— The winds of Boston and vicinity, part II. *Ann. Astron. Obs. of Harvard College.* 83 Part 1:28-47.

1918

McAdie, A. *A manual of aerography for the United States Navy.* Washington: Govt. Pr. Office. 165pp.

—— Blue Hill Observatory. *Rept. of the President of Harvard College, 1917-1918*: 226-227.

—— Meteorology and the national welfare. *Sci. Mon.* 6:176-187.

—— Nova Albion—1579. Reprinted from *Proc. Amer. Antiquarian Soc.* October, 12 pp.

—— Observations and investigations made at the Blue Hill Meteorological Observatory in the year 1917. *Ann. Astron. Obs. of Harvard College.* 83 Part 2:65-92.

—— Observations and investigations made at the Blue Hill Meteorological Observatory in the year 1918. *Ann. Astron. Obs. of Harvard College.* 83 Part 3:97-125.

1919

McAdie, A. A quick method of measuring cloud heights and velocities. *Ann. Astron. Obs. of Harvard College.* 83 Part 4:162-168

—— Atmospheric pollution. *Mon. Wea. Rev.* 47:806-807.

—— Blue Hill Observatory. *Rept. of the President of Harvard College, 1918-1919:* 182-183.

—— Franklin's electrical experiment. *Sci. Amer.* 121:128.

—— Review of the annual report of the Committee for the Investigation of Atmospheric Pollution. *Sci.* 50:501.

—— The freedom of the skies. Some of the problems that will have to be solved as the human race takes to the air. *Sci. Amer.* 122:84.

—— "The great fleet," (Rev) 1914-1916 by Admiral Viscount Jellicoe of Tamps. *Sci.* 50:21-23.

—— The work of the aerographic section of the Navy. *Mon. Wea. Rev.* 47:225-226.

—— Uniformity in aerographic notation. *Ann. Astron. Obs. of Harvard College.* 83 Part 4:169-177.

—— Uniformity in aerological records. The desirability of universal scientific units. *Sci. Amer. Suppl.* 87:15-16.

——, and G. P. Payne. Uniformity in symbols (in meteorology). *Sci.* 50:411-412.

1920

McAdie, A. An awkward unit. *Nature.* 106:179.

—— Blue Hill Observatory. *Rept. of the President of Harvard College. 1919-20*:229-230.

—— Further discussion of the kilograd scale. *Bull. Amer. Meteor. Soc.* 1:96-97.

—— Gravity and aerostatic pressure on fast ships and airplanes. (Effect on barometer readings.) *Sci.* n.s. 51:144-145.

—— Muggy days and thirsty air. The mechanics that keep us cool and why it fails to work in damp weather. *Sci. Amer.* 123, 220:234-235.

—— Muggy days and thirsty air. *Tycos.* 11:9-11.

—— Observations and investigations made at the Blue Hill Meteorological Observatory in the year 1919. *Ann. Astron. Obs. of Harvard College.* 83 Part 4:131-161.

—— Records of night cloudiness for astronomers. *Pub. of the Astron. Soc. of the Pacific.* 32:300-306.

—— The attainment of high levels in the atmosphere. *Sci.* 51:348.

—— The Einstein theory of relativity. Prize essay. *Sci. Amer.*

—— The Kelvin-kilograd thermometer scale. *Bull. Amer. Meteor. Soc.* 1:84.

—— Wandering storms. *Geog. Rev.* 10:37-41.

1921

McAdie, A. Blue Hill Observatory. *Rept. of the President of Harvard College. 1920-1921:* 264-265.

—— Observations and investigations made at the Blue Hill Meteorological Observatory in the year 1920. *Ann. Astron. Obs. of Harvard College.* 83 Part 5:183-217.

—— Relativity and the absurdities of Alice. *Atl. Mon.* June, 811-814.

—— Review of Jellicoe's "Crisis of the naval war." *Sci.* 53:501.

—— Review of the annual report of the committee for the investigation of atmospheric pollution. *Sci.* 53:389-392.

—— Symbols in science. *Ann. Astron. Obs. of Harvard College.* 83 Part 5:218-220.

—— The energy of cyclones. Quoted on, in *Mon. Wea. Rev.* 49:5.

——, and H. Lyman. Atmospheric pollution. *Mon. Wea. Rev.* 49:159-160.

1922

Fisher, W. J. Low sun phenomena. *Ann. Astron. Obs. of Harvard College.* 86 Part 1:38-45.

McAdie, A. Blue Hill Observatory. *Rept. of the President of Harvard College. 1921-22:*210-211.

—— C. G. S. units in aerography-kilobar, kilocal, kilograd. *Ann. Astron. Obs. of Harvard College.* 86 Part 1:46-56.

—— General adoption of the centesimal system of angular measurement with application to anenometers and nephoscopes. *Trans. Amer. Geophys. Union.* 3:50-53.

—— Gravitation versus relativity. *N. Y. Times Book Rev.* 31 December.

—— Height and velocity of low clouds in relation to surface humidity and wind. *Ann. Astron. Obs. of Harvard College.* 86 Part 1:33-37.

—— Kilobar, kilocal, kilograd. *Sci.* 57:207-208.

—— Monsoon and trade winds as rainmakers and desertmakers. *Geog. Rev.* 12:412-419.

—— New methods of observing winds at flying levels over the ocean. *Bull. Nat. Res. Council.* 3 Part 2:94-101.

—— Observations and investigations made at the Blue Hill Meteorological Observatory in the year 1921. *Ann. Astron. Obs. of Harvard College.* 86 Part 1:5-32.

—— Review of the annual report of the committee for the investigation of atmospheric pollution. *Sci.* 55:596.

—— The monsoon as rainmaker.

—— The steering line of hurricanes. *Bull. Nat. Res. Council.* 3 Part 2:102-108.

—— Visibility vs. victory. *Harvard Grad. Mag.* 30:163-168. (Reprinted as "Where weather won" in War Weather Vignettes).

—— *Wind and Weather.* New York: The Macmillan Co. 82 p.

Paine, G. F. The measurement of atmospheric humidity. *Ann. Astron. Obs. of Harvard College.* 86 Part 1:57-61.

1923

McAdie, A. *A Cloud Atlas.* New York: Rand McNally and Co. 57 p.

—— Alice in Wonderland as a relativist. *N. Y. Times.* 11 March.

—— Aviation and the navy (A letter from A. McAdie). *Aviation Mag.* 18:709.

—— Beneath the fog. *Harvard Grad. Mag.* 31:230.

—— Blue Hill Observatory, *Rept. of the President of Harvard College, 1922-1923,* 249

—— Dampness indoors and out. *Bull. Amer. Meteor. Soc.* 4:38-40.

—— *Making the Weather.* New York: The Macmillan Co. 88 pp.

—— Observations and investigations made at the Blue Hill Meteorological Observatory in the year 1922. *Ann. Astron. Obs. of Harvard College.* 86 Part 2:67-94.

—— Paradoxical rainfall data. *Nature.* 17 March, 362.

—— Review of "The aviator. New explorations of the air." *N. Y. Times Book Rev.* 26 July.

—— Review of the annual report of the committee for the investigation of atmospheric pollution. *Sci.* 58:330-331.

—— The depreciation of the pound. *Sci.* 57:235.

—— The truth about rainmaking. *Harvard Alumni Bull.* 25:957-959.

—— Units and constants in aerography. *Ann. Astron. Obs. of Harvard College.* 86 Part 2:95-106.

—— Velocity of sound in free air. *Ann. Astron. Obs. of Harvard College.* 86 Part 2:107-117.

—— Velocity of sound in free air and type of air structure. *Phys. Rev.* 21:377-378.

—— Wanted, warming pans. *Harvard Grad. Mag.* 30:503.

1924

McAdie, A. A hundred pounds. *Sci.* 58:142.

—— Air power. *Aviation*. 19:478.

—— Blue Hill Observatory, *Rept. of the President of Harvard College, 1923-1924:* 251-252.

—— Drams, scruples and king's noses. *Atl. Mon.* 820-822.

—— Dry and wet seasons. *Ann. Astron. Obs. of Harvard College*. 86 Part 4:227-232.

—— Gilpin in the world war. *Harvard Grad. Mag.* 32:229. (Reprinted as "John Gilpin aloft" in War Weather Vignettes.)

—— How big is an acre? *Atl. Mon.* July 78-79.

—— Observations and investigations made at the Blue Hill Meteorological Observatory in the year 1923. *Ann. Astron. Obs. of Harvard College*. 86 Part 4:199-226.

—— Present status of metric system in the Unites States. *Sci.* 59:528-529.

—— Review of the annual report of the committee for the investigation of atmospheric pollution. *Sci.* 60:136-137.

—— The wind gods at Gallipoli. *Harvard Grad. Mag.* 31:575.

—— Uniformity in weights and measures. *Sci.* 57:38.

1925

McAdie, A. *An account of the work at the Blue Hill Observatory*. Cambridge (Mass.): Harvard University. 17 pp.

—— Blue Hill Observatory, *Rept. of the President of Harvard College, 1924-1925,* 240-241.

—— Franklin and lightning. *Atl. Mon.* July, 67-73.

—— Franklin's kite experiment and modern kite work. *Bull. Amer. Meteor. Soc.* 6:38-39.

—— Observations and investigations made at the Blue Hill Meteorological Observatory in the year 1924. *Ann. Astron. Obs. of Harvard College*. 87 Part 1:5-32.

—— Smithsonian weather forecasts. *Sci.* 62:418-419.

—— Special maps for charting storm movements. *Assoc. Amer. Geogrs.* 15:41.

—— The Kennelly-Heaviside layer. Discussion and correspondence. *Sci.* 61:540.

—— War and water. *Harvard Grad. Mag.* 32:580.

—— *War Weather Vignettes*. New York: Macmillan. 62 p.

—— War without end. *Harvard Grad. Mag.* 33:201.

Paine, G. P. Aerodynamics of the psychrometer. *Ann. Astron. Obs. of Harvard College*. 87 Part 1:33-126.

1926

McAdie, A. Blue Hill Observatory, *Rept. of the President of Harvard College, 1925-1926:* 251-252.

—— Discussion "The metric movement." *Sci.* 64:379-380.

—— Forty years of Blue Hill. *Harvard Alumni Bull.* 29: Oct. 28.

—— Historic dark days. *Tycos.* Oct. 153-154.

—— How high is the sky?. *Atl. Mon.* October, 489-493.

—— *Man and Weather.* Cambridge (Mass.): Harvard University Press. 99 pp.

—— Phoenix at Skagerak. *Harvard Grad. Mag.* 34:33-37.

—— Review of "Conditions of validity of Macallums microchemical test for calcium.*Sci.* 64:Dec.

—— Review of the annual report of the committee for the investigation of atmospheric pollution. *Sci.* 63:310.

—— The flier's aspect of aerography. *Aviation.* The structure of air. 21:702; The water vapor in free air. 21:841-842; Dispelling the clouds. 21:954-955.

—— The wind and flood hazard (in "Investment lessons of the great Florida hurricane"). *The Annalist, New York Times,* Oct. 1 1926.

—— When fleets foozle. *Harvard Grad. Mag.* 33:532.

1927

McAdie, A. Aerography. *Tech. Engrg. News.* 314.

—— Beyond the trades. *Harvard Grad. Mag.* 34 March.

—— Blue Hill Observatory, *Rept. of the President of Harvard College, 1926-1927:* 260-261.

—— Dissipating fog. *Atl. Mon.,* May 672-676.

—— Every home an Observatory. *Bull. Amer. Meteor. Soc.* 8:27.

—— Hyperbar and infrabar. *Bull. Amer. Meteor. Soc.* 8:68-69.

—— Icarus and melting wax. *Sci.* 66:280.

—— Lewis Carrol and relativity. *N. Y. Times.*

—— Lightning discharges. *Meteor. Mag.* 62:106-108.

—— Long-range weather forecasts. *Sci.* 65:374.

—— Observations and investigations at the Blue Hill Meteorological Observatory in the years 1925 and 1926. *Ann. Astrom Obs. of Harvard College.* 87 Part 2:131-174.

—— Pi(e) for donkeys. *School and Soc.* 15:110.

—— Review of "Aeronautical meteorology" by W. R. Gregg. *Sci.* 60:15-16.

—— Review of the smoke problem. *Geog. Rev.* 17:350.

—— That bewitching vegetable. Contributor's Club. *Atl. Mon.* February:276-278.

—— The flier's aspect of aerography. *Aviation.* The Shenandoah and the squall. 22:215-216. Electric storms present certain hazards. Few cases of planes being struck - one on record proved inconsequential. 22:571-573. Forecasting (the various instruments and methods used by pilots and their relative value in weather forecasting) 22:1043-1044.

—— The second siege of Troy. *Harvard Grad. Mag.*, 35, September.

1928

McAdie, A. A Harvard graduate who beat the Dutch. *Harvard Grad. Mag.* 36:183-188.

—— Airgraphy of aerography. *Sci.* 66:16.

—— Blue Hill Observatory. *Rept. of the President of Harvard College, 1927-1928*:281-282.

—— Cloud classification. *Obs. and Invest. Blue Hill Meteorological Observatory in the year 1927.* Cambridge, Mass. 18-26 plus plates.

—— *Cloud nomenclature.* Cambridge, Memoir: Harvard University Press.

—— Fog and sub-cooled water vapor (ice-loading). *D. Guggenheim fund for promotion of aerography.*

—— Franklin's kite experiment and the energy of lightning. *Mon. Wea. Rev.* 56:216-219.

—— Hazards of lightning. *Bull. D. Guggenheim Fund for the Promotion of Aerography.* No. 10, N.Y. Sept.

—— New data on weather maps. *Ann. Astron. Obs. of Harvard College.* 17:33-34.

—— *Observations and investigations made at the Blue Hill Meteorological Observatory in the year 1927.* Blue Hill Meteorological Observatory. 5-17.

—— Phenomena preceeding lightning. *Mon. Wea. Rev.* 56:219-220.

—— Shakespeare and his crowd apparently never smoked. *New York Times.* 3 June.

1929

McAdie, A. An unusual lightning flash. *Mon. Wea. Rev.* 57, 197-198.

—— Blue Hill Observatory. *Rept. of the President of Harvard College, 1928-1929*, 254-256.

—— Cloud formations as hazards in aviation. *Obs. and Invest. at the Blue Hill Observatory in the year 1928*. Cambridge, Mass. 19-20 plus plates.

—— Daily maps of world weather. An attempt at three dimensional geography. *Geog. Rev.* 19:87-93.

—— George Downing, a Puritan politician. (Reprinted from *Proc. Amer. Antiquarian Soc.* April.) 11 p.

—— Inversions and grass minimum temperatures. *Meteor. Mag.* 64:110-113.

—— New forms of dry and wet bulb thermometers. *Sci.* 70:172-173.

—— *Observations and investigations made at the Blue Hill Meteorological Observatory in the year 1928.* Cambridge: Blue Hills Meteorological Observatory. 5-18.

—— The hazard of sub-cooled fog and ice storms in aviation. *Bull. Amer. Meteor. Soc.* 10:37-39. (Abstract.)

—— The hazard of thunderstorms in aviation. *Bull. Amer. Meteor. Soc.* 10:36-37. (Abstract.).

—— Weather hazards in aviation. *Sci. Mon.* 29:66-71.

1930

McAdie, A. Aerographics. *Observations and Investigations made at the Blue Hill Observatory in the year 1929*. Cambridge, Mass; Blue Hill Meteorological Observatory. 19-25

—— Blue Hill Observatory. *Rept. of the President of Harvard College, 1929-1930*, 268-269.

—— Clouds and the airman. *Nat. Aeron. Mag.* Dec. 49-59.

—— *Clouds.* Mass.: Harvard Univ. Press. 22 p.

—— *Fog.* New York: The Macmillan Co. 23 p.

—— Ice warning semaphore. *Observations and investigations at the Blue Hill Meteorological Observatory in the year 1929.* Cambridge: Blue Hill Meteorological Observatory. 26-30.

—— *Observations and investigations Made at the Blue Hill Meteorological Observatory in the year 1929.* Cambridge, Blue Hill Meteorological Observatory. 5-18.

—— Shakespeare on smoking. *Harvard Grad. Mag.* 37:281-286

1931

McAdie, A. Blue Hill Observatory. *Rept. of the President of Harvard College, 1930-1931*, 280-281.

—— Meteorological conditions during the air raid on London. Oct. 19-20, 1917. *Nature*. 127:198.

—— Terramotum! Quid bonum?. *Harvard Alumni Bull*. 33:1150-1153.

—— The clouds. *Harvard Alumni Bull*. 33:478-479.

—— The commercial importance of fog control. *Ann. Assoc. Amer. Geogr*. 21:91-100.

—— Thomas Jefferson at home. *Proc. Amer. Antiquarian Soc*. April, 22 p.

1932

Brooks, C. F. A recent tour of some geographic centers of northwest Europe and sea-temperature by bucket on express liners. *Trans. Amer. Geophys. Union*. Repts. and papers, Oceanography, 1932, Thirteenth Ann. Meet. 255-258 and 258-263.

—— Geographical record. Obituary of Robert DeCourcy Ward. *Geog. Rev*. 22:161.

—— Robert DeCourcy Ward, Climatologist. *Ann Assoc. Amer. Geograph*. 22:33-43.

—— The Blue Hill Observatory. *Rept. of the President of Harvard College, 1931-1932*: 279-282.

—— The weather profiles across the North Atlantic (Author's summary). *Bull. Amer. Meteor. Soc*. 13:169-170.

—— What the atmosphere does to solar radiation. *Bull. Amer. Meteor. Soc*. 13:217-220.

Fassig, O. L. Fifty years of North American rainfall. *Bull. Amer. Meteor. Soc*. 13:205-206.

Fergusson, S. P. A symposium on climatic cycles. Presented before Nat. Acad. Sci. at Ann. Meet., Washington, D.C., April 26, 1932. *Bull. Amer. Meteor. Soc.*. 13:121-126.

——, and C. F. Brooks. A program for the observation of weather-changes during the total solar eclipse of August 31, 1932. *Trans. Amer. Geophys. Union*. Repts. and papers - meteorology, 1932, Thirteenth Ann. Meet., 117-120.

1933

Brooks, C. F. A cross-section of modern meteorology. *Bull. Amer. Meteor. Soc*. 14:291-295.

—— A meteorological cross section: The Köppen birthday volumes. *Geo. Rev*. 23:472-478.

—— Rainstorm of September 16-17, 1932 in New England. *Bull. Amer. Meteor. Soc*. 14:25-27.

—— The Blue Hill Observatory. *Rept. of the President of Harvard University, 1932-1933*, 285-288.

—— The value of cloud-observations. Abstract. *Trans, Amer. Geophys. Union*. 14th Ann. Meeting, 1933:84-85.

—— William Gardiner Reed, 1884-1932. (Excerpts from memoir read at Washington meeting of the Association of Amer. Geog., Dec 29, 1932.) *Bull. Geograph. Soc.* Phila. 21:96-98.

Fassig, O. L. The trade-winds of the Eastern Caribbean. *Trans. Amer. Geophys. Union.* 14th Ann. Meeting, 1933:69-78.

Fergusson, S. P. The early history of aerology in the United States. *Bull. Amer. Meteor. Soc.* 14:252-256.

Harwood, E. M. An 18 degree halo. *Mon. Wea. Rev.* 61:327.

——, and C. F. Brooks. Monthly sequences of sea-surface temperature on the New York-San Juan steamship route. *Trans. Amer. Geophys. Union.* 14 Ann. Meeting, 1933:173-180.

Haurwitz, B. Investigations of atmospheric periodicities at the Geophysical Institute, Leipzig, Germany. *Mon. Wea. Rev.* 61:219-221.

—— The recent theory of Giao concerning the formation of precipitation in relation to the polar-front theory. *Trans. Amer. Geophys. Union.* 14 Ann. Meeting, 1933:89-91.

Kimball, H. H. Recent advances in the science of meteorology and in its practical applications. (Presidential address, Atlantic City Meet., Dec. 27, 1933). *Bull. Amer. Meteor. Soc.* 14:3-6.

—— Solar radiation measurements obtained at the Blue Hill Meteorological Observatory of Harvard University during the Second International Polar Year, August 1932 to August 1933. *Mon. Wea. Rev.* 61:230-232.

—— The Atlantic City meeting. *Bull Amer. Meteor. Soc.* 14:7-8.

—— The use of glass color-screens in the study of atmospheric depletion of solar radiation. *Mon. Wea. Rev.* 61:80-83.

Stone, R. G. Melting hailstones cause fog at New London. *Bull. Amer. Meteor. Soc.* 14:242.

Wexler, H. A. A comparison of the Linke and Angstrom measures of atmospheric turbidity and their application to North American air-masses. *Trans. Amer. Geophys. Union.* 14 Ann. Meeting, 1933:91-99

1934

Bjerknes, J., C. F. Brooks, S. P. Fergusson, et al. For mountain observatories: Memorandum on the importance of meteorological mountain observatories with special reference to the new Observatory on Mount Washington. *Bull. Amer. Meteor. Soc.* 15:93-94.

Brooks, C. F. The Blue Hill Observatory. *Rept. of the President of Harvard College, 1933-1934:*297-300.

—— William Morris Davis: A biographical note and annotated bibliography. *Bull. Amer. Meteor. Soc.* 15:56-61.

——, and E. M. Harwood. A preliminary comparison of monthly weather with ocean-temperature sequences in the Western Atlantic region. *Trans Amer. Geophys. Union.* 15 Ann. Meeting, 1934:211-214.

——, and C.H. Pierce. Dust Storms. *Bull. Amer. Meteor. Soc.*, 15:107-109

Fergusson, S. P. Aerological studies on Mount Washington. *Trans. Amer. Geophys. Union.* 15 Ann. meeting, 1934:114-117.

—— An international comparison of standard barometers. *Mon. Wea. Rev.* 62:364-366.

—— The sensitiveness of anemometers. *Bull. Amer. Meteor. Soc.* 15:95-99.

—— Daytime radiation at Blue Hill Meteorological Observatory in 1933 with application to turbidity in American air masses. *Harvard Meteor. Studies.* No. 1 Cambridge: Harvard University Press. 31 p.

Haurwitz, B. Concerning atmospheric turbidity. *Bull. Amer. Meteor. Soc.* 15:19-20. (Abstract.)

Kimball, H. H. Turbidity and water-vapor determinations from solar radiation measurements at Blue Hill, and relations to air-mass types. *Mon. Wea. Rev.* 62:330-333.

——, and H. Wexler. Meteorologischen. *Zeitschrift.* 6:237-238.

Kimball, H. H., and I. F. Hand. The use of glass color-screens in the study of atmospheric depletion of solar radiation. Abstract. *Bull. Amer. Meteor.* Soc. 15:18-19.

Namias, J. An introduction to the study of air mass analysis. *Bull. Amer. Meteor. Soc.* 15:184-190.

—— Some aspects of the surface of subsidence. *Trans. Amer. Geophys. Union.* 15th Ann. Meeting, 1934:105-114.

—— Subsidence within the atmosphere. *Harvard Meteor. Studies.* No. 2 Harvard Univ. Press.

Pierce, C. H. The cold winter of 1933-34. *Bull. Amer. Meteor. Soc.* 15:61-78.

Pierce, A. H. The dust storm of November 12 and 13, 1933. *Bull. Amer. Meteor. Soc.* 15:31-35.

Schell, I. Differences between temperatures, humidities, and winds on the White Mountains and in the free air. *Trans Amer. Geophys. Union.* 15th Ann. meeting; 1934:118-124.

Sise, A. F., and C. F. Brooks. An investigation of ultra-high frequency radio transmissions in relation to weather. *Bull. Amer. Meteor. Soc.* 15:238-239.

Stone, R. G. Controlled weather: fog dissipation and rain making. *Bull. Amer. Meteor. Soc.* 15:205-206.

—— Die Entwicklung der amerikanischen Bergobservatorien und das derzeitige Netz von Bergstationen in den verinigten Staaten von Amerika. *Jahresbericht des Sonnblick, Veriennes.* 1934:11-30.

—— Dust storms of April 10 and 11, 1934, at Sioux City, Iowa. *Bull. Amer. Meteor. Soc.* 15:196-198.

—— Snow for skiing in New Hampshire. *U.S. Eastern Ski Ann.* 8 pp.

—— The history of mountain meteorology in the United States and the Mount Washington Observatory. *Trans. Amer. Geophys. Union.* 15th Ann. meeting 1934:124-133.

Wexler, H. Remarks concerning an extension of Angstrom's coefficient of atmospheric turbidity. *Bull. Amer. Meteor. Soc.* 15:20-21.

1935

Baker, R. F. Preliminary measurements of ultra-violet at Blue Hill Meteorological Observatory. *Mon. Wea. Rev.* 63:221-222.

Bent, A. E. A radio sounding at Blue Hill Meteorological Observatory. *Bull. Amer. Meteor. Soc.* 16:265-267.

—— Ultra-high frequency radio activities at Blue Hill Meteorological Observatory. *Bull. Amer. Meteor Soc.* 16:59-60.

—— Use of ultra-high frequencies in tracking meteorological balloons. *Trans. Amer. Geophys. Union.* 16th Ann. meeting 1935:141-144.

—— Use of ultra-high radio frequencies in meteorology. *Bull. Amer. Meteor. Soc.* 16:23-24.

Brooks, C. F. An early morning weather profile from Cape Cod to central Massachusetts. *Bull. Amer. Meteor. Soc.* 16:93-94.

—— A quartet of complex halos and the weather which followed them. *Bull. Amer. Meteor. Soc.* 16:305-306.

—— Blue Hill Observatory, fifty years old. *Harvard Alumni Bull.* 37:11 pp.

—— Microbarometric oscillations at Blue Hill, Part III. Air waves at Blue Hill, Dec. 17-18, 1931. *Bull. Amer. Meteor. Soc.* 16:159.

—— The Blue Hill Meteorological Observatory. *Rept. of the President of Harvard University. 1934-1935:*306-308.

—— The Robert deCourcy Ward climatological collection. *Bull. Amer. Meteor. Soc.* 16:319-320.

——, and E. M. Harwood. Cloud observations in short-term forecasting of snow-storms. *Trans. Amer. Geophys. Union.* 16th Ann. meeting, 1935:110-114.

Clayton, H. H. Meteorological periods and solar periods. *Trans. Amer. Geophys. Union.* 16th Ann. meeting 1935:158-160.

—— On developing long-range weather forecasting. *Bull. Amer. Meteor. Soc.* 16:236-237.

—— Rating weather forecasts. *Bull. Amer. Meteor. Soc.* 15:279-281.

Fergusson, S. P. Some recent comparisons of anemometers. *Bull. Amer. Meteor. Soc.* 16:303-304.

Haurwitz, B. A theoretical study of wind-velocity and wind-direction in curved air-currents. *Trans. Amer. Geophys. Union.* 16th Ann. meeting 1935:124-126.

——— Microbarometric oscillations at Blue Hill. Part 1. Waves of presure and wind at the top of a ground inversion. *Bull. Amer. Meteor. Soc.* 16:153-157.

——— On the change of wind with elevation under the influence of viscosity in curved air currents. *Gerlands Bietr. Geophys.* 45:243-267.

——— The height of tropical cyclones and of the "eye" of the storm. *Mon. Wea. Rev.* 63:45-49.

Hull, R. A. Air-mass conditions and the bending of ultra-high frequency waves. *QST.* June 1935.

Kimball, H. H. Intensity of solar radiation at the surface of the earth, and its variations with latitude, altitude, season and time of day. *Mon. Wea. Rev.* 63:1-4.

Lange, K. O. Radio-meteorographs, Part 1. *Bull. Amer. Meteor. Soc.* 16:233-236.

——— Radio-meteorographs. Part II. The Moltchanoff and Askanna Radio-meteorographs (A) the "Kammgerat" of Moltchanoff. *Bull. Amer. Meteor. Soc.* 16:267-271.

——— Radio-meteorographs. Parts III-V. *Bull. Amer. Meteor. Soc.* 16:297-299.

——— The radio-meteorograph project of Blue Hill Meteorological Observatory, Harvard University—a preliminary report. *Trans. Amer. Geophys. Union.* 16th Ann. meeting, 1935:144-147.

McKenzie, A. An interesting use of ultra-high frequency radio in a meteorological study. *Bull. Amer. Meteor. Soc.* 16:317-318.

Millar, F. G. Rapid methods of calculating the Rossby diagram. *Bull. Amer. Meteor. Soc.* 16:229-233.

Pickard, G. W., A. E. Bent and C. F. Brooks. A report of progress in the investigation of ultra-high frequency radio transmission, November 22 and 23, 1934. *Bull. Amer. Meteor. Soc.* 16:60-62.

Schell, I. I. Free-air temperature from observations on mountain peaks with application to Mount Washington. *Trans. Amer. Geophys. Union.* 16th Ann. meeting, 1935:126-141.

——— Translation of Chromov's rules for forecasting synoptic situations. *Bull. Amer. Meteor. Soc.* 16:21-22

Spilhaus, A. F., and S. P. Fergusson. Two contributions to anemometry. The cup anemometer in an unsteady wind, by Spilhaus. Some recent comparisons of anemometers, by Fergusson. *Bull. Amer. Meteor. Soc.* 15:301-304.

Stone, R. G. A modern classification of weather types for synoptic purposes. *Bull. Amer. Meteor. Soc.* 16:324-326.

—— Climatological data for air-conditioning. *Bull. Amer. Meteor. Soc.* 16:241-242.

—— Meteorology and climatology (1935). *Amer. Yearbook.* 667-672.

—— Microbarometric oscillations at Blue Hill, Part II. H. Helm Clayton's statistical study of wave-like pressure oscillations observed at Blue Hill between 1885 and 1893. *Bull Amer. Meteor. Soc.* 16:158-159.

—— Quayle on modification of climate in Australia. *Bull. Amer. Meteor. Soc.* 16:148-150.

—— Report on sites for mountain observatories in the Southern Appalachians. *Ms. Rept. to U.S. Weather Bureau,* Aug. 25 pp.

—— The dates of closing of Lake Champlain by ice, 1816-1935. *Bull. Amer. Meteor. Soc.* 16:112.

—— The great Pacific Northwest storm of October 21, 1934. *Bull. Amer. Meteor. Soc.* 16:165.

Wexler, H. Turbidities of American air masses and conclusions regarding the seasonal variation in atmospheric dust content. *Mon. Wea. Rev.* 62:397-402.

1936

Baker, R. F. Measurement of Schott glass filter temperatures. *Mon. Wea. Rev.,* 64:5-6

Baldwin, H. I., and C. F. Brooks. Forests and floods in New Hampshire. *New Engl. Regional Planning Comm.* Publication No. 47. 28 p.

Brooks, C. F. Blue Hill Meteorological Observatory, 1885, Milton Mass. In *Harvard Univ. Handbook, 1936.* Cambridge: Harvard University Press 247-248.

—— Blue Hill Meteorological Observatory. *Rept. of the President of Harvard University, 1935-1936,* 323-327.

——, and A. J. Connor. *Climatic Maps of North America.* Cambridge, Mass.

——, and R.G. Stone. Charles Fitzhugh Talman, 1874-1936. *Bull. Amer. Meteor. Soc.* 17:243-245.

Church, P. E. A geographical study of New England temperature. *Geograph. Rev.* 26:283-291.

Fergusson, S. P. The exhibit of meteorological apparatus and accessories. *Ann. Sci. Exhib. of the A. A. A. S.* 1936:4 p.

—— C. F. Brooks and B. Haurwitz. Eclipse-meteorology, with special reference to the total solar eclipse of 1932. *Trans. Amer. Geophys. Union.* 17th Ann. meeting, 1936:129-134.

Haurwitz, B. On the vertical wind distribution in anticyclones, extratropical and tropical cyclones under the influence of eddy viscosity. *Gerlands Bietr. Geophys.* 47:206-214.

Kimball, H. H. Determination of atmospheric turbidity and water vapor content. *Mon. Wea. Rev.* 64:1-5.

——, and I. F. Hand. *The intensity of solar radiation as received at the surface of the earth and its variation with latitude, altitude, the season of the year and the time of the year and the time of day, biological effects of radiation.* New York: McGraw-Hill Book Co. 66 p.

Lange, K. O. Radio-meteorographs. The 1935 radio-meteorographs of Blue Hill Meteorological Observatory. *Bull. Amer. Meteor. Soc.* 17:1-24.

Namias, J. Glossary of elementary terms used in articles I to VII of the series, "An introduction to the study of air mass analysis." *Bull. Amer. Meteor. Soc.* 17:20-23.

Pear, C. B. An automatic temperature reporting station. *Bull. Amer. Meteor. Soc.* 17:265.

Schell, I. I. On the diurnal variation of wind velocity. *Gerlunds Bietr. Geophys.* 47:60-72.

—— On the vertical distribution of wind velocity over mountain summits. *Bull. Amer. Meteor. Soc.* 17:295-300.

Stone, R. G. Discussion of "Some observations on the climate and 'northeaster' storms of the Kodiak Archipelago" by W. B. Marriam. *Bull. Amer. Meteor. Soc.* 17:353-354.

—— Fog in the United States and adjacent regions. *Geog. Rev.* 26:111-134.

—— Mean annual snowfall for the United States. A map in, Climatic Maps of North America. By C. F. Brooks, A. J. Connor and others. Cambridge.

—— Meteorology and climatology. *Amer. Yearbook.* 1936, 680-686.

—— The great rains in Panama, Nov. 1935. (co-author). *Bull. Amer. Meteor. Soc.* 17:391-392.

1937

Bent, A. E. Mount Washington Observatory carries on. *Appalachia.* 21:553-557.

—— Radio equipment for the Harvard Radio-meteorographs. *Bull. Amer. Meteor. Soc.* 18:99-107.

Brooks, C. F. A five-year program of research and instruction in aerology and aeronautical meteorology. *Trans. Amer. Geophys. Union.* 18th Ann. meeting 1937:1 pp.

—— Blue Hill Meteorological Observatory. *Rept. of the President of Harvard University, 1936-1937*:335-340.

—— W1XFW. *Harvard Alumni Bull.* 2 July 1085-1091.

——, and H. I. Baldwin. How forests retard floods. *Amer. Forests.* June 1937:6 pp.

——, and Oliver L. Fassig, 1860-1936. *Bull. Amer. Meteor. Soc.* 18:28-30.

——, and A. F. Thiessen. The meteorology of the great floods in the Eastern United States. *Geog. Rev.* 27:269-290.

——, and R. Wexler. Seasonal variation of the ultra violet on Great Blue Hill, near Boston, Mass. *Bull. Amer. Meteor. Soc.* 18:298-302.

Church, P. E. Temperatures of the Western North Atlantic from Thermograph records. *Assoc. D'Oceanographic Phy.* Pub. Sci. No. 4 1937:32 pp.

Fergusson, S. P. The measurement of humidity at low temperatures. *Bull. Amer. Meteor. Soc.* 18:380-381.

Haurwitz, B. The Norwegian wave-theory of cyclones. *Bull. Amer. Meteor. Soc.* 18:193-201.

—— The physical state of the upper atmosphere. *J. Roy. Astron. Soc. of Canada.* Oct 1936-Feb 1937, 96 p.

—— Total solar and sky radiation on Mount Washington. N. H. *Mon. Wea. Rev.* 65:97-99.

Lange, K. O. Comparative measurements between Harvard radio-meteorographs and conventional meteorographs. *Beit. Phys. Freien Atmos.* 24:243-254.

—— The 1936 radio-meteorographs of Blue Hill Meteorological Observatory. *Bull. Amer. Meteor. Soc.* 18:107-126.

——, C. Harmantas, C. P. Pear and D. P. Keily. A series of 31 radio soundings at Cambridge, Massachusetts, in February and March 1937. *Trans. Amer. Geophys. Union.* 18th Ann. meeting 1937:1p.

Pagliuca, S. Icing measurements on Mount Washington. *J. Aero. Sci.* 4:399-402.

—— The problem of forecasting sleet for highway and industrial purposes. *Trans. Amer. Geophys. Union.* 18th Ann. meeting 1937:551-554.

Pear, C. B. Balloons for weather forecasting. *Sci. Digest.* 3:5-7. Move complete in *Electronics.* 10.

Stone, R. G. Bibliography of Oliver Fassig. *Bull. Amer. Meteor. Soc.* 18:31-34.

—— Climate and disease, by Guy Hinsdale. (Notes and editing.) *Bull. Amer. Meteor. Soc.* 6 Chapters between 1937 and 1939.

—— Meteorology and climatology. *American Yearbook,* 716-720.

—— On the causes of deaths during the heat wave of July 1936 at Detroit. *Bull. Amer. Meteor. Soc.* 18:233-236.

—— Spring snow survey reports and stream flow forecasts. *Bull. Amer. Meteor. Soc.* 18:178-179.

—— The classification, nomenclature and definitions of the forms of ice and snow according to Seligman. *Bull. Amer. Meteor. Soc.* 18:5-10, 47-53.

——, and C. F. Brooks. Wilhelm Schmidt, 1882-1936. *Bull. Amer. Meteor. Soc.* 18:34.

1938

Arenberg, D. L. A year of ultra-violet measured with a dosimeter at Blue Hill. *Trans. Amer. Geophys. Union.* 19th Ann. meeting 1938:140-143.

—— The triple point of water and the icing of airplanes. *Bull. Amer. Meteor. Soc.* 19:383-384.

Brooks, C. F. An example of Arctic control of our seasonal weather. *Bull. Amer. Meteor. Soc.* 19:395-396.

—— Blue Hill Meteorological Observatory. *Rept. of the President of Harvard University, 1937-1938:* 387-390.

—— Need for universal standards for measuring precipitation, snowfall and snow-cover. Trans. of the meetings of the Int. Commiss. of Snow and of Glaciers, Edinburgh. *Int. Assoc. Hydro. Bull.* 23 1938:7-52.

—— (Ed.) Some American papers and notes on climatic variations. *Bull. Amer. Meteor. Soc.* 19:161-224.

—— West Indian hurricanes that blast New England. *Middletown Scientific Assoc,* 570th Regular meeting, Dec. 3 p.

—— Wind-shields for precipitation-gages. *Trans. Amer. Geophys. Union.* 19 Ann. meeting 1938:540-542.

Chaney, R. W. A summary of the climatic data in the papers on cenazoic paleontology of Western North America. Blue Hill Obs., for the. *Amer. Committee of the Int. Commission of Climatic Variations.* Amsterdam July:1938.

Clayton, H. H. Centers of action and long period weather changes. *Bull. Amer. Meteor. Soc.* 19:76-77.

Dorsey, H. G. Local forecasting of heavy winter precipitation at Blue Hill. *Bull. Amer. Meteor. Soc.* 19:281-297 and 335-339.

—— Some local reversals of pressure gradient at Blue Hill. *Bull. Amer. Meteor. Soc.* 19:453-454.

Lange, K. O. The application of the Harvard radio-meteorograph to a study of icing conditions. *J. Aero. Sci.* 6:59-63.

Pear, C. B. The design and performance of radio equipment for radio-metrographs. *Bull. Amer. Meteor. Soc.* 19:299-309

Schulman, E. Douglas on climatic cycles and tree growth. *Bull. Amer. Meteor. Soc.* 19:204-211

—— Nineteen Centuries of rainfall history in the Southwest. *Bull. Amer. Meteor. Soc.* 19:211-214

—— On Angstrom's turbidity coefficient. *Bull. Amer. Meteor. Soc.* 19:398-399

Slocum, G., and D. L. Arenberg. The hailstorm of April 29, 1938 in Washington, D. C. *Mon. Wea. Rev.* 66:275-277.

Stone, R. G. Note on the Cheyenne type of fog regime. *Bull. Amer. Meteor. Soc.* 19:72-73.

—— On the relation of weather and snow reports to winter sports. *Trans. Amer. Geophys. Union.* 19th Ann. meeting 1938 Part II:720-724.

—— Pacific Slope winter sports conditions broadcast weekly. *Bull. Amer. Meteor. Soc.* 19:87-88.

—— Paulcke on alpine snow deposits. *Bull. Amer. Meteor. Soc.* 19:80-83.

—— Remarks on Kidson's and Weightman's "The cyclone series in the Caribbean Sea, Oct 17-24, 1935." *Bull. Amer. Meteor. Soc.* 19:256-259.

—— Small tornado at Silver Lake, N.H., July 17, 1937. *Bull. Amer. Meteor. Soc.* 19:93-94.

—— The distribution of the average depth of snow-on-the-ground in New York and New England: method of study. *Trans. Amer. Geophys. Union.* 19th Ann. meeting 1938:486-492.

—— The great blizzards of December 1937 at Buffalo. *Bull. Amer. Meteor. Soc.* 19:88-90.

—— The great snowfall of early February 1938 in the Sierra Nevada. *Bull. Amer. Meteor. Soc.* 19:86-87.

1939

Arenberg, D. L. Turbulence as the major factor in the growth of cloud crops. *Bull. Amer. Meteor. Soc.* 20:444-448.

Brooks, C. F. Blue Hill Meteorological Observatory. *Rept. of the President of Harvard College and Repts. of Departments.* 1938-39, 428-435.

—— Forecasting the weather. *Harvard Alumni Bull.* 41:150-153.

—— Hurricanes into New England, meteorology of the storm of September 21, 1938. *Geo. Rev.* 29.

—— The hurricane-meteorological postscript. *Harvard Alumni Bull.* 41:1165-1168.

——, and I. I. Schell. Certain symnoptic antecedents of severe cold waves in Southern New England. *Bull. Amer. Meteor. Soc.* 20:439-444.

——, and A. H. Thiessen. The meteorology of the great floods in the eastern United States. *The Smithsonian Rept. for 1938,* 325-348.

Conrad, V. The frequency of various wind velocities on a high isolated summit. *Bull. Amer. Meteor. Soc.* 20:373-375.

Fergusson, S. P. Experimental studies of anemometers. *Harvard Meteor. Studies.* No. 4. Cambridge: Harvard Univ. Press.

—— Experimental studies of anemometers. (Abstract.) *Bull. Amer. Meteor. Soc.* 20:307-309.

—— Sensitive open-scale instruments for the detection of minor disturbances of the atmosphere. *Bull. Amer. Meteor. Soc.* 20:135-141.

Friend, A. W. Continuous determination of air-mass boundaries by radio. Preprint. *Bull. Amer. Meteor. Soc.* May:1939.

Haurwitz, B. Pressure and temperature variations in the free atmosphere and their efect on the life history of cyclones. *Bull. Amer. Meteor. Soc.* 20:282-287.

—— The interaction between the polar front and the tropopause. *Int. Geod. and Geophys.* Washington Assembly. Sept.:1939.

——, and E. Haurwitz. Pressure and temperature variations in the free atmosphere over Boston. *Harvard Meteor. Studies.* No. 3:74p. Cambridge: Harvard University Press.

Lange, K. O. Vereisungsmessungen mit der Harvard-Radiosonde. *Sonderdruck aus Beitrage zur Phys. der Freien Atmos.* 25:241-250.

Schell, I. I. Some long-range forecasting methods and activities in Europe. *Bull. Amer. Meteor. Soc.* 20:44-48.

—— The interdependence of the atmospheric circulations of the northern and southern hemisphere. *Bull. Amer. Meteor. Soc.* 20:332-333.

Stone, R. G. The lapse rates in different air masses at Dickson Island, West Siberia. *Bull. Amer. Meteor. Soc.* 20:376-378.

—— Tropical climatology and physiology in relation to the acclimatization of white settlers. In *White Settlers in the Tropics*, by A. G. Price: 275-301.

1940

Brooks, C. F. Blue Hill Meteorological Observatory. *Rept. of the President of Harvard University and Repts. of Departments, 1939-40.* 5 p.

—— Further experience with shielded precipitation-gages on Blue Hill and Mount Washington. *Trans. Amer. Geophys. Union.* 1940:482-485.

—— Hubert on the African origin of the hurricane of September 1938. *Trans. Amer. Geophys. Union.* 1940:251-253.

—— Hurricanes into New England, meteorology of the storm of September 21, 1938. *The Smithsonian Rept.* 241-251.

—— Oliver Lanard Fassig (1860-1936). *Proc. Amer. Acad. Arts and Sci.* 74:118-120.

—— Snow favors cold and fair weather. Blue Hill Meteorological Observatory. *Blue Hill Notes.* No. 6:15. 2 p.

—— Some North American connections of Caribbean climate. *Proc. 8th Amer. Sci. Conf. Phy. and Chem. Sci.* 297-311.

—— The worst weather in the world. *Appalachia.* 23:194-202.

Friend, A. W. Developments in meteorological soundings by radio waves. *J. Aero. Sci.* 7:347-352.

Pear, C. B. Radio equipment for an unmanned weather station. *Bull. Amer. Meteor. Soc.* 21:107-110.

Schell, I. I. Foreshadowing the severity of the iceberg season south of Newfoundland. *Bull. Amer. Meteor. Soc.* 21:7-10.

Stone, R.G. A note on the distortion of the hurricane San Felipe (II) by the mountains of Puerto Rico. *Trans Amer. Geophys. Union.* 1940:254-255.

—— The distribution of the average depth of snow on the ground in New York and New England (II): Curves of Average depth and variability. *Trans. Amer. Geophys. Union.* 1940:672-692.

1941

Ackerman, E. The Köppen classification of climates in North America. *Geogr. Rev.* 31:105-111.

Arenberg, D. L. The formation of small hail and snow pellets. *Bull. Amer. Meteor. Soc.* 22:113-116.

Brooks, C. F. Apparent rate of progress as affected by changing intensity of hurricane San Felipe (II) while crossing Puerto Rico. *Trans Amer. Geophys. Union.* 1941:426-428.

—— Clouds in aerology and forecasting (I). *Bull. Amer. Meteor. Soc.* 22:335-345.

—— Henry Helm Clayton at 80. *Bull. Amer. Meteor. Soc.* 22:375-377.

—— Spreading cirrus clouds formed during airplane maneuvers over eastern Massachusetts. *Bull. Amer. Meteor. Soc.* 22:234.

—— The measurement of snowfall with shielded gages. *Proc. Central Snow Conf.* East Lansing, Mich. 1:193-195.

—— Two winter storms encountered by Columbus in 1493 near the Azores. *Bull. Amer. Meteor. Soc.* 22:303-309.

——, S. P. Fergusson, H. H. Kimball, B. Haurwitz, E. S. Brooks, J. Namias, C. H. Pierce, H. Wexler and E. M. Brooks. Eclipse meteorology with special reference to the solar eclipse of August 31, 1932. *Harvard Univ. Meteor. Studies.* No. 5. Cambridge: Harvard University Press. 109 p.

Burthoe, R. W. Bibliographic tools for meteorological research. *Bull. Amer. Meteor. Soc.* 22:357-361.

Conover, J. H. Isallobars and wind directions as indicators of the direction of movement of secondary cyclones on the middle Atlantic Coast in winter. *Bull. Amer. Meteor. Soc.* 22:276-278.

Conrad, V. Structure of the weather on Mount Washington. *Bull. Amer. Meteor. Soc.* 22:297-298.

—— The variability of precipitation. *Mon. Wea. Rev.* 69:5-11.

Elsasser, W. M. A statistical analysis of the earth's internal magnetic field. *Phy. Rev.* 60:876-883.

Friend, A. W. Further comparisons of meteorological soundings by radio waves with radiosonde data. *Bull. Amer. Meteor. Soc.* 22:53-61.

Schell, I. I. Foreshadowing Montana's winter precipitation (verification). *Trans. Amer. Geophys. Union.* 1941:428-429.

—— Foreshadowing the winter's precipitation in Montana. *Bull. Amer. Meteor. Soc.* 22:24.

—— Foreshadowing this winter's precipitation in Montana and Florida. *Bull. Amer. Meteor. Soc.* 22:424.

Stone, R. G. Health in tropical Climates. *Climate and Man., Yearbook of Agriculture.* 247-261.

1942

Arenberg, D. L., and R. W. Elsner. The microscopic structure of rime. *Bull. Amer. Meteor. Soc.* 23:276-280.

Brooks, C. F. Blue Hill Meteorological Observatory. *Rept. of the President of Harvard College and Repts. of Departments, 1940-1941.*

—— Clouds. *Sky and Telescope.* 1:3 pp.

—— Clouds in aerology and forecasting (I). *Bull. Amer. Meteor. Soc.* 22:335-345.

Conrad, V. *Fundamentals of physical climatology.* Cambridge: Harvard University Press. 121 p.

—— The interiurnal variability of temperature on Mount Washington. *Trans. Amer. Geophys. Union.* 23rd Ann. meeting 1942:279-283.

—— World climate and world war. *Bull. Amer. Meteor. Soc.* 23:264-273.

Elsasser, W. M. Heat transfer by infrared radiation in the atmosphere. *Harvard Meteor. Studies.* No. 6. Harvard Univ. Press. 107 p.

Eustis, R. S. The winds over New England in relation to topography. *Bull. Amer. Meteor. Soc.* 23:383-387.

Schell, I. I. Foreshadowing this winter's precipitation in Montana and Florida. *Bull. Amer. Meteor. Soc.* 23:131, 182-183.

—— On the use of climatically coherent areas in seasonal forecasting. *Bull. Amer. Meteor. Soc.* 23:182-183.

Stetson, H. T., and C. F. Brooks. Auroras observed at the Blue Hill Meteorological Observatory, 1885-1940. *Terres. Magnet. and Atmos. Elect.* March, 21-29.

Stone, R. G. On the mean circulation of the atmosphere over the Caribbean. *Bull. Amer. Meteor. Soc.* 23:4-16.

1943

Arenberg, D. L. Determination of icing-conditions for airplanes. *Trans. Amer. Geophys. Union.* 99-122.

Brooks, C. F. Blue Hill Meteorological Observatory. *Rept. of the President of Harvard College and Repts. of the Departments. 1941-1942.* 434-441.

—— Hail schells. *Bull. Amer. Meteor. Soc.* 24:53.

—— March weather. *Sky and Telescope.* 2:5.

—— Non-instrumental determination at the ground of ceiling, icing layers, turbulence and winds aloft. *Bull. Amer. Meteor. Soc.* 24:20-21.

—— Some North American connections of Caribbean climate. *Proc. 8th Amer. Sci. Cong.*

—— Waves, wind and weather. In *Science from Shipboard* by H. Wolff, New York. 11-52.

Conrad, V. Interhourly variability of temperature. *Trans. Amer. Geophys. Union.* 24 Ann. meeting 1943:122-125.

—— The climate of the Mediterranean region. *Bull. Amer. Meteor. Soc.* 24:127-145.

Fergusson, S. P. Gustiness under various weather conditions. *Bull. Amer. Meteor. Soc.* 24:22-29.

Hand, I., J. H. Conover and W. A. Boland. Simultaneous pyreheliometric measurements at different heights on Mount Washington, N. H. *Mon. Wea. Rev.* 71:65-69.

Liverence, W. F., and C. F. Brooks. Cloudiness and sunshine in New England. *Bull. Amer. Meteor. Soc.* 24:263-274.

Schell, I. I. The sun's spottedness as a possible factor in terrestrial pressure. *Bull. Amer. Meteor. Soc.* 24:85-93.

Schulman, E. Filter measurements of solar radiation at Blue Hill Meteorological Observatory. *Harvard Meteor. Studies.* No. 7. Harvard Univ. Press. 68 p.

Stone, R. G. On the practical evaluation and interpretation of the cooling power in bioclimatology. *Bull. Amer. Meteor. Soc.* 24:295-306 and 327-339.

1944

Brooks, C. F. Blue Hill Meteorological Observatory. *Repts. of the President of Harvard College and Repts. of the Departments, 1942-1943.* 8 p.

—— International cloud-nomenclature and coding. *Trans. Amer. Geophys. Union.* 1944:494-502.

—— Spot forecasting approach to synoptic forecasting. *Bull. Amer. Meteor. Soc.* 25:35.

Chapman, C., and C. F. Brooks. Meteorology in the warring forties. *Geog. Rev.* 34:466-475.

Conrad, V. *Methods in Climatology.* Cambridge: Harvard University Press.

—— Radiation and cloudiness. *Trans. Amer. Geophys Union.* 445-455.

Fergusson, S. P. On the improvement of certain meteorological instruments, exposure and techniques. *Bull. Amer. Meteor. Soc.* 25:289-298.

Stone, R. G. The average length of the season with snow-cover of various depths in New England. *Trans. Amer. Geophys. Union.* 1944:875-881.

1945

Brooks, C. F. Blue Hill Meteorological Observatory. *Rept. of the President of Harvard College and Repts. of Departments, 1943-44.* 8 p.

—— Minimum water content of hail deposits. *Bull. Amer. Meteor. Soc.* 26:122.

—— The significance of rapid rises in temperature on Mount Washington in severe cold waves. *Bull Amer. Meteor. Soc.* 26:292-293.

——, and C. Chapman. The New England hurricane of September 1944. *Geograph. Rev.* 35:132-136 also Smithsonian Rept. 1945:235-246.

——, J. H. Conover and others. *A Guide to Cloud Coding.* U.S. Weather Bureau. Mimeo. 10 pp.

Burhoe, R. W. Note on finding a representative snowfall measurement. *Bull. Amer. Meteor. Soc.* 26:341-342.

Conover, J. H. Seasonal weather summaries. *Blue Hill Meteorological Observatory* Quarterly. (mimeo)

Conrad, V. Earthquakes, atmospheric pressure, geological structure. *Trans. Amer.Geophys. Union.* 26:201-202.

—— Some remarks upon the destructive effects of the hurricane, September 14-15, 1944, observed at Hyannis, Cape Cod, Massachusetts. *Trans. Amer. Geophys. Union.* 26:217-219.

Cowles, J. Report of the committee to visit Blue Hill Meteorological Observatory.

Friend, A. W. A summary and interpretation of ultra-short wave propagation data collected by the late Ross A. Hull. *Proc. of the Inst. of Radio Engs.* 33:358-373.

1946

Brooks, C. F. Alexander George McAdie. *Harvard Univ. Gazette.* 41:99-100.

—— Blue Hill Meteorological Observatory. *Rept. of the President of Harvard College and Repts. of the Departments, 1944-1945,* 346-352.

—— Convergence in the Lake Region as a factor contributing to snowfall in an arctic air mass. *Bull. Amer. Meteor. Soc.* 27:8.

—— Fluffiest snow. *Bull. Amer. Meteor. Soc.* 27:91.

—— On the growth of hailstones. *Bull. Amer. Meteor. Soc.* 27:358.

—— and E. S. Brooks. Sunshine recorders: a comparative study of the burning-glass and thermometric systems. *Ann. Meeting, Amer. Meteor. Soc.* 28 Dec. 1946.

——, and C. Chapman. The New England hurricane of September 1944. *Smithsonian Rept. for 1945.* 235-246.

——, S. P. Fergusson and L. G. Gratin. Minute on the life and services of Alexander McAdie. *Harvard Univ. Gazette* February.

Conover, J. H. "Baseball" hail from towering cumulonimbus. *Weather.* 217.

—— Seasonal weather summaries. *Blue Hill Meteorological Observatory* Quarterly. Mimeo.

—— Tests of the Priez Aerovane in the natural wind at Blue Hill Meteorological Observatory. *Bull. Amer. Meteor. Soc.* 27:523-531.

Conrad, V. Diurnal variations of precipitation and forecasting. *Trans. Amer. Geophys. Union.* 27:35-40.

—— Earthquakes, atmospheric pressure tendency, geological structure. *Bull. Seismo. Soc. Amer.* 36:5-16.

—— Polygon nets and their physical development. *Amer. J. Sci.* 244:277-296.

—— Preliminary cloud statistics. *Mon. Res. Bull., Mt. Wash. Obs.* 2:17 pp.

—— Problems of health resort climatology. *Bull. Amer. Meteor. Soc.* 27:152-154.

—— Statistical investigation of the Mount Washington series of icing observations. *The Mt. Wash. Obs. Monthly Res. Bull.* 2 Oct. 17 pp.

—— Trigger causes of earthquakes and geological structure. *Bull. Amer. Meteor. Soc.* 36:357-362.

—— Usual formulas of continentality and their limits of validity. *Trans. Amer. Geophys. Union.* 27:663-664.

Haurwitz, B. Insolation in relation to cloudiness and cloud density. *J. Meteor.* 3:154-166.

—— Insolation in relation to cloud type. *J. Meteor.* 3:123-124.

—— Relations between solar activity and the lower atmosphere. *Trans. Amer. Geophys. Union.* 27:161-163.

Schell, I. I. Foreshadowing the winter precipitation in Montana and Florida. *Bull. Amer. Meteor. Soc.* 27:33-34.

—— Foreshadowing this winter's (1945/46) precipitation in the Northern Rocky Mountains and North Pacific States Region and verification. *Bull. Amer. Meteor. Soc.* 27:131-132, 348-349.

—— Single-radiosonde analysis in local 8-hour forecasting of precipitation. *Bull. Amer. Meteor. Soc.* 27:164-168.

1947

Blue Hill pocket cards, 2nd Ed. *Blue Hill Meteor. Obs., Harvard Univ.* 4 pp.

Brooks, C. F. Blue Hill Meteorological Observatory. *Rept. of the President of Harvard College and Repts. of Departments, 1945-46.* 9 p.

—— Recommended climatological networks based on the representativeness of climatic stations for different elements. *Trans. Amer. Geophys. Union.* 28:845-846.

——, and E. S. Brooks. Sunshine recorders: a comparative study of the burning-glass and thermometric systems. *J. Meteor.* 4:105-115.

Conover, J. H. Seasonal weather summaries. *Blue Hill Meteorological Observatory Quarterly.* (mimeo.)

Fergusson, S. P., and C. F. Brooks. Henry Helm Clayton, 1861-1941. *Sci.* 105:247-248.

Howell, W. E. Evaluation of errors in multi-cylinder icing observations due to variations between two adjacent sites, inclination to the wind, and length of exposure. *Memorandum Rept. Harvard-Mt. Washington Icing Res. Proj.*

Mitchell, J. M. Jr. Comparisons of the Stewart anemometer plastic-cup wheel in the natural wind at Blue Hill Meteorological Observatory. *Bull. Amer. Meteor. Soc.* 28:249-251.

Schell, I. I. Dynamic persistence and its applications to long range foreshadowing. *Harvard Meteor. Studies.* No. 8. 80 p.

1948

Brooks, C. F. Blue Hill Meteorological Observatory. *Rept. of the President of Harvard College and Repts. of Departments, 1946-47:* 10 pp.

—— The climatic record: its content, limitations and geographic value. *Ann. Assoc. Amer. Geograph.* 38:153-168.

—— Brooks, C. F., W. E. Howell, and J. H. Conover. Harvard University Blue Hill Meteorological Observatory. *Blue Hill Meteorological Observatory,* First quarterly report, 1948-49, to the U.S. Weather Bureau, contract CWB 8099. Typed, 31p.

——, and W. E. Howell. Harvard–Mt. Washington icing researches 1946-47. *Mt. Wash. Obs. Bull.* 3-4.

Conover, J. H. Observations and photographs of a cold front made from an airplane. *Bull. Amer. Meteor. Soc.* 29:313-318.

——, and S. H. Wollaston. Seasonal weather summaries for Blue Hill Meteorological Observatory. *Blue Hill Meteorological Observatory* Quarterly. (Mimeo.)

Conrad, V. A new criterion of relative homogeneity of climatological series. *Arch. Meteor. Geophys. Bioklim.* Ser. B, Bd. 1, Heft 1:1-8.

Harvard University and Mt. Washington Observatory, Inc. *Harvard–Mt. Washington Icing Research Rept. 1946-47.* U.S. Air Force, Air Material Comm., Wright-Patterson AF Base Dayton, Ohio. Contract W35-038-ac-15827.

Haurwitz, B. Insolation in relation to cloud type. *J. Meteor.* 5:110-113.

Howell, W. B. On the climatic description of physiographic regions. *Ass. Assoc. Amer. Geograph.* 39:12-25.

Howell, W. E. *Harvard–Mt. Washington Icing research Report 1946-47.* AF Tech. Rept. 5676, U.S. Air Force, Air Material Com. Dayton Ohio. 802 pp.

—— The growth of cloud drops in uniformly lifted air. *Doctoral Dis.* Massachusetts Institute of Technology.

Mitchell, J. M. An example of micro-oscillations in pressure, wind and temperature at Blue Hill Meteorological Observatory. *Bull. Amer. Meteor. Soc.* 30:105.

1949

Brooks, C. F. Blue Hill Meteorological Observatory. *Rept. of the President of Harvard College and Repts. of Departments. 1947-48.* 6 p.

—— The revised cloud codes CL, CM, CH. *Bull. Amer. Meteor. Soc.* 30:31-33.

—— W. E. Howell and J. H. Conover. Mt. Washington and Blue Hill data aids to snowstorm forecasting for Boston. *Blue Hill Meteor. Obs.* Report to the U.S. Weather Bureau. Contract CWB 8099. 2nd quarerly rept. 44 p. typed.

Brooks, C. F., W. E. Howell, and J. H. Conover. Mt. Washington and Blue Hill data aids to snowstorm forecasting for Boston. *Blue Hill Meteorological Observatory.* Third quarterly rept. Contract CWB 8099 June 29:33p. plus appendix. Typed.

Brooks, C. F., V. Conrad, W. E. Howell, and J. H. Conover. Mt. Washington and Blue Hill data aids to snowstorm forecasting for Boston. *Blue Hill Meteorological Observatory.* Final report 1948-49. Contract CWB 8099. Typed 47p.

Chang, P. K. Methods of evaluating and comparing cold waves, with special reference to New York City. *Bull. Amer. Meteor. Soc.* 30:107-108.

Conover, J. H. and S. H. Wollaston. Seasonal weather summmaries Blue Hill Meteorological Observatory. *Blue Hill Meteorological Observatory.* Quarterly. (Mimeo.)

——, and S. H. Wollaston. Cloud systems of a winter cyclone. *J. Meteor.* 6:249-260.

Howell, W. E. A comparison of three multicylinder icing meters and a critique of the multicylinder method. *Mt. Wash. Obs. Res. Rept.* (ozalid). 46 p.

—— Axial-oscillation icing rate meter. ozalid. 5 p.

—— Cone device for drop size measurement. ozalid. 7 p.

—— R. Wexler and S. Braun. Contributions to the theory of constitution of clouds. *Mt. Wash. Obs. Res. Rept.* Part I, 43 p, Part II, 203 p.

—— Disk type icing meters. 5 pp. ozalid.

—— On the climatic description of physiographic regions. *Ann. of the Assoc. of Amer. Geog.* 39:12-25.

—— The growth of cloud drops in uniformly cooled air. *J. Meteor.* 6:134-149.

—— Universal decimal classification numbers for meteorology. (Large sheet.) *Blue Hill Meteorological Observatory, Harvard University.* 1 p.

Local climatological data for Blue Hill Meteorological Observatory. *U.S. Dept. of Commerce, U.S. W. B.* Monthly beginning July 1949.

Schell, I. I. The sun's spottedness as a possible factor in the frequency of anticyclones in Northwestern North America. *Bull. Amer. Meteor. Soc.* 40:292-294.

1950

Bibliography of meteorological instruments. Blue Hill Meteorological Observatory. Report to the U.S. Weather Bureau. Contract. CWB 8120. 165 p.

Brooks, C. F. Blue Hill Meteorological Observatory. *Rept. of the President of Harvard College and Repts. of Departments, 1948-49,* 289-300.

—— Interdiurnal variability of temperature extremes. *Bull. Amer. Meteor. Soc.* 32:192.

—— Thirty years of the American Meteorological Society. *Bull. Amer. Meteor. Soc.* 31:210-214.

—— Will New York rainmaking affect New England rainfall? *Weatherwise.* 3:30-31.

——, and I. I. Schell. Forecasting heavy snowstorms at Blue Hill. I. Tracks of the cyclones producing heavy snows in the winters of 1938/9 to 1946/7. II. Pressure, wind and temperature on Mt. Washington during the 24 hours prior to the onset of heavy snow at Blue Hill. III. Precipitation index at Washington, D. C. and Buffalo, N. Y., associated with heavy winter snowstorms at Blue Hill. *Bull. Amer. Meteor. Soc.* 31:131-133. 31:163-167, 31:219-222.

——, W. E. Howell and J. H. Conover. Mt. Washington and Blue Hill data aids to snowstorm forecasting for Boston. *Blue Hill Meteor. Obs.* Report to the U.S Weather Bureau. Contract CWB 8120. 2nd Quart. Rept. 31 p typed.

——, V. Conrad, W. E. Howell and J. H. Conover. Mt. Washington and Blue Hill data aids to snowstorm forecasting for Boston. *Blue Hill Meteor. Obs.* Report to the U.S Weather Bureau. Contract CWB 8120. Final Rept. 53 p typed.

——, V. Conrad, W. E. Howell and J. H. Conover. Mt. Washingston and Blue Hill data aids to snowstorm forecasting for Boston. *Blue Hill Meteor. Obs.* Report to the U.S. Weather Bureau. Contract CWB 8120. 1st Quart. Rept. 23 p typed.

Conover, J. H. Tests and adaption of the Foxboro dew-point recorder for weather observatory use. *Bull. Amer. Meteor. Soc.* 31:13-22.

——, and S. H. Wollaston. Seasonal weather summaries for Blue Hill Meteorological Observatory. *Blue Hill Meteorological Observatory.* Quarterly.

Conrad, V., and L. W. Pollak. Methods in Climatology. Cambridge. Harvard University Press. 2nd ed. rev. and enl. 459 p.

Fergusson, S. P. Standard measures and the economical production of graphs and figures. *Sci.* 112:403-404.

Howell, W. E. Contributions to the theory of the constitution of clouds. *Mt. Wash. Obs. Res. Rept.* 53 p.

—— More rain for New Yorkers. *Weatherwise.* 3:27-29.

——, and R. Geiger. *Climate near the ground.* Cambridge, Mass. 482 p.

Local climatological data for Blue Hill Meteorological Observatory. *U.S. Dept. of Comm. U.S.W.B.* monthly and annual.

1951

Brooks, C. F. Blue Hill Meteorological Observatory. *Rept. of the President of Harvard College and Repts. of Departments, 1949-50.* 13 p.

—— The thunderstorm vs. the airline. *Bull. Amer. Meteor. Soc.* 32:145.

—— Brooks, C. F. The use of clouds in forecasting. *Compendium of Meteorology,* Amer. Meteor. Soc., Boston, MA. 1167-1178.

——, V. Conrad, W. E. Howell and J. H. Conover. Mt. Washington and Blue Hill data aids to snowstorm forecsting for Boston. Report to the Weather Bureau. Contract CWB 8135.

Conover, J.H. A new combination anemometer, wind vane and wind direction recorder for Mt. Washington Obs. *Bull. Amer. Meteor. Soc.* 32:386-390.

—— Are New England winters getting milder? *Weatherwise.* 4:5-9.

—— The snowstorm of February 21-23, 1950. *Blue Hill Meteor. Obs.* Report to the U.S. Weather Bureau. Contract CWB 8135. 17 p typed.

——, and W. L. Gates. Tests of various types of maximum and minimum thermometers. *Blue Hill Meteorological Observatory.* Harvard Univ. Contract CWB 8120. 32 p. typed.

——, and T. G. Nastos. Tests of Stewart and Victor precipitation gages. *Blue Hill Meteorological Observatory.* Contract CWB 8120. 7 p typed.

——, C. F. Brooks and V. Conrad. Snowstorm forecasting for Boston including manual for objective forecasting. *Blue Hill Meteor, Obs.* Report to the U.S. Weather Bureau. Contract CWB 8135. Final Rept. 1950-51, 101 p.

——, S. H. Wollaston and P. C. Dalrymple. Seasonal Weather Summaries at Blue Hill Meteorological Observatory. *Blue Hill Meteorological Observatory,* Quarterly. Mimeo.

Conrad, V. A method of estimating the periodic constituent of a geophysical series. *Arch. Meteor. Geophys. Bioklim.* Ser. A Bd. 4:324-337.

Fergusson, S. P. Nephoscopes I and II. *Bull. Amer. Meteor. Soc.* 32:259-266 and 308-313.

Howell, W. E. A comparison of icing conditions on Mount Washington with those encountered in flight. *Trans. Amer. Geophys. Union.* 32:179-188.

—— The classificaton of cloud forms. In *Compendium of Meteorology.* Boston, MA: American Meteorological Society. 1161-1166.

Karapiperis, P. P. The climate of Blue Hill according to air masses and winds. *Harvard Meteor. Studies.* No. 9. 105 p.

—— The influence of ground condition and cloudiness on the diurnal march of surface vapor pressure at Blue Hill, Milton, Mass. *Trans. Amer. Geophys. Union.* 32:547-551.

—— The tower of the winds. *Weatherwise.* 4:112-113.

Local climatological data for Blue Hill Meteorological Observatory. *U.S. Dept. of Comm. U.S.W.B.* Monthly and annual.

1952

Brooks, C. F. Blue Hill Meteorological Observatory. *Rept. of the President of Harvard College and Repts. of the Departments.* 1950-51, 225-235.

—— Current dew research in Israel. *Bull. Amer. Meteor. Soc.* 33:55.

Conover, J. H., and C. F. Brooks. *Blue Hill Meteorological Observatory.* Harv. Univ. Semi-annual progress rept. 1951-52 Contract CWB 8163 20 July. 20 p typed.

Conrad, V. Climatic changes or cycles. *Arch. Meteor. Geophys. Bioklim.* Ser. B. Bd. 4 Heft 2:109-121.

—— Temperature, wind direction and speed in the tropics. *Berichte des Deutschen Wetterdienstes in der US-Zone.* No. 42:3-5.

Howell, W. E., and P. P. Karapiperis. Interpretation of the rainfall climate of Marathon, Greece. *Proc. Athens Acad.* 27:106-117.

Karapiperis, P. P. Interhourly variability of temperature at Blue Hill, Mass. *Arch. Meteor. Geophys. Bioklim.* Ser. B. 4:52-56.

—— Interdiurnal variability of temperature at Blue Hill. II, *Arch. Meteor. Geophys. Bioklim.* Ser. B:4:57-64.

—— The diurnal march of vapor pressure on sea-breeze days at Athens, Greece. *Quart. J. Royal Meteor. Soc.* 78:82-84.

U.S. Weather Bureau. Local climatological data for Blue Hill Meteorological Observatory. *U.S. Dept. of Comm. U.S.W.B.* Monthly and annual.

Wexler, R. Precipitation growth in stratiform clouds. *Quart. J. Royal Meteor Soc.* 78:363-371.

—— Theory of the radar upper band. *Quart. J. Royal Meteor. Soc.* 78:372-376.

Wollaston, S. H. Seasonal weather summaries. *Blue Hill Meteorological Observatory*, Quarterly. Mimeo.

1953

Brooks, C. F. Baur's formula indicates a cold winter for New England. *Bull. Amer. Meteor. Soc.* 34:41.

—— Blue Hill Meteorological Observatory. *Rept. of the President of Harvard College and Repts. of the Departments, 1951-1952,* 331-347.

—— Completion of the Köeppen–Geiger Handbuch der Klimatologie. *Prof. Geograph.* 5:21-22.

—— Current scientific projects at the Observatory. *Mount Washington Obs. News Bull.* No. 23:1.

—— What of the winter? Text of broadcasts over WGBH-FM. Blue Hill Observatory. *Why the Weather Series.* 4 and 11:Dec. Mimeo. 3 p. mimeo.

——, and J. P. Webber. The tornados of June 9, 1953. Some remarks on wind, pressure, fall of debris, hail, clouds, and steering. *Bull. Amer. Meteor. Soc.* 34:379. (Abstract.)

Conover, J. H. Climatic Changes as interpreted from Meteorological Data, in *Climate Change*, Harlow Shapley, ed.Cambridge, Mass: Harvard University Press. 221-230.

—— Blue Hill Meteorological Observatory. Harvard Univ.. Final report, 1951-52, to the United States Weather Bureau. Contract CWB 8163.

—— J. P. Webber and C. F. Brooks. *Blue Hill Meteorological Observatory.* Harvard Univ. semi annual prog. rept. 1952-1953. Contract CWB 8310. 19 p. typed.

Conover, J. H., J. P., Webber, and C. F. Brooks. *Blue Hill Meteorological Observatory.* Harvard University. Final report, 1951-52, to the United States Weather Bureau. Contract CWB 8163. Typed. 55p.

Conrad, V. The components of the diurnal variation of temperature on Mt. Washington and other summits. *Arch. Meteor. Geophys. Bioklim.* Ser. A. Bd. 6, Heft 1:78-107.

—— The semidiurmal temperature wave and peculiarities of the diurnal variations of temperature on mountain summits. *Geofis. Pura Appl.* 26:88-96.

Gross, R. E. *Report of the committee to visit Blue Hill Observatory, 1953.*

Karapiperis, P. P. Predicting minimum temperature (especially frost) by the evening wet-bulb at Blue Hill, Mass. in spring. *Geophys. Pura E Applicata-Milano.* 24:2-8.

—— The structure of the weather at Blue Hill, as indicated by the hours at which the diurnal extremes of temperature occur. *Arch. Meteor. Geophys. Bioklim.* Ser. B. Bd. 4 Heft 3:275-280.

Local climatological data for Blue Hill Observatory. *U.S. Dept. of Commerce, U.S.W.B.*, Monthly and annual.

Mitchell, J. M. On the causes of instrumentally observed secular temperature trends. *J. Meteor.* 10:244-261.

Wexler, R. Radar echoes from a growing thunderstorm. *J. Meteor.* 10:285-290.

—— The physics of tropical rain. Chap 6. in *Tropical Meteor.* New York: McGraw-Hill. 155-176

Wollaston, S. H. Seasonal weather summaries, *Blue Hill Meteorological Observatory.* Quarterly Mimeo.

1954

Brooks, C. F. Blue Hill Meteorological Observatory. *Report of the President of Harvard College and Repts. of the Departments, 1952-53.* 16 p.

—— (contributions to the methodology of observations at sea) *World Meteor. Organ.* Tech. Note No. 2, Methods of observation at sea. Part I—Sea surface temperature, WMO-No. 16, TP 8:35 pp. Geneva; Observations made during transatlantic voyages pp 2-4; Analysis of observations at different depths pp.5-15: Part II—Air temperature and humidity, atmospheric pressure, cloud height, wind rainfall and visibility, WMO-No. 40 TP:15, 35 pp. Geneva: Temperature and humidity, pp. 1-3; wind. p. 20: rainfall, pp. 27-29.

—— Indications for a mild winter in New England. *Bull. Amer. Meteor. Soc.* 35:132-133.

—— Note on the homogenity of climatological records. *Blue Hill Meteorological Observatory.* 3 pp.

—— and A. E. Bent. Winter report. *Mount Washington Obs. News Bull.* No. 24:1.

—— J. P. Webber, P. H. Putnins and F. B. Burhoe. Current library accessions. *Blue Hill Meteorological Observatory and American Meteorological Society.* Mimeo. 755-873.

Conover, J. H. Strain type remote recording precipitation gage. *In Precipitation measurement study,* final rept. of Illinois State Water Survey, Met. Lab. at the Univ. Of Ill., Urbana Ill. Report to U.S. Army on Contract DA-36-039 SC-15484. 21-23. (Abstract in Bull. of *Amer. Meteor. Soc.* 34:430.)

—— J. P. Webber, R. J. Boucher and C. F. Brooks. *Blue Hill Meteorological Observatory.* Harvard Univ.; Semi-annual progress report, 1953-54. Contract CWB 8381. July 27. 19 p typed.

Local climatological data for Blue Hill Observatory. *U.S. Dept. of Commerce; U.S.W.B.,* Monthly and annual.

Webber, J. P., J. H. Conover, R. Wexler, C. F. Brooks, E. S. Brooks, J. Honig and R. J. Reed. *Blue Hill Meteorological Observatory.* Harvard Univ.; Final report 1952-53 to the United States Weather Bureau on Contract CWB 8310. Typed. 66 pp.

Wexler, R., R. J. Reed and J. Honig. Atmospheric cooling by melting snow. *Bull. Amer. Meteor. Soc.* 35:48-51.

Wollaston, S. H. Seasonal weather summaries, *Blue Hill Meteorological Observatory.* Quarterly. Mimeo.

U.S. Weather Bureau. Local climatological data for Blue Hill Meteorological Observatory.

1955

Boucher, R. J., P. H. Putnins, J. P. Webber, R. Wexler and C. F. Brooks. Radar-synoptic investigation of the rainstorms of November 23 to 25, 1953, in the Boston area. *Blue Hill*

Meteorological Observatory Meteor. Radar. Studies No. 1. Harvard Univ.; Air. Res. and Dev. Command: Cont. AF 19 (604)-950:67 pp.

Brooks, C. F. Blue Hill Meteorological Observatory. *Rept. of the president of Harvard College and repts. of the departments, 1953-54.* 25 pp.

—— Forecasting from snow crystals. *Mt. Wash. News Bull.* 28:9-10.

Conover, J. H. Spiral precipitation patterns in extatropical cyclones. *Blue Hill Meteorological Observatory, Meteor. Radar Studies No. 2.* Harvard Univ.: Air Res. and Dev. Command. Cont. AF 19(604)-950: 11 p.

——, R. J. Boucher, P. H. Putnins, R. Wexler and C. F. Brooks. *Blue Hill Meteorological Observatory.* Harvard Univ.:semi-annual report 1954-55. Contract CWB 8523. 17 p. typed.

——, ——, ——, ——, ——, Jen-hu Chang and E. S. Brooks. *Blue Hill Meteorological Observatory.* Harvard Univ.: Final report 1954-55 to the United States Weather Bureau on contract CWB 8523. 26 p. typed.

——, ——, J. P. Webber and R. Wexler. *Blue Hill Meteorological Observatory.* Harvard Univ.:Final report 1953-54 to the United States Weather Bureau on Contract CWB 8381. 36 p. typed.

Local climatological data for Blue Hill Meteorological Observatory. *U.S. Dept. of Commerce. U.S.W.B.* Monthly and annual.

Wexler, R. Radar analysis of precipitation streamers observed 25 February 1954. *J. Meteor.* 12:391-393.

—— The melting layer. *Blue Hill Meteorological Observatory Meteor. Radar. Studies No. 3.* Harvard Univ.: Air Res. and Dev. Command:Cont. AF 19 (603)-950. 18 p.

Wollaston, S. H. Seasonal weather summaries, *Blue Hill Meteorological Observatory.* Quarterly. Mimeo.

1956

Blue Hill pocket cards. *Blue Hill Meteorological Observatory.* Harvard Univ.:4p.

Brooks, C. F. Another busy winter ahead. *Mt. Wash. Obs. News Bull.* No. 30:1-2.

—— Blue Hill Meteorological Observatory. *Rept. of the President of Harvard College and Repts. of the Departments.* 1954-55, 442-461.

—— Note on the climate of Mount Washington. *Mt. Wash. Obs. News Bull.* No. 30:6-7.

——, J. H. Conover, R. J. Boucher, P. Putnins, Jen-hu Chang, E. S. Brooks and R. E. Ashley. *Blue Hill Meteorological Observatory.* Harvard Univ.: Final report 1955-56 to the United States Weather Bureau on Contract CWB 8608. 51 p. typed.

Conover, J. H., R. J. Boucher, P. H. Putnins and C. F. Brooks. *Blue Hill Meteorological Observatory*. Harvard Univ.: semi-annual report 1955-56. Contract CWB 8608. 12 April. 45 p. typed.

Conrad, V. On thermal springs, a contribution to the knowledge of their nature. *Arch. Meteor. Geophys. Bioklim*. Series A Bd. 9 Heft 3:371-405.

Local climatological data for Blue Hill Meteorological Observatory. *U.S. Dept. of Comm. U.S.W.B*. Monthly and annual.

Putnins, P. The paradox of the "inhomogeneity" of annual temperatures at Blue Hill. *Arch. Meteor. Geophys. Bioklim*. Series B. Bd. 7 Heft 3-4:305-316.

Wexler, D. R., and D. Atlas. Factors influencing radar-echo intensities in the melting layer. *Quart. J. Royal Meteor Soc*. 82:349-351.

Wollaston, S. H. Seasonal weather summaries, *Blue Hill Meteorological Observatory*. Quarterly. Mimeo.

1957

Boucher, R. J. Synoptic–dynamic implications of 1.25 cm vertical beam radar echoes. Proc. 6th Wea. Radar Conf. *Amer. Meteor Soc.*, 179-188.

—— Synoptic–physical implications of 1.25 cm vertical beam radar echoes. *Blue Hill Meteorological Observatory Meteor. Radar Studies No. 5*. Harvard Univ. Air Res. and Dev. Command. Cont. AF 19(604)-950. 24 p.

Brooks, C. F. Blue Hill Meteorological Observatory. *Rept. of the President of Harvard College and Repts. of the Departments. 1955-1956:*700-713.

—— Our twenty-fifth year. *Mt. Wash. Obs., News Bull*. No. 31:1.

—— The new international definitions of hydrometeors. *Proceedings Conference on the Physics of Cloud and Precipitation Particles*, Sept. 1955. Pergamon Press. 415-420.

—— Project snowflake. *Mt. Wash. Obs. News Bull*. No. 31:1-2.

—— Proposed new tower. *Mt. Wash. Obs. News Bull*. No. 31:8.

——, R. J. Boucher, J. H. Conover, R. Wexler, A. Glaser, V. Conrad, E. S. Brooks and R. E. Ashley. E. *Blue Hill Meteorological Observatory*. Harvard Univ.: Semi-annual report 1956-57. Contract CWB 9006. 25 p. typed.

Chang, J. H. Global distribution of the annual range in soil temperature. *Trans. Amer. Geophys. Union*. 38:718-723.

—— World patterns of monthly soil temperature distribution. *Ann. Assoc. Amer. Geograph*. 47:241-249.

Conrad, V. Brief note on estimating the significance of a correlation coefficient. *Arch. Meteor. Geophys. Bioklim.* Ser. B; Bd.8 Heft 2:257-259.

—— Höhenlage und Sonnenacheindauer an Neiderhlagstagen. *Wetter Leben.* 9:1-4.

Glaser, A. H., and J. H. Conover. Meteorological utilization of images of the earth's surface transmitted from a satellite vehicle. *Blue Hill Meteorological Observatory.* Harvard Univ.; Milton Mass. Air. Res. and Devel. Command. Contract AF 19(604)-1589 phase II. 145 p.

L'hermitte, R. I. Display of precipitation echo contours. Proc. 5th Wea. Radar Conf. Suppl. Boston: *Amer. Meteor. Soc.* 23-27

Local climatological data for Blue Hill Observatory. *U.S. Dept. of Commerce; U.S.W.B.* Monthly and annual.

Wexler, R. Advection and the melting layer. *Blue Hill Meteorological Observatory, Meteor. Radar Studies No. 4.* Harvard Univ. Air Res. and Dev. Command;Cont. AF19(604)-950. 12 p.

—— Particle size distribution in rain and snow inferred from Z-R relations. *Blue Hill Meteorological Observatory, Meteor. Radar Studies No. 7.* Harvard Univ. Air. Res. and Dev. Command. Cont. AF 19(604) 950. 7 p.

—— The lapse rate in the melting layer. In Proc. 6th Weather Radar Conference. Boston, *Amer. Meteor. Soc.*:83-87.

—— and D. Atlas. Moisture supply and growth of stratiform precipitation. *J. Meteor.* 15:531-538. Proc. 6th Weather Radar Conference, Boston. *American Meteor Society.* 69-76.

——, and D. Atlas. Moisture supply and growth of stratiform precipitation. *BHMO Meteor. Radar Studies, No. 6.* Harvard Univ. Air Res. and Dev. Command. Cont. AF 19(604) 950. 19 p.

Wollaston, S. H. Seasonal weather summaries, *Blue Hill Meteorological Observatory.* Quarterly. Mimeo.

1958

Ashley, R., R. J. Boucher and J. H. Conover. *Blue Hill Meteorological Observatory.* Harvard University Semi-annual progress report 1957-58. Contract CWB 9271. 5 p typed.

Blue Hill Meteorological Observatory. Study of synoptic–dynamic influences on the nature of cloud and precipitation echoes. *BHO Radar Studies.* Harvard Univ. Oct. 1953-February 1958: Air Res. and Dev. Command. Cont. AF 19 (604) 950. 36 p.

Boucher, R. J. Synoptic and meso-scale aspects of severe local storms in New England. *Blue Hill Meteorological Observatory* Harvard University. U.S. Weather Bur. Cont. CWB 9271. 36 p. Append. 56 p.

——The development and growth of precipitation in three New England coastal cyclones—a radar symoptic analysis. *Blue Hill Meteorological Observatory Meteor. Radar Studies* No. 9. Harvard Univ. 44 pp. Air Res. and Dev. Command. Cont. AF 19(604) 950. 44 p.

Brooks, C. F., and E. S. Brooks. The accuracy of wind-speed estimates at sea. *Trans. Amer. Geophys. Union.* 39:52-57.

——, and J. H. Conover. Blue Hill Observatory. *Rept. of the President of Harvard College, 1956-1957.* 612-625.

——, and S. J. Richardson. The Blue Hill Meteorological Library. *Harvard Library Bull.* 12:271-281.

Chang, J. H. Ground temperature. *Blue Hill Meteorological Observatory. Harvard University.* Quartermaster Res. and Dev. Command. Cont. DA 19-129-QM-348. Vol. 1:text 300 pp.: Vol. 2:tables 196 pp.

Conover, J. H. Blue Hill Meteorological Observatory. *Rept. of the President of Harvard College and Repts. of the Departments 1957-1958.* 527-532.

—— *Blue Hill Meteorological Observatory.* Harvard Univ. final report 1957-58 to the United States Weather Bureau on contract CWB 9271. Typed 4 pp.

—— Charles Franklin Brooks, Meteorologist. *Sci.* 863-864.

——, and R. J. Boucher. *Blue Hill Meteorological Observatory.* Harvard Univ. Final report 1956-57 to the U.S. W. B. on contract CWB 9006. 19 p. typed.

Local climatological data for Blue Hill Observatory. *U.S. Dept. of Comm. U.S.W.B.* Monthly and annual.

Shackford, C. R. Radar observation of cold front thunderstorms of July 14, 1956. *Blue Hill Meteorological Observatory, Meteor. Radar Studies* No. 11. Air REs. and Dev. Command:Cont. AF 19(604)- 950. 39 p.

Thompson, G. K. A system for recording eight meteorological elements simultaneously on moving film (Weatherlog). *Blue Hill Meteorological Observatory Meteor. Radar Studies* No.8. Air Res. and Dev. Command: Cont. AF 19 (604) 950. 60 p.

Wexler, R. and D. Atlas. Vertical structure of continuous streamer-form precipitation. *Blue Hill Meteorological Observatory, Meteor. Radar Studies* No. 10. Air Res. and Dev. Command Cont. AF 19(604)-950. 38 p.

——, and D. Atlas. Moisture supply and growth of stratiform precipitation. *J. Meteor.* 15:531-538.

Wollaston, S. H. Seasonal weather summaries, *Blue Hill Meteorological Observatory.* Quarterly. Mimeo.

1959

Boucher, R. J. Synoptic–physical implications of 1.25 cm vertical-beam radar echoes. *J. Meteor.* 16:312-326.

Brooks, C. F. The society's first quarter century. *Weatherwise.* 12:223-230.

Conover, J. H. Cloud patterns and related air motions derived by photography. *Blue Hill Meteorological Observatory.* Final report for Cont. AF 19(604)-1589. 268 pp.

Goody, R. M. Blue Hill Meteorological Observatory. *Reports of the President of Harvard College and Repts. of the Departments, 1958-1959:*509-512.

Local climatological data for Blue Hill Meteorological Observatory. *U.S. Dept. of Comm. U.S.W.B.* Monthly and annual.

Thompson, G. K. Services aiding the study and recording of meteorological phenomena in correlation with radar reflectivity patterns of precipitation as observed at or adjacent to the Blue Hill Meteorological Observatory, Milton, Mass. Final Rept. Cont. AF19 (604)-4942 for period 1 Nov. 1958-31 Oct. 1959. Includes raindrop spectrograph. *Harvard University, Blue Hill Meteorological Observatory.* 9 p. typed.

1960

Conover, J. H. Cirrus patterns and related air motions near the jet stream as derived by photography. *J. Meteor.* 17:532-546.

Goody, R. M. Blue Hill Meteorological Observatory. *Rept. of the President of Harvard College and Repts. of the Departments, 1959-1960,* 623-625.

Local climatological data for the Blue Hill Meteorological Observatory. *U.S. Dept. of Commerce, U.S. W. B.* Monthly and annual.

Volz, F. Einige Beobachtungen Über ungewohnliche Regenbogen und Hinweise auf wünschenswerete Beobachtungen. *Meteor. Rundsch.* 13:117-118.

—— Some aspects of the physics of rain and the optics of the rainbow. *Geophys. Monogr.* 5:280-286.

——, and R. M. Goody. Twilight intensity at 20° elevation. *Blue Hill Meteorological Observatory Harvard University Sci. Rept. 1.* Contr. AF (604)-4546:46 pp.

1961

Goody, R. M. Blue Hill Meteorological Observatory. *Rept. of the President of Harvard College and Repts. of the Departments 1960-1961.* 527-529.

Local climatological data for the Blue Hill Meteorological Observatory. *U.S. Dept. of Comm. U.S.W.B.* Monthly and annual.

Volz, F. Der Regenbogen. *Hand. geophys.* 8:Physik d. Atmosph. 1. Kap. 14:3 Berlin:943-1020.

—— Twilight intensity at 20° elevation. Results of observations. *Blue Hill Meteorological Observatory*, Harvard University. Rept. 2 on Contract AF 19 (604)–4546. 23 pp.

1962

Goody, R. M. Blue Hill Meteorological Observatory. *Report of the President of Harvard College and Repts. of the Departments. 1961-1962*, 626-628.

Local climatological data for Blue Hill Meteorological Observatory. *U.S. Dept. of Comm. U.S.W.B.* Monthly and annual.

Noxon, J. F., and R. M. Goody. Observations of day airglow emission. *J. Atmos. Sci.* 19:342-343.

——, and V. R. Jones. Observations of the (o,o) level of the ($^1\Delta g - ^3\bar{\Sigma}g$) system of oxygen and twilight airglow. *Nature.* 196;157.

Volz, F., and R. M. Goody. The intensity of twilight and upper atmospheric dust. *J. Atmos. Sci.* 19:385-406.

—— Twilight intensity at 20° elevation. Analysis and discussion of observations. *Blue Hill Meteor. Obs.* Final Report. Cont. AF 19 (604)- 4546. 47 pp.

1963

Goody, R. M. A note on the refraction of a twilight ray. *J. Atmos. Sci.* 20:502-505.

—— Blue Hill Meteorological Observatory. *Rept. of the President of Harvard College and Repts. of the Departments, 1962-1963:*493-495. .

Local climatological data for the Blue Hill Meteorological Observatory. *U.S. Dept. of Comm. U.S.W.B.* Monthly and annual.

Noxon, J. F. Observations of daytime aurora. *J. Atmos. Terr. Phys.* 25:637-645.

——, and T. P. Markham. Airglow emission during a solar eclipse. *J. Geophys. Res.* 68:6059-6061.

1964

Goody, R. M. Blue Hill Meteorological Observatory. *Rept. of the President of Harvard College and Repts. of the Departments 1963-64.* 431-432.

Local climatological data for Blue Hill Observatory. *U.S. Dept. of Comm. U.S.W.B.* Monthly and annual.

Noxon, J. F. A study of the 6300A oxygen line in the day airglow. *J. Geophys. Res.* 69:3245-3255.

—— The latitude dependence of OH rotational temperature in the night airglow. *J. Geophys. Res.* 69:4087-4092.

1965

Goody, R. M. Blue Hill Meteorological Observatory *Report of the president of Harvard College and Repts. of the Departments 1964-1965.* 487-490.

——, and J. F. Noxon. Frequency incoherent scattering in skylight. *Izv. Acad. Sci. USSR: Atmos. Ocean. Phys.* 3:275-281.

Local climatological data for Blue Hill Meteorological Observatory. *U.S. Dept. of Comm. U.S.W.B.* Monthly and annual.

Noxon, J. F. Noncoherent scattering of skylight. *Atmos. and Oceanic Phys. Series.* 1:275. (P.E.)

1966

Goody, R. M. Blue Hill Meteorological Observatory. *Rept. of the President of Harvard College and Repts. of the Departments, 1965-1966.* 477-480.

——, and H. McClees. High frequency electromagnetic pressure modulator. *Rev. Sci. Inst.* 37:1273-1274.

Local climatological data for Blue Hill Meteorological Observatory. *U.S. Dept. of Comm. U.S.W.B.* Monthly and annual.

Noxon, J. F. A method for observation of coronal emission lines. *Astrophys.* J. 145:400-410.

1967

Goody, R. M. Blue Hill Meteorological Observatory. *Rept. of the President of Harvard College and Repts. of the Departments, 1966-1967,* 496-499.

Local climatological data for Blue Hill Observatory. *U.S. Dept. of Comm. U.S.W.B.* Monthly and annual.

Noxon, J. F. Dayglow observations. In *Aurora and Airglow.* New York Reinhold Pub. Corp. New York 123-131.

—— Interpretation of the dayglow. In *Aurora and Airglow.* Reinhold Pub. Corp. New York: 315-321.

—— Oxygen spectra in dayglow, twilight, and during an eclipse. *Nature.* 213:350-352.

1968

Goody, R. M. Blue Hill Meteorological Observatory. *Rept. of the President of Harvard College and Repts. of the Departments 1967-1968.* 448-450.

—— Cross-correlating spectrometer. *J. Opt. Soc.* 58:900-908.

——, and T. McCord. Continued search for the Venus airglow. *Planet. Space Sci.* 16:343-351.

Local climatological data for Blue Hill Meteorological Observatory. *U.S. Dept. of Comm. U.S.W.B.* Monthly and annual.

Noxon, J. F. Day airglow. *Space Sci. Rev.* 8:91-134.

1969

Carleton, N. P., R. M. Goody, et al. Measurement of the abundance of CO_2 in the Martian atmosphere. *Astrophy. J.* 155:323-331.

Goody, R. M. Blue Hill Meteorological Observatory. *Rept. of the President of Harvard College and Repts. of the Departments 1968-1969.* 450-453.

—— Time variations in atmospheric N_2O in eastern Massachusetts. *Planet. Space Sci.* 17:1319-1320.

Local climatological data for Blue Hill Meteorological Observatory. *U.S. Dept. of Comm. ESSA.* Monthly and annual.

1970

Goody, R. M. Blue Hill Meteorological Observatory. *Rept. of the President of Harvard College and Repts. of the Departments, 1969-1970,* 481-483.

Local climatological data for Blue Hill Meteorological Observatory. *U.S. Dept. of Comm. ESSA.* Monthly and annual.

Noxon, J. F. Auroral emission from O_2 ($^1\Delta g$). *J. Geophys. Res.* 75:1879-1891.

—— Metastable oxygen: origin of atmospheric absorbtion near 50 km. *Sci.* 168:1120-1121.

—— Noxon, J. F. Optical emission from $O(^1D)$ and $O_2(b^1\Sigma g)$ in ultraviolet photolysis of O_2 and CO_2. *J. Chem. Phys.,* 52:1852

——, and A. E. Johanson. Effect of magnetically conjugate photoelectrons on 0I(6300A). *Planet. Space Sci.* 18:1367-1379.

1971

Goody, R. M. Blue Hill Meteorological Observatory. *Rept. of the President of Harvard College and Repts. of the Departments 1970-1971.* 453-454.

Local climatological data for Blue Hill Meteorological Observatory. *U.S. Dept. of Comm. NOAA.* Monthly and annual.

Noxon, J. F. Interpretation of F. region nightglow. The Radiating Atmosphere, ed. by. B. M. McCormac and A. V. Jones. Reidel Dordrecht-Holland. 64-

1972

Local climatological data for Blue Hill Meteorological Observatory. *U.S. Dept. of Comm. NOAA.* Monthly and annual.

Mullaney, H., M. D. Papagiannis and J. F. Noxon. Parallel study of 6300A airglow emission and ionospheric scintillation. *Planet. Space Sci.* 20:41-46.

Noxon, J. F. Recent high latitude optical observations. In *Magnetospheric Ionosphere Interactions.* ed. by K. Folksted; Universitetsforlaget Oslo.

——, and A. E. Johanson. Changes in thermospheric molecular oxygen abundance inferred from twilight 6300A airglow. *Planet. Space Sci.* 20:2125-2151.

1973

Local climatological data for Blue Hill Meteorological Observatory. *U.S. Dept. of Comm. NOAA.* Monthly and annual.

Noxon, J. F. OI emission. *Physics and Chemistry of Upper Atmospheres.* ed. by B. M. McCormac. Dordrecht-Holland: D. Reidel. 213-218.

1974

National Oceanic and Atmospheric Adminstration. Local climatological data for Blue Hill Meteorological Observatory. *U.S. Dept. of Comm. NOAA.* Monthly and annual.

1975

Local climatological data for Blue Hill Meteorological Observatory. *U.S. Dept. of Comm. NOAA.* Monthly and annual.

Markham, T. P., J. Buchau, R. E. Antcil and J. F. Noxon. Airborne study of equatorial 6300A nightglow. *J. Atmos. Terrest. Phys.* 37:65-74.

1976

Local climatological data for Blue Hill Meteorological Observatory. *U.S. Dept. of Comm. NOAA.* Monthly and annual.

Noxon, J. F., and J. V. Evans. Simultaneous optical and incoherent scatter observations of two low-latitude auroras. *Planet. Space Sci.* 24:425-442.

Roble, R. G., J. F. Noxon and J. V. Evans. The intensity variation of the atomic oxygen red line during morning and evening twilight on 9-10 April 1969. *Planet. Space Sci.* 24:327-340.

1977

Local climatological data for Blue Hill Meteorological Observatory. *U.S. Dept. of Comm. NOAA.* Monthly and annual.

1978

Local climatological data for Blue Hill Meteorological Observatory. *U.S. Dept. of Comm. NOAA.* Monthly and annual.

1979

Local climatological data for Blue Hill Meteorological Observatory. *U.S. Dept. of Comm. NOAA.* Monthly and annual.

1980

Local climatological data for Blue Hill Meteorological Observatory. *U.S. Dept. of Comm. NOAA.* Monthly and annual.

1981

Blue Hill Meteorological Observatory 1981 Hurricane plotting chart. *Blue Hill Observatory.* 3 p.

Blue Hill Weather Club Bull. 1:1. 2 p.

Local climatological data for Blue Hill Meteorological Observatory. *U.S. Dept. of Comm. NOAA.* Monthly and annual.

1982

Blue Hill Observatory 1982 Hurricane Plotting Chart. Blue Hill Observatory. Houghton Mifflin Co. 3 p.

Blue Hill Weather Club Bull. 2:1-4. 2-4 p.

Local climatological data for Blue Hill Meteorological Observatory. *U.S. Dept. of Comm. NOAA.* Monthly and annual.

1983

Blue Hill Observatory 1983 Hurricane Plotting Chart. *Blue Hill Observatory*. The Foxboro Co. 2 p.

Blue Hill Weather Club Bull. 3:1-5. 2-4 p.

Local climatological data for Blue Hill Meteorological Observatory. *U.S. Dept. of Comm. NOAA.* Monthly and annual.

1984

Blue Hill Weather Club Bull. 4:1-4. 2-4 p.

Blue Hill Observatory 1984 Hurricane Plotting Chart. *Blue Hill Observatory*. The Foxboro Co. 2 p.

Conover, J. H. The Blue Hill Observatory. *Weatherwise*. 37:296-303.

Local climatological data for Blue Hill Meteorological Observatory. *U.S. Dept. of Comm. NOAA.* Monthly and annual.

1985

Blue Hill Weather Club Bull. 5:1. 4 p.

Conover, J. H. Highlights of the history of the Blue Hill Meteorological Observatory and the early days of the American Meteorological Society. *Bull. Amer. Meteor. Soc.* 66:30-37.

Local climatological data for the Blue Hill Meteorological Observatory. *U.S. Dept. of Comm. NOAA.* Monthly and annual.

Appendix D

Chronology of Some Important and Interesting Events

1884	Rotch conceives the idea of the Observatory.
1885	The Observatory opened and observations began, 1 February.
1885	A. McAdie made atmospheric electricity measurements from the tower with a kite.
1886	Base station opened, 1 July.
1888	Valley station opened, 1 July.
1888	Rotch was appointed assistant in meteorology at Harvard College.
1889	East wing library opened.
1890	Cloud height measurements by double theodolite began.
1891	Government weather agency shifted from the Signal Service to the Weather Bureau under the Department of Agriculture.
1893	Metropolitan Park Commission took Blue Hills for parkland.
1894-97	Clayton derived cloud model and wind patterns aloft in respect to cyclones and anticyclones. Undertook the climatology of cloud heights and motions.
1894	The first kite ascent in the world carrying a thermograph took place, 4 August.
1896	A lease was negotiated between Harvard University and Park Commission.
1897	Rotch purchased Denny estate on Canton Avenue which became his permanent house.
1899	First radio experiments began, 16 June.
1899	Clayton proposed extension of the Dewey decimal system for meteorological subjects.
1900	Acetylene gas was used for illumination, 12 July.
1900	Highest kite sounding made from Blue Hill, 4,815 m MSL, 19 July.
1900	Clayton published kite sounding data as related to cyclones and anticyclones.
1901	First kite flights over the ocean were made, 22 August.
1901	Kite flights from a transatlantic ship passage commenced, 28 August.
1902	West wing completed.
1903	New library opened.

1903	Trolley line extended past the base.
1904	First balloonsonde in U.S. Launch by Fergusson from St. Louis, Missouri, 15 September.
1905	Concrete wall and iron fence were erected around the Observatory.
1905	Clayton and colleague of de Bort's made soundings over tropical Atlantic Ocean.
1905	Park Commission erected refractory at base.
1906	Rotch was appointed professor of meteorology at Harvard, 1 September.
1906-07	Fergusson made kite soundings from Twin Mountain, N.H. to compare with Mount Washington summit data.
1907	Clayton piloted balloon with others on record flight: St. Louis to Asbury Park, N.J., 21-23 October.
1908	New three-story concrete tower was completed, 4 June.
1908	Balloonsondes were made from Pittsfield, Massachusetts.
1909	Rotch instructed his first graduate student at Harvard.
1909	The first double theodolite observation of a pilot balloon was made in America, 7 July.
1912	Abbott Lawrence Rotch died, 7 April.
1912	Water pipe was laid to base for water supply.
1913	Observatory was officially taken over by Harvard with $50,000 endowment from Rotch.
1913	Broken kite wire stopped a train.
1913	Lightning came down kite wire, 6 March.
1913	Alexander George McAdie was appointed director, 1 October.
1914	Electric lights were put in service, 16 January.
1914	McAdie offered two courses at Harvard.
1917	McAdie started to teach rudiments of meteorology and forecasting to U.S. naval officers.
1918	McAdie was commissioned Lt. Commander of Naval Reserve Force was granted six months leave.
1918	McAdie organized Aerograph Section of U.S. Navy.
1918	In recognition of his teaching, McAdie was appointed Abbott Lawrence Rotch Professor of Meteorology at Harvard.
1922	Valley station closed, 1 July.
1924	South gale tore part of roof off three successive times, January.
1926	Fund drive netted $85,000 to place endowment at $176,000.
1930	Observations were not published for the first time.
1931	McAdie retired.

1931	Charles Franklin Brooks was appointed director of the observatory and professor of meteorology at Harvard.
1931	Observational program was expanded to include solar radiation. Fergusson started restoration of instruments.
1931	Weather station was opened atop the Geographic Institute building at Harvard.
1932	Climatological library was moved to Geographic Institute, Cambridge.
1932	Hazen thermometer shelter was moved from backyard to enclosure and reactivated.
1932	Mount Washington Observatory reopened, October.
1933	Meteorograph was set up on Mt. Wachusett.
1933	Al Sise and Arthur Bent made radio contact on 5 m with Mt. Washington, 10 May, a world distance record on that wavelength. G.W. Pickard and H.S. Shaw become involved with radio experiments.
1933	Daily radio contacts with Mt. Washington commenced on 60.6 MHz. Call letters WIXW were assigned to the Observatory, 7 November.
1934	The great wind on Mt. Washington was timed by radio at Blue Hill, 12 April.
1934	Northwest gale carried all but two masts from the tower, 27 December.
1935	Observatory's 50th anniversary. Brooks proposed the development of the radio-meteorograph.
1935	First successful radio-meteorograph transmission, airplane to Blue Hill, occurred 17 April.
1935	First successful radio transmission from a free balloon occurred, 23 October.
1935	First successful radio-meteorograph flight was made, 23 December.
1936	Balloon launcher was completed, 14 August.
1936	Probably the first remote transmission of temperature in U.S., at Harris Hill, Elmira, N.Y. was performed by Lange and Pear, June.
1937	Metropolitan District Commission police radio was set up on tower.
1938	A hurricane on 21 September knocked out power for seven days.
1939	Automatic weather station was developed.
1940	U.S. Weather Bureau Solar Radiation Supervising Station, headed by Irving F. Hand, was moved to Blue Hill, November.

1940   District Weather Bureau forecast office opened at Boston under in-fluence of Brooks and others.

1941   After Pearl Harbor, U.S. Army men moved in to spot airplanes and make ballistic wind soundings.

1942   Brooks was appointed as consultant to the National Defense Re-search Committee.

1942   Radio-meteorograph development concluded.

1943-44 Brooks and Conover taught weather observers for the Weather Bureau.

1943   New Metropolitan District Commission and state police radio trans-mitters were placed on northeast and southwest sides of sum-mit.

1946   A steel platform was added to the south side of tower.

1946   Balloon ascents by MIT carried cosmic ray counters.

1946   Henry Helm Clayton died, 26 October.

1948   Harvard University cut off the use of unrestricted funds. Closure became a possibility.

1948   First government contract. Observatory began snowstorm forecast-ing with the Weather Bureau.

1950   Cruft Laboratory places a van in backyard for measurement of clouds of electrons at high levels.

1951   Overseers planned to close Observatory in five years.

1951   WGBH-FM went on the air, October. Transmitter was located in west wing basement. Antenna northwest of observatory.

1952   The road was plowed regularly in winter.

1952   Demands for books and information from library made it one of the most active branches of the Observatory.

1953   Debris from Worcester tornado fell on the hill.

1953   Ground was broken for WGBH-TV and FM building.

1954   WGBH-TV tower was completed.

1954   Brooks retired, after 35 years, as secretary of the American Me-teorological Society.

1954   Weather radar of the Air Force was placed on tower at the center of the hill. Console located inside a Jamesway Hut in backyard.

1955   Air Force weather radar personnel moved into the west wing base-ment.

1955   Weather radar reports were transmitted twenty-four hours per day by Air Force personnel.

1955   WGBH-TV went on the air.

1955   Jamesway Hut was moved to Air Force compound on center of hill.

1957   Temporary building in the backyard was opened for use.

1957   Sixth Weather Radar Conference was sponsored by Observatory, Air Force Cambridge Research Center, Massachusetts Institute of Technology and American Meteorological Society.

1957   Dr. Brooks retired 1 September and party was held for him at the Rotch Estate.

1957   Thomas Cabot succeeded in a compromise with Harvard University and a search for a new director was made.

1957   John H. Conover was made acting director, September 1957–June 1958.

1958   Brooks died suddenly, 8 January.

1958   Base house was sold.

1958   Base station was closed.

1958   Cambridge climatological station was closed.

1958   Existing contractual work of cloud and weather radar studies, ground temperatures, instrument development and library services were terminated. Library was broken up, with a part going to the Gordon McKay Library in Cambridge.

1958   Richard M. Goody was appointed to the Abbott Lawrence Rotch chair of Dynamic Meteorology and director of the Observatory, 1 July. New programs at the Observatory and courses were offered at Harvard University.

1959   U.S. Weather Bureau took over the climatological observations, 1 July.

1959   Solar Radiation Supervisory Station was closed.

1959   Temporary building in the backyard was removed.

1959   Observers room was moved from the first to the second floor of tower and new instruments were installed.

1959   Sterling Price Fergusson died, 16 November.

1961   Air Force Weather Radar Group left.

1962   Renovations were completed.

1962   Noxon and Goody make first observation of radiation by atomic oxygen at 6300 A and 6364 A in the day airglow.

1962   Siderostat completed outside west wing.

1963   Summit road surfaced and drainage system restored. Vehicle and plow garaged in backyard.

1966   Successful measurements were made of quantity of nitrous oxide in atmosphere. Pressure-modulated radiometer was developed.

1967   Professor Goody was appointed Mallinckrodt Professor of
         Planetary Physics.

1968   Rotch monument was moved inside the Observatory yard.

1968   Goody had plans to close the Observatory. National Oceanic and At-
         mospheric Administration declined to continue involvement.

1971   Conover persuades National Oceanic and Atmospheric Administra-
         tion to reverse decision and Metropolitan District Commission
         acquires the building. Weather Service continues observa-
         tions. Metropolitan District Commission caretakers occupied
         the remainder of the building.

1976   Climatology Station was made a "reference station."

1981   Blue Hill Observatory Weather Club and Museum was established
         by William Minsinger.

1984   Weather Club received $10,000 from Harvard to aid in the restora-
         tion of Observatory.

1985   Hundredth anniversary was celebrated with fireworks, speeches,
         and a dinner, 1 February.